Mechatronic Systems, Applications

Edited by
Annalisa Milella, Donato Di Paola
and Grazia Cicirelli

Mechatronic Systems, Applications

Edited by Annalisa Milella, Donato Di Paola and Grazia Cicirelli

Republished by InTech

Janeza Trdine 9, 51000 Rijeka, Croatia

Technical Editor Sonja Mujacic

Cover Designer Dino Smrekar

Additional hard copies can be obtained from orders@intechopen.com

Mechatronic Systems, Applications,
Edited by Annalisa Milella, Donato Di Paola and Grazia Cicirelli
p. cm.
ISBN 978-953-307-040-7

For EU product safety concerns:
IN TECH d.o.o., Prolaz Marije Krucifikse Kozulić 3, 51000 Rijeka, Croatia,
info@intechopen.com or visit our website at intechopen.com.

Meet the editors

Annalisa Milella received the Laurea (summa cum laude) and Research Doctorate degrees from the Politecnico of Bari, Italy, in 2002 and 2006, respectively, both in Mechanical Engineering. In 2005, I was a visiting PhD student at the EPFL Autonomous Systems Laboratory. Currently, I am a researcher at the Institute of Intelligent Systems for Automation (ISSIA), National Research Council (CNR) of Bari, Italy. My main research interests include: - computer vision applied to robotics and intelligent systems - self-localization methods for mobile robots - robotic non-destructive inspection - robotic surveillance systems

Grazia Cicirelli received the Laurea degree (summa cum laude) in Computer Science from the University of Bari (Italy) in 1994. Until 2001 she held grants from the Italian National Research Council (CNR) for research activities in Robotics and Image Processing. From 2001 she is a Technologist Researcher at the Institute of Intelligent Systems for Automation (ISSIA) of CNR in Bari. Her principal interests include pattern recognition, artificial intelligence, image processing for robotic applications and intelligent systems for video-surveillance. She has worked on and directed numerous research projects in different research areas such as Quality Control, Intelligent Transportation Systems, Autonomous Mobile Robotics. She is author of numerous research papers published in International Conference Proceedings, National and International Journals. She is a co-inventor of 1 international patent on the development of a visual system for event detection in a sport context. Dr. Cicirelli regularly serves as reviewer on various Conferences and International Journals including: Industrial Robot: An International Journal International Journal of Advanced Robotics Systems IEEE Transactions on Intelligent Transportation Systems. She received the 2013 Award for Excellence as Outstanding Reviewer to "Industrial Robot: An International Journal" (Emerald) for the significant contribution made throughout 2012. She is a member of the Editorial Board at the International Journal of Advanced Robotic Systems (IJARS). Furthermore, she is Editor of the permanent Topic "Robot Navigation" and Associate Editor of the permanent Topic "Vision Systems" at the same journal.

Preface

Mechatronics, the synergistic blend of mechanics, electronics, and computer science, has evolved over the past twenty-five years, leading to a novel stage of engineering design. By integrating the best design practices with the most advanced technologies, mechatronics aims at realizing highquality products, guaranteeing, at the same time, a substantial reduction of time and costs of manufacturing. Mechatronic systems are manifold, and range from machine components, motion generators, and power producing machines to more complex devices, such as robotic systems and transportation vehicles. This book is concerned with applications of mechatronic systems in various fields, like robotics, medical and assistive technology, human-machine interaction, unmanned vehicles, manufacturing, and education. The Editors would like to thank all the authors who have invested a great deal of time to write such interesting chapters, which we are sure will be valuable to the readers.

A brief description of every chapter follows. Chapters 1 to 6 deal with applications of mechatronics for the development of robotic systems. Chapter 1 presents the design and realization of a novel bio-inspired climbing caterpillar robot. The climbing technology is combined with bio-inspired research to create a novel robotic prototype, which has a cognitive potential, and can climb and move flexibly in its working environment. Chapter 2 introduces two novel fuzzy logic-based methods to estimate the location of passive RFID tags using a mobile robot equipped with RF reader and antennas, and a laser rangefinder. It is shown that both approaches are effective in supporting mobile robot navigation and environment mapping for robotic surveillance tasks. Chapter 3 deals with the design of a contact sensor for robotic applications. The main contributions of the chapter are the design of the contact sensor, and the use of a neural network for force vector identification based on measures of sensor body deformation. In Chapter 4, the authors develop an intelligent home security system, consisting of a multisensor fire-fighting robot and a remote control system. The robot is able to navigate autonomously, avoid obstacles, detect fire source and fight it. It can also transmit the environment status to a distant user. Users can both receive information and control the robot remotely. Chapter 5 presents the design and integration of a power detection and diagnosis module to measure the residual power of an autonomous robot. The detection, isolation and diagnosis algorithm use a multilevel multisensory fusion method. The module is integrated in the software architecture of the robot and can transmit the detection and diagnosis status to the main controller. The design and implementation of smart environments with applications to mobile robot navigation is the focus of Chapter 6. The authors develop a so-called Intelligent Space (iSpace), where distributed sensor devices including mobile robots can cooperate in order to provide advanced services to the users.

Medical and assistive technologies and human-machine interaction systems are the topic of chapters 7 to 13. Chapter 7 presents some robotic systems for upper and lower limbs rehabilitation. Then, it focuses on the application of mechatronics to rehabilitation for functional assessment and movement analysis. Finally, it discusses open issues in the field of robotics and mechatronic systems for rehabilitation. Chapter 8 is concerned with the design of a wearable sensor system, which includes body-mounted motion sensors and a wearable force sensor for measuring lower limb orientations, 3D ground reaction forces, and joint moments in human dynamics analysis. In Chapter 9, the authors describe a new navigation system that is able to autonomously handle a laparoscope, with a view to reducing latency, allowing real-time adjustment of the visual perspective. The system consists of an intuitive mechatronic device with three degrees of freedom and a single active articulation. It is shown that this new mechatronic system allows surgeons to perform solo surgery. Furthermore, downtime for cleaning and positioning is reduced. Chapter 10 presents a model-based fault detection and isolation (FDI) method for a powered wheelchair. Faults of three internal sensors (two wheel resolvers and one gyro), one external sensor (Laser Range Sensor), and two wheel motors are handled. Interacting-Multiple-Model estimator and Kalman Filter are applied to FDI of the internal sensors, whereas FDI of external sensor is detected considering the errors related to scan matching. Different experiments are carried out in order to prove the robustness of the proposed approach. Several projects concerning the use of virtual reality for electric wheelchair driving learning are described in Chapter 11. In Chapter 12, a magnetorheological technology for human-machine interaction through haptic interfaces is introduced. A novel operating mode that reduces uncontrolled forces, as well as the inertia of moving parts, is proposed. Modelling and experimental characterization of the system is presented using two haptic interfaces: a haptic interface for musical keyboards and a novel Human Machine Interface for automotive cockpits. Chapter 13 describes and validates experimentally a new impedance control scheme for a two-DOF Continuous Passive Motion (CPM) device for an elbow joint.

Chapters 14 and 15 concern mechatronic systems for autonomous vehicles. Specifically, Chapter 14 presents a road sign recognition technique to be used for the development of Intelligent Transport Systems (ITS), while Chapter 15 is focused on the development of an Unmanned Ground Vehicle (UGV) for task-oriented military applications.

Chapters 16-19 deal with mechatronics in manufacturing contexts. Chapter 16 analyses the dynamics of microparts along a sawtooth surface with horizontal and symmetric vibrations, and presents experimental results obtained with a micropart feeder using bimorph piezoelectric actuators and 0603 capacitors. Chapter 17 presents a combination of an optimized pallet pattern generation algorithm, an industrial robot simulator, and a modified trajectory optimization algorithm. The focus of Chapter 18 is on the development of an automated measurement and grading system for the High Brightness-LED dies in the fabrication section based on machine vision. Chapter 19 presents selected results of two extensive surveys targeted on adoption and utilization of advanced manufacturing technology.

Chapter 20 concludes the book, describing a method for the installation of mechatronics education in schools.

Annalisa Milella, Donato Di Paola and Grazia Cicirelli

Contents

1

A Bio-Inspired Small-Sized Wall-Climbing Caterpillar Robot

Houxiang Zhang[1], Wei Wang[2],
Juan Gonzalez-Gomez[3] and Jianwei Zhang[1]
1. University of Hamburg
Germany
2. Beijing University of Aeronautics and Astronautics
China
3. School of Engineering, Universidad Autonoma de Madrid
Spain

1. Introduction

Climbing robots work in a special vertical environment and use mobility against gravity (Zhang, 2007). They are a special potential sub-group of mobile technology. In the recent 15 years, there have been considerable achievements in climbing robot research worldwide by exploring potential applications in hazardous and unmanned environments (Virk, 2005). The typical application of climbing robots includes reliable non-destructive evaluation and diagnosis in the nuclear industry, the chemical industry and the power generation industry (Longo, et al., 2004), welding and manipulation in the construction industry (Armada, et al., 1998), cleaning and maintenance for high-rise buildings in the service industry (Elkmann, et al., 2002) and urban search and rescue in military and civil applications (Wu, et al., 2006).

However, until now, there are few successful prototypes that are both small enough and move flexibly enough to negotiate surfaces with a complex structure. It is common to design rather big and heavy climbing robots. The difficulties of developing a flexible and small climbing robot with full locomotion capabilities include not only the weight reduction of the mechanism but also the miniaturization of the flexible construction. An additional problem is the fact that the intelligent technology in many climbing robotic prototypes is not developed enough.

The purpose of this paper is to present a novel bio-inspired climbing caterpillar robot which is currently under construction in our consortium. We combine the climbing technology with bio-inspired research to create a novel robotic prototype which has a cognitive potential and can climb and move flexibly in its working environment. This paper only concentrates on the design and realization of the current climbing robotic prototype. Other details such as gaits, motion kinematics and dynamics will be discussed in other publications.

This paper is organized as follows. First the related work on climbing robots and the biologically inspired mobile robotic system will be introduced systematically in section 2. At the beginning of section 3, we investigate the climbing locomotion mechanism adopted by caterpillars. Based on this, our on-going climbing robotic project will be introduced. Different aspects including system design, mechanical implementation and control realization will be presented in detail. Although we designed two climbing caterpillar robotic configurations, the simpler inchworm configuration is the focus for discussion in this paper. After pointing out future work, our conclusions are given in the end.

2. Related research in literature

2.1 Climbing mechanism of caterpillars

Climbing robots are a kind of mobile robots. There are two important issues for designing a successful climbing robotic prototype. The first one is the adhesion principle, the second one is mechanical kinematics.

Many climbing robots use legged structures with two (biped) to eight legs, where more limbs inherently provide redundant support during walking and can increase the load capacity and safety. The robots with multiple-leg kinematics are complex due to several degrees of freedom. This kind of robots which use vacuum suckers and grasping grippers for attachment to buildings are too big, too heavy and too complex. As the simplest kinematical model in this class, bipeds vary most significantly in the style of their middle joints. Robots by Nishi (Nishi, 1992) and the robot ROBIN (Pack, et al., 1997) use a revolute middle joint. A prismatic middle joint is used by ROSTAM IV (Bahr, et al., 1996), while the robot by Yano (Yano, et al., 1997) does not have a middle joint but simply a rigid central body. ROSTAM IV, the smallest robot in this class built to date, weighs approximately only 4 kg, but the reliability and safety of its movement is not satisfying.

The robot ROMA (Abderrahim, et al., 1991) is a multifunctional, self-supporting climbing robot which can travel into a complex metallic-based environment and self-support its locomotion system for 3D movements. Generally, construction and control of these robots is relatively complicated. The other problem is that the climbing robots based on the grasping method often work in a specialized environment such as metal-based buildings. In order to realize a climbing movement, the mechanical structure of the robots is not designed modularly.

Inspired by gecko bristles, the last few years have witnessed a strong interest in using molecular force as a new attachment method for climbing robots. Flexible climbing prototypes with multi-legs (Sitti, et al., 2003) and with wheels (Murphy, et al., 2006) have been emerging. From the locomotion viewpoint, there is no difference to the other climbing prototypes.

The prototypes with a wheeled and chain-track vehicle are usually portable. The adhesion used by this kind of robot is negative pressure or propellers, therefore the robots can move continuously. A smart mobile robot was proposed as a flexible mobile climbing platform carrying a CCD camera and other sensors. It uses a negative pressure chamber to attach to vertical surfaces. Even if this kind suction is not sensitive to a leakage of air, the negative pressure is not good enough for safe and reliable attachment to a vertical surface when the robot crosses window frames. An improved smart structure with two linked-track vehicles was proposed, which can be reconfigured so that the robot can move between surfaces standing at an angle of 0 - 90 degrees due to the pitching DOF actuated by the joint to increase the flexibility (Wang, et al., 1999). Recently, many similar climbing prototypes with wheels and chain-tracks have been presented worldwide.

With sliding frames, a climbing robot can be made simpler and lighter from the kinematic point of view, which is one of the most important specifications for devices working off ground. This kind of climbing robots features pneumatic actuation, which can effect a linear sliding movement better than electric motor systems. In 1992, a pneumatic climbing robot with a sliding frame was developed for cleaning the glass surface of the Canadian Embassy in Japan (Nishigami, et al., 1992). However, the robot cannot move sideways. Since 1996, our group has been developing a family of Sky Cleaner autonomous climbing robots with sliding frames for glass-wall cleaning (Zhang, et al., 2005). The first two prototypes are mainly used for research, but the last one is a semi-commercial product designed for cleaning the glass surface of the Shanghai Science and Technology Museum. The benefits of this locomotion principle are offset by nonlinear control methods and difficulties of the pneumatic systems. As a conclusion, it can only be used for specialized environments such as glass curtain walls.

Some limbless robots are also capable of climbing. However, using friction, snake-like prototypes can only climb up and down a tube with a suitable diameter (Granosik, et al., 2005). The robot has to have a shape that allows as much contact as possible with the tube's inner surface. The other example of these kinds of limbless climbing robots is the Modsnake (Wright et al. 2007) developed at the CMU's Biorobotics Laboratory. This robot consists of 16 modules and it is capable of climbing on the inside or outside of a tube. Actually, these are pipe robots rather than climbing robots.

2.2 Bio-inspired mobile robots and control methods

The last few years have witnessed an increasing interest in implementing biological approaches for mobile robotic design and research. A lot of impressive work including multi-legged robots, snake-like robots, and robotic fish has been done on bio-inspired mobile robotic technology recently.

For example, the robot RiES (Spenko, et al., 2008) with 4-6 legs can climb glass surfaces using nano material and walk on wall surfaces using metal nails. This robot adapts to the cockroach's locomotion model, and its design implements the modular approach. At the Boston Dynamic Institute, two world-renowned bio-inspired mobile robots have been developed. The Littledog robot (Pongas, et al., 2007) with four legs is designed for research on learning locomotion to probe the fundamental relationships among motor learning, dynamic control, perception of the environment, and rough terrain locomotion. Then there is the BigDog robot (Raibert, et al., 2008), which is the alpha male of the Boston Dynamics family of robots. It is a quadruped robot that walks, runs, and climbs on rough terrain and carries heavy loads. These two mobile prototypes are not only well designed from the mechanical point of view, but also concerning their high level of intelligence.

Snake-like robots, also called limbless robots, make up the other big group in the bio-spired mobile robotic family. The snake-like robots were first studied by Hirose, who developed the Active Cord Mechanism (ACM) (Hirose, 1993). Recently some new versions have been developed in his group (Togawa, et. al., 2000). S. Ma et al. in Japan and his Chinese colleagues at the Robotics Laboratory of Shenyang Institute of Automation also developed their own yaw-connecting robot and studied the creeping motion on a plane and on a slope

(Chen, et. al., 2004). Other prototypes are SES-2 (Ute, et al., 2002), S5 (Miller, 2002), WormBot (Conradt, et al., 2003) and swimming Amphibot I (Crespi, 2005).
The classical approach to controlling limbless robots is based on the inverse kinematics. The joint's angles are obtained from the desired trajectory of the supporting points or the center of mass. The limbless robots can be considered as hyper-redundant manipulators, formed by infinite joints. Chirjkjian employed functions to describe the shapes that the manipulator must adopt, and got the angular expressions (Chirjkjian, et al., 1995). Lipkin found much success in applying a three-dimensional variation of this approach to generate crawling, climbing and swimming gaits for the Modsnake robot (Lipkin et al., 2007). Goldman examined the kinematics of climbing a pole and calculated the joint angles for fitting a snake robot body to a helical backbone curve (Goldman et al., 2007).
In nature the vertebrates and invertebrates have special neurons called Central Pattern Generators (CPGs). These centers oscillate and produce rhythms that control muscle activity to carry out actions such as breathing, bowel movements, masticating, locomotion, etc. Based on biological studies, mathematical models are constructed from these oscillators which are then applied to robots to control locomotion. One of the pioneers in applying CPG models to robotics is the EPFL's Bio-inspired Robotics Laboratory (Ijspeert, 1998). In 2004, together with Crespi, they implemented the first prototype of Amphibot, demonstrating the viability of his bio-inspired model for robot locomotion. All their research work on the use of CPGs for locomotion control in Robots is reviewed in (Ijpeert, 2008). In the Biological neuron-computation group of the Autonomous University of Madrid, Herrero-Carron modeled and implemented CPGs based on Rulkov's model to control an eight segment caterpillar robot (Herrero-Carron, 2007).

3. A Bio-inspired climbing caterpillar robot

3.1 Climbing mechanism of the caterpillars

Caterpillars are among the most successful climbers and can maneuver in complex three-dimensional environments, burrow, and hold on to the substrate using a very effective passive grasping system (Mezoff, et al., 2004). They consist of a head and neck part, a body with several segments and a tail end part, as shown in Fig. 1 and Fig. 2. Their movement depends mainly on the muscle's expansion and contraction. Caterpillars use passive grip to secure themselves to complex branched substrates and can effect multidimensional movements. They are able to bend, twist and crumple in ways that are not possible with a rigid skeleton. The prolegs provide astonishing fault-tolerant maneuvering ability and stable, passive attachment.
Caterpillars have the following advantages of climbing compared with other animals from the system design viewpoint.
1) Good length to pitch-back moment ratio (Spenko, et al., 2008): A big length to pitch-back moment ratio of the robotic mechanical design is better to realize a reliable attachment and to decrease the danger of the climbing movement.
2) Distributed modular design: Caterpillars are with several segments which are similar to identical modules so that the mass of the body is distributed into all segments. During the climbing movement many segments are attached to the surface while only some numbers of segments are moving, thus makes the robot safer than other climbing kinematics principles.

There are two kinds of typical locomotion modes adapted by different caterpillars. The corresponding representative worms are the inchworm (Fig. 1) and Manduca sexta larvae (Fig. 2) respectively. Caterpillar kinematics models are also presented in two figures. In order to analyze the kinematics of caterpillars, an adhesion module is indicated as "△" and an active rotating joint module is indicated as "–○–" in our discussion.

Fig. 1. Inchworm and its locomotion mechanism

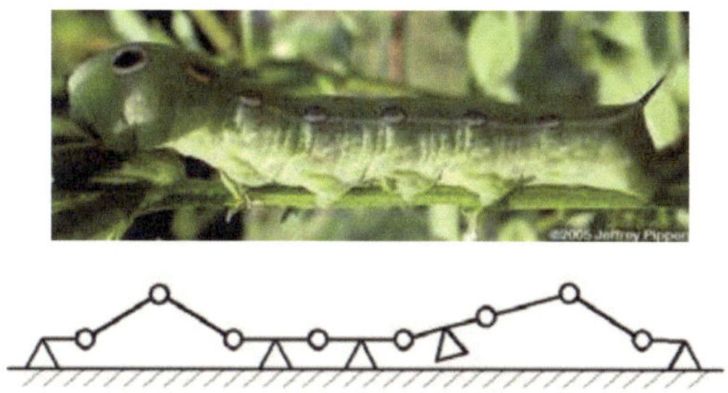

Fig. 2. Manduca sexta larvae and its locomotion mechanism

In this paper, we only concentrate on the inchworm configuration due to the following reasons.

1) The inchworm performs a gait different from that of normal caterpillars. Although the inchworm also consists of three limbs, the head, the tail and the trunk, the body limb is totally different since it possesses no proleg at all.

2) Due to its simple body structure, the inchworm has to adopt a simple gait to move. While crawling, it lifts the tail first, contracts the trunk, and then drops the tail a short distance from its original position in the forward direction. At this time, the inchworm is like a bow. Then, it lifts its head, stretches the trunk and drops the head. A gait is completed and a certain distance has been covered in a forward movement.

3) There is only one adhesion module adhering to the wall during motion; during the motion, the body delivers an incomplete wave.

In the following part of this chapter, we are going to present our climbing caterpillar robot design with an inchworm configuration. Different aspects will be introduced in detail.

3.2 Prototype design

Based on the investigation of natural climbing caterpillars, the most important requirement for our robotic system moving on a slope with different materials is extraordinary motion capabilities. Two mechanical units are definitely necessary for designing a light mechanical climbing robot: structure with flexible locomotion capabilities and a safe and reliable attachment device.

On the other hand, as a bio-inspired robot imitating a natural caterpillar, the proposed climbing caterpillar robot should be as intelligent as possible. In order to move freely, it is also important for the mobile robot not to be wired or otherwise connected to the environment. The robot should carry all devices it requires: onboard power, the controller, and wireless communication.

In this project, we combine climbing techniques with a modular approach to realize a novel prototype as a flexible wall climbing robotic platform featuring an easy-to-build mechanical structure, a low-frequency vibrating passive attachment principle and various locomotion capabilities (Zhang, et al., 2007). This multifunctional bio-inspired modular climbing caterpillar will:

1) be capable of walking and climbing not only in different environments but also on the vertical surfaces and ceilings on the inside of buildings;

2) possess locomotion capacities including pitching, yawing, lateral shift, and rotating;

3) feature sensor-servo-based active perception of the environment.

Fig. 3 shows pictures taken from a 3D-animation of the planned robotic caterpillar on a vertical wall. This system, which is currently under development at the University of Hamburg, is based on the technology that already exists in our consortium.

Fig. 3. 3D-animation of the planned robotic caterpillar

Another feature of this prototype lies in a new attachment principle. Currently, it is noted that four attachment principles are valid for climbing robot design. First there is no possibility of using magnetic force on general lightweight climbing robots except for some special cases that work on ferromagnetic surfaces (Akinfiev, et al., 2002). Molecular force (Sitti, et al., 2003) as a new attachment method for climbing robots is very promising; however, the benefits of this novel adhesive principle are offset by expensive manufacturing prices and difficulties. Based on the current technological level, real industrial application will still take some time.

The grasping gripper is relatively prevalent for designing a reliable attachment unit for a climbing robot. The climbing robots using grippers generally work in a specialized environment, such as metal-based buildings (Abderrahim, et al., 1991). As a result, we cannot implement this idea on our prototype either. This not the attachment mechanism adopted by natural climbing caterpillars.

Actually, natural caterpillars use the passive suckers on their prologs for moving and climbing, as mentioned above. The vacuum in these suckers is usually established by vacuum ejectors or vacuum pumps which are easy to control. These advantages are offset by the long air tube or relatively heavy devices that need to be added to the climbing robots, which limit the application of this adsorption method to smart wall-climbing robots, such as our proposed caterpillar robot.

A new low-frequency vibrating passive suction method is presented in order to keep the merits and eliminate the shortcomings of using normal active vacuum suckers (Zhang, et. al., 2007.). This passive idea also comes from the natural caterpillars' adhesion principle. Application of a new low-frequency vibrating passive suction method makes it possible to forego the conventional heavy vacuum ejectors and realize an effective simple adsorption, furthermore to improve the inspired technological level and flexibility of the locomotion capability.

3.3 Mechanical module design and Inchworm configuration realization

As we mentioned above, this paper only concentrates on robotic inchworm design and realization, as shown in Fig. 4. Actually, the mechanical modules are uniform and identical, making it possible to use them to build a general caterpillar robotic configuration. For two reasons, the inchworm configuration is the first milestone in this long-term project.

1) It is the basic and simplest configuration as a robotic caterpillar; the other configuration can be based on it from the mechanical viewpoint. For example, a larvae robot can be created by connecting two inchworm prototypes.

2) The gaits of an inchworm robot are relatively simple. However, it is more challenging for us to validate our design, especially concerning the new passive attachment. All results will be valuable for our future research.

Fig. 4. Robotic inchworm design in CAD and real prototype

The inchworm robot consists of three serially connected modules for moving. Actually, the modular design is identical. Though it is possible to construct the robot with three uniform modules, in order to realize a mimic inchworm configuration, the first and the last modules feature passive suckers while the middle on has no attachment unit.

One body module includes an active joint and an attachment module, as shown in Fig. 5. The joint module consists of two brackets with some holes, an RC servo, a shaft, and a flange (Wang, et al., 2008). As a result of actuation by the servo, one DOF active rotating joint within ±90 degrees enables two brackets to adopt pitching movements. Brackets 1 and 2 are fixed to the shell and axis of the servo motor respectively. When the motor is running, these two brackets rotate around the shaft in the middle. The mechanical interfaces on the outside plates of the brackets allow for the joint modules to be assembled either in parallel axes or perpendicular axes.

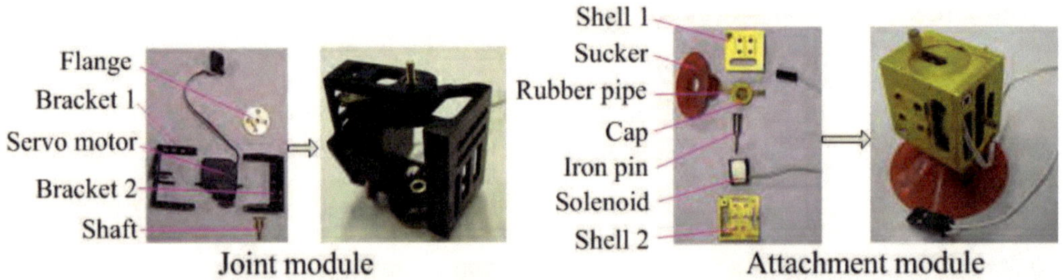

Fig. 5. Mechanical design of the joint module and attachment module

The attachment module without any embedded DOF consists of two shells, a passive sucker, a solenoid valve, and other small parts. The vacuum in the sucker is generated only by the distortion of the sucker. A simple mechanism driven by a solenoid is used to release the vacuum in the passive sucker. When the solenoid is not actuated, a rubber pipe connecting the inner side of the sucker to the outside air is shut off by an iron pin and cap under the force of a spring. The sucker can be attached to flat surfaces. If the solenoid is actuated, the iron pin will be withdrawn from the cap to connect the inside of the sucker with the outside air through the pipe. The vacuum in the sucker is released, and the sucker can be lifted. On the two shells of the attachment module, the mechanical interfaces are the same as those on the joint module. Thus the attachment module can be directly connected to the joint module. The basic performances of two modules are listed in Table 1, in which the elastic coefficient of the passive sucker deserves special attention, because its influence is important for the gait realization. In order to lighten the weight, all mechanical parts are manufactured from aluminum.

Joint Module	Performances	Adhesion Module	Performances
Size: length×width×height (mm^3)	35×37×30	Size: length×width×height (mm^3)	26×32×20
Weight (g)	19.2	Weight (g)	27.8
Max Output Torque (Nm)	0.2	Max Attaching Force, *Fn* (N)	40
Output Angle (Degrees)	0-180	Max Sliding Force, *Fq* (N)	15
		Max Turning Torque, *M* (Nm)	0.4
		Attaching force-to-weight Ratio	100

Table 1. Performance of the Demonstrated Modules

3.4 Control realization

The inchworm robot should own enough intelligence to imitate a natural creature. First, the robot should carry onboard power, the controller, and wireless communication units. Second, as we mentioned before, the system should be low-cost to be used for different applications such as locomotion analysis or bio-inspired investigation. As a result, to ensure its ability of performing different gaits, there is enough space in each module for sensors, the onboard controller, and batteries. Considerable stress is laid on weight reduction as well as on construction stiffness to achieve a dexterous movement mechanism.

Fig. 6 shows the principle of a distributed control system. Each module has embedded intelligent capabilities with an independent onboard controller with two layers. On the one hand, each module is equal from the control view point. According to their different locomotion functions, various programs will run on the single module respectively. All motion commands can be sent to a certain module individually or broadcast to all modules through the I^2C bus according to the task requirements. On the other hand, any module can be nominated as a master control which is in charge of high-level control functions such as path planning, navigation, localization. At the moment, a PC is used as a consoler to set up the parameter configuration and generate locomotion gaits. It can also be directly connected to the bus through RS232. In this way, the PC can be considered as a virtual module in the robot system and plays the role of the master or a graphic user interface (GUI).

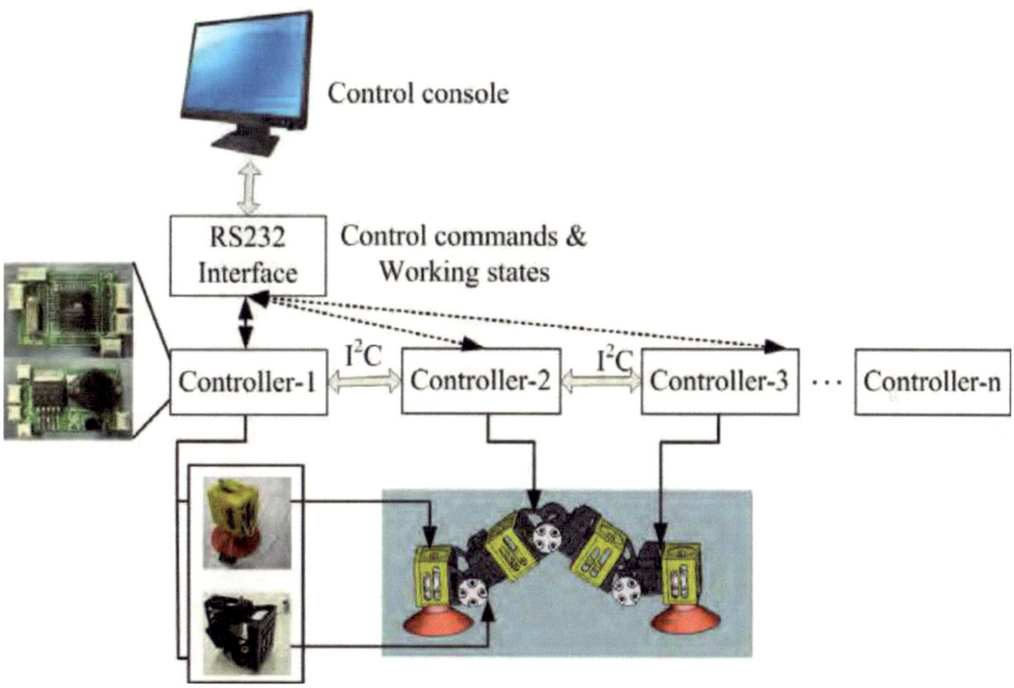

Fig. 6. Control realization

Each controller has one channel of Pulse Width Modulation (PWM) output to control the servo motor, one on-off output to control the solenoid valve, three digital or analog sensor inputs to collect sensor data, one I^2C bus and one RS232 serial port. The number of the controllers in a caterpillar robot is determined by the number of the modules. The controllers can communicate with each other by the I^2C bus and receive the orders from a console through the RS232 serial port. While the robot works, the information about its working state and the sensor data will be sent back to the console at the same time.

4. Locomotion and on-site experiments

4.1 Locomotion control

In order to control the climbing of the inchworm robot, an unsymmetrical phase method (UPM) is proposed. That means, the movement of attaching the suckers to the wall is faster than that of lifting the sucker from the wall. Fig. 7 shows five typical steps in the gait of an inchworm robot climbing on a flat wall, as well as the angle of each joint and the state of each sucker in one control cycle. At the beginning and end steps t_0 and t_4, the angle values of three joints are all zero.

The state of the sucker is controlled by a corresponding solenoid which has only two states, on or off. The high level means that the solenoid is actuated and the sucker is released. The low level denotes the inverse state.

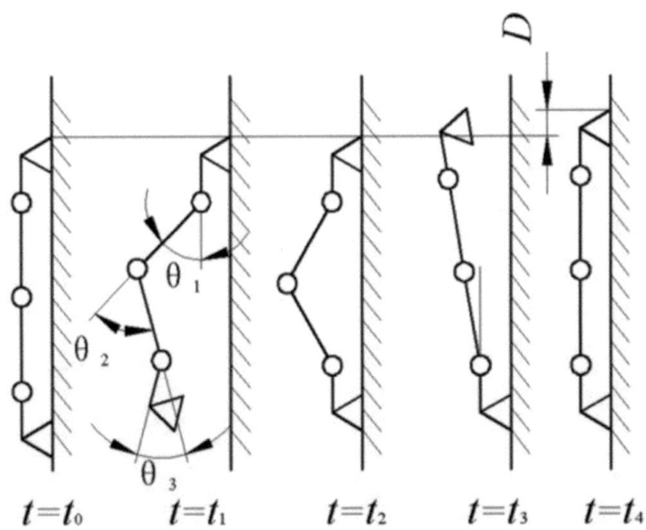

Fig. 7. The locomotion gaits of inchworm robot

At time t_1, the inchworm robot lifts the lower sucker when the angle values of its three joints fulfill the relationship shown in equation (1). Where $\Delta\theta_L$ is a constant which is named the impact angle and defined by experiments.

$$\theta_1+\Delta\theta_L=\theta_3=-(1/2)\theta_2=\theta_L \tag{1}$$

According to (1), this sucker moves not only forward but also up the wall. During the time between t_1 and t_2, the robot puts down the sucker by turning Joint 1. As mentioned above, the time between t_0 and t_1 is much longer than the time between t_1 and t_2, so the control phases are unsymmetrical during two periods. The inchworm robot uses the impact force between the sucker and wall produced by UPM to compress the passive sucker well and to attach firmly and reliably to the wall.

In the lowering period of UPM, when the sucker makes contact with the wall, the force F acting on the sucker can be expressed by (2).

$$F=F_1+F_2=(M+I\omega/\delta_t)/A \tag{2}$$

Where:

F_1 is the force produced by the joint driver whose output torque is M;
F_2 is the force introduced by the impulse acting on the sucker;
I is the turning inertia of all moving parts;
ω is the joint velocity;
A is the distance between the unattached sucker and rotating joint;
δ_t is the impulse time.

The values of some parameters in (2) are shown below. At step t_1, A is equal to 0.13 m, M is 0.2 Nm and I is 1.62×10^{-4} kg m^2. The values of ω and t can be attained in real experiments, ω is 5.2 rad/s and δt is 2×10^{-3}s. As a result, F_1 is equal to 1.5N and F_2 is equal to 3.2N. That means that the compression distortion values of the sucker produced by F_1 and F_2 are 0.9mm and 1.8mm respectively, according to the compression elastic coefficient of the sucker.

The joint trajectories in Fig. 7 are denoted by equations (3) - (5), which are loaded in the controllers to calculate the joint angles in real time. Details can be found in (Wang, et.al., 2009).

$$\theta_1(t)=\begin{cases} \dfrac{\theta_L+\Delta\theta_L}{t_1-t_0}t, & t\in[t_0,t_1] \\ \theta_L+\Delta\theta_L-\dfrac{\Delta\theta_L}{t_2-t_1}t, & t\in(t_1,t_2] \\ \theta_L(1-\dfrac{t}{t_3-t_2}), & t\in(t_2,t_3] \\ 0, & t\in(t_3,t_4] \end{cases} \tag{3}$$

$$\theta_2(t)=\begin{cases} -\dfrac{2\theta_L}{t_1-t_0}t, & t\in[t_0,t_1] \\ -2\theta_L, & t\in(t_1,t_2] \\ -2\theta_L(1-\dfrac{t}{t_3-t_2}), & t\in(t_2,t_3] \\ 0, & t\in(t_3,t_4] \end{cases} \tag{4}$$

$$\theta_3(t)=\begin{cases}\dfrac{\theta_L}{t_1-t_0}t, & t\in[t_0,t_1]\\ \theta_L, & t\in(t_1,t_2]\\ \theta_L-\dfrac{\theta_L-\Delta\theta_L}{t_3-t_2}t, & t\in(t_2,t_3]\\ \Delta\theta_L(1-\dfrac{t}{t_4-t_3}), & t\in(t_3,t_4]\end{cases} \tag{5}$$

4.2 Climbing tests

Recently, a series of on-site tests have been made to confirm our design and to find out the appropriate ω in the equation (2). In these tests, first one sucker of the inchworm robot is fixed on a glass wall by a clamp, and another one is lowered and lifted repeatedly. The compression value of the free sucker is recorded. To compress the sucker and lower it by 1.5mm, ω should be nearly 2.8rad/s; while for the maximum compression value of up to 3mm, ω should reach 6rad/s. Because a too-large joint velocity will interfere with the stability of the attached sucker when the robot is climbing the wall, 5.2 rad/s is taken as the joint velocity during the lowering motion.

After that step, we made a climbing test on a glass surface. The inchworm robot realizes continuous motion on the vertical wall successfully with the gait presented in Fig. 7. Fig. 8 shows the procedure of the inchworm robot climbing up for the course of one gait; the maximal step length is 5mm, and the time of one step is 1.8s.

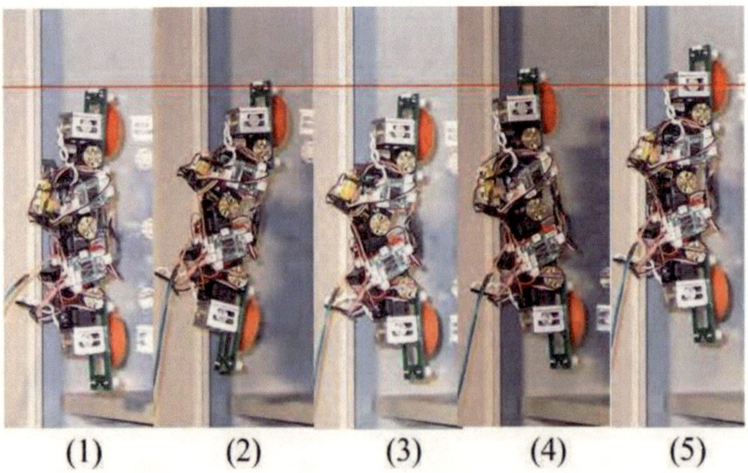

Fig. 8. Climbing testing on glass

5. Conclusions and future work

This paper presents a novel, bio-inspired, small climbing caterpillar robot. The discussion is focused on the inchworm configuration since it is the simplest and basic structure compared to the other climbing mechanisms adopted by natural caterpillars. All related aspects including design motivations, system integration, mechanical module design, control hardware, and locomotion realization are introduced in detail. In contrast to conventional theoretical research, the project introduced in this project successfully implements the following innovations:

1. It proposes a climbing robot based on a modular reconfiguration concept. The robot features a simple, light mechanical structure and a novel passive attachment.
2. The distributed control system completes the modular design. A UPM locomotion method enables the robot to climb vertical surfaces reliably.
3. Related tests have shown that the inchworm robot can implement safe climbing in a certain locomotion gait. This implies the mechanical feasibility, the rationality of the design and the flexible movement adaptability of the robot.

There are still a lot of technical problems for us to uncover, such as vibration during movement, evaluation of different locomotion parameters, even if the inchworm configuration is the simplest one. We should improve the gait control methods to diminish the internal force in the caterpillar robot, and realize more reliable motions, such as climbing between two surfaces in different planes, crossing a barrier on the wall etc.

Second, currently all of the locomotion capabilities are pre-programmed, as we mentioned. In future, our research will focus on the realization of real autonomy.

Third, other caterpillar robotic configurations will be designed and tested soon. Future research will include finding the rules of constructing a reasonable configuration with the passive joint and active joint, testing the feasibility of passive joints in the caterpillar model and selecting the safest climbing gait.

6. Reference

Zhang, H. (2007) Edited Book, Climbing & Walking Robots Towards New Applications, 2007, I-Tech Education and Publishing.

Virk, G. (2005) The CLAWAR Project-Developments in the Oldest Robotics Thematic Network, 2005 IEEE Robotics & Automation Magazine, June, pp.14-20.

Longo, D., Muscato, G. (2004) A Modular Approach for the Design of the Alicia3 Climbing Robot for Industrial Inspection, Industrial Robot: An International Journal, Vol.31, No.2, pp.148-158.

Armada, M., Gonzalez de Santos, P., Prieto, P., Grieco, J. (1998). REST: A Sixlegged Climbing Robot, European Mechanics Colloquium, Euromech 375, Biology and Technology of Walking, pp. 159-164.

Elkmann,N., Felsch, T., Sack, M., Saenz, J., Hortig, J.(2002). Innovative Service Robot Systems for Facade Cleaning of Difficult-to-Access Areas, Proceedings of the 2002 IEEE/RSJ International Conference on Intelligent Robots and Systems EPFL, Lausanne, Switzerland, October, pp.756-762.

Wu, S., Li, M., Xiao, S., Li, Y.(2006). A Wireless Distributed all Climbing Robotic System for Reconnaissance Purpose, Proceeding of the 2006 IEEE International Conference on Mechatronics and Automation, Luoyang, China, June 25-28, pp.1308-1312.

Nishi, A. (1992). A Biped Walking Robot Capable of Moving on a Vertical Wall, Mechatronics, Vol. 2, No. 5, pp. 543-554,

Pack, R., Christopher, J., Kawamura, K. (1997). A Rubbertuator-based Structure-climbing Inspection Robot, Proc. IEEE Int. Conf. Robotics and Automation, Vol. 3, pp. 1869-1874.

Bahr, B., Li, Y., Najafi, M. (1996). Design and Suction Cup Analysis of a Wall Climbing Robot, Comput. Electr. Eng., Vol. 22, No. 3, pp. 193-209.

Yano, T., Suwa, T., Murakami, M., Yamamoto, T. (1997). Development of a Semi Self-contained Wall Climbing Robot with Scanning Type Suction Cups, Proc. IEEE Int. Conf. Intelligent Robots and Systems, Vol. 2, pp. 900-905.

Abderrahim, M., Balaguer, C., Giménez, A., Pastor J.M., Padron, V.M. (1991). ROMA: A Climbing Robot for Inspection Operations, Proceedings 1999 IEEE International conference on Robotics and Automation, Detroit, Michigan, U.S.A., May, pp.2303-2308.

Sitti, M., Fearing, R. (2003). Synthetic Gecko Foot-robots, Proceeding of IEEE international Conference on Robotics and Automation, Vol. pp.1164-110.

Murphy, M. P., Tso, W., Tanzini,M., Sitti, M. (2006). Waalbot: An Agile Small-Scale Wall Climbing Robot Utilizing Pressure Sensitive Adhesives, Proceeding of the 2006 IEEE/RSJ International Conference on Intelligent Robots and System, Beijing, China, October 9-15, pp.3411-3416.

Wang, W., Zong, G. (1999). Analysis on The Mechanics Stability for a New Kind of Robot, Journal of Robot, Vol.21, No.7, pp.642-648.

Nishigami, M., Mizushima, T. (1992). Glass Roof Cleaning Robot System 'Canadian Crab', Journal of the robotics society of Japan, Vol.10, No.5, pp.40-42.

Zhang, H., Zhang, J., Liu, R., Zong, G. (2005). Realization of a Service Robot for Cleaning Spherical Surfaces, International Journal of Advanced Robotic Systems, Vol.2, No.1, pp. 53-58.

Granosik, G., Borenstein, J. (2005). Integrated Joint Actuator for Serpentine Robots, IEEE/ASME Transactions on Mechatronics, Vol. 10, pp. 473-481.

Wright, C., Johnson, A., Peck, A., McCord, Z. Naaktgeboren, A., Gianfortoni, P., Gonzalez-Rivero, M., Hatton, R., and Choset. H. (2007). Design of a modular snake robot. Proceedings of the IEEE/RSJ International Conference on Intelligent Robots and Systems, San Diego, U.S.A., October, 2007, pp. 2609–2614.

Spenko, M. J., Haynes, G. C., Saunders, J. A., Cutkosky, M. R., Rizzi, A. A., Full, R. J. (2008). Biologically Inspired Climbing with a Hexapedal Robot, Journal of Field Robotics, Vol. 25, No.4-5, pp. 223-242.

Pongas, D.; Mistry, M. Schaal, S.(2007). A Robust Quadruped Walking Gait for Traversing Rough Terrain, Proceeding of 2007 IEEE International Conference on Robotics and Automation, Los Angeles, CA, U.S.A., 10-14 April, 2007, pp.1474-1479.

Raibert, M., Blankespoor, K.; Nelson, G.; Playter, R. et al., (2008), BigDog, the Rough-Terrain Quadruped Robot, Proceedings of the 17th World Congress The International Federation of Automatic Control Seoul, Korea, July 6-11, 2008, pp.10822-10825.

Hirose, S. (1993). Biologically Inspired Robots (Snake-like Locomotor and Manupultor), Oxford University Press, 1993.

Togawa, K., Mori, M., Hirose, S. (2000). Study on Three-dimensional Active Cord Mechanism: Development of ACM-R2, Proceedings of the 2000 IEEE/RSJ International Conference on Intelligent Robots and Systems, Takamatsu, Japan, Oct.31-Nov.5, 2000, pp.2242-2247.

Chen, L., Wang, Y., Ma, S.(2004). Studies on Lateral Rolling Locomotion of a Snake Robot, Proceedings of the 2004 IEEE/RSJ International Conference on Intelligent Robots and Systems, Sendai, Japan, Sept.28-Oct.2, pp. 5070-5074, 2004.

Ute, J., Ono, K.(2002). Fast and Efficient Locomotion of a Snake Robot Based on Self-excitation Principle, Proceedings of 7th International Workshop on Advanced Motion Control, July 3-5, pp. 532- 539, 2002.

Miller, P. (2002). Snake Robots for Search and Rescue, Neurotechnology for Biomimetic Robots, MIT PRESS, pp. 271-284, 2002.

Conradt, J., Varshavskaya, P., (2003). Distributed Central Pattern Generator Control for a Serpentine Robot. Proceedings of the 13th International Conference on Artificial Neural Networks (ICANN), Istanbul, Turkey, June, 2003, pp.338-341.

Crespi, A., Badertscher, A., Guignard, A., Ijspeert, A. (2005). Swimming and Crawling with an Amphibious Snake Robot, Proceedings of the 2005 IEEE International Conference on Robotics and Automation, Barcelona, Spain, April 18-22, pp.3024-3028, 2005.

Chirjkjian, G., and Burdick, J. (1995). Kinematically Optimal Hyperredundant Manipulator Configurations. IEEE transactions on Robotics and Automation, Vol. 11, pp.794–806.

Lipkin, K., Brown, I., Peck, A., Choset, H., Rembisz, J., Gianfortoni, P., Naaktgeboren, A. (2007). Differentiable and Piecewise Differentiable Gaits for Snake Robot, Proceedings of IEEE/RSJ Intl. Conference on Intelligent Robots and Systems, San Diego, CA, USA, Oct. 29-Nov. 2, pp.1864–1869.

Goldman, G. and Hong, D. (2007). Determination of Joint Angles for Fitting a Serpentine Robot to a Helical Backbone Curve, Proceeding of 4th International Conference on Ubiquitous Robots and Ambient Intelligence, Pohang, S. Korea, Nov. 22-24, 2007.

Ijspeert, A.(1998). Design of Artificial Neural Oscillatory Circuits for the Control of Lamprey- and Salamander-like Locomotion Using Evolutionary Algorithms. PhD thesis, University of Edimburg, 1998.

Ijspeert, A.(2008). Central Pattern Generators for Locomotion Control in Animals and Robotics, Neural Networks, Vol. 21, pp. 642-653, 2008

Herrero Carrón, F.(2007). Study and Application of Central Patter Generators to the Control of a Modular Robot. Master's thesis, Escuela Politécnica Superior, Universidad Autónoma de Madrid, 30. August, 2007

Mezoff, S., Papastathis, N., Takesianm, A., Trimmer, B.A (2004). The Biomechanical and Neural Control of Hydrostatic Limb Movements in Manduca Sexta, J Exp Biol, Vol.207, pp. 3043-3054.

Zhang, H., Gonzalez-Gomez, J., Chen, S., Wang, W., Liu, R., Li, D., Zhang, J. (2007). A Novel Module Climbing Caterpillar Using Low-frequency Vibrating Passive Suckers, Proceeding of 2007 IEEE/ASME International Conference on Advanced Intelligent Mechatronics, ETH Zurich, Switzerland, pp.85-90.

Akinfiev, T., Armada, M.(2002). The Influence of Gravity on Trajectory Planning for Climbing Robots with Non-Rigid Legs, Journal of Intelligent and Robotic Systems, Vol.35, No.3, pp.309-326, 2002.

Wang, W., Wang, Y., Zhang, H., Wang, K., Zhang, J. (2008). Analysis of the Kinematics of Modular Climbing Caterpillar Robots, Proceeding of 2008 IEEE/ASME International Conference on Advanced Intelligent Mechatronics, Xi'an, China, 4-7 June, pp.84-89.

Wang, W., Zhang, H., Zhang, J., Zong, G. (2009). Crawling Gait Realization of Modular Climbing Caterpillar Robot, Submit to IROS2009, U.S.A.

2

RFID Technology for Mobile Robot Surveillance

Annalisa Milella, Donato Di Paola and Grazia Cicirelli
Institute of Intelligent Systems for Automation, National Research Council
Italy

1. Introduction

The increasing need for automated surveillance systems in indoor environments such as airports, warehouses, production plants, etc. has stimulated the development of intelligent systems based on mobile sensors. Differently from traditional non-mobile surveillance systems, those based on mobile robots are still in their initial stage of development, and many issues are currently open for investigation (Everett, 2003; Dehual et al., 2007).
The use of robots significantly expands the potential of surveillance systems, which can evolve from the traditional passive role in which the system can only detect events and trigger alarms, to active surveillance in which a robot can be used to interact with the environment, with humans or with other robots for more complex cooperative actions (Burgard et al., 2000; Vig & Adams, 2007).
A major challenge in surveillance tasks using mobile robots is that of providing the robot with a suitable knowledge of the environment to both navigate safely and perform inspection tasks. Furthermore, in order to effectively exploit mobility and multi-functionality, it is important to develop integrated control systems, capable of addressing simultaneously a number of problems, such as task planning, dynamic task sequencing, resolution of conflicts for shared resources, event-based feedback control.
These issues are part of our current research concerning the development of a multi-sensor Surveillance Mobile Robot (SMR). The SMR consists of a commercial mobile robot (see Fig. 1), which is equipped with various sensors including a laser rangefinder and a Radio Frequency IDentification (RFID) device, and takes advantage of a hybrid control architecture to implement both high-level functions, like mission execution monitoring, and low-level reactive control. So far, several behaviours and tasks have been implemented and experimented. In this chapter, we focus on environment mapping and exploration using RFID technology.
An RFID device typically consists of radio frequency (RF) tags, a reader with one or more antennas, and software to process the tag readings. The reader interrogates the tags, receiving their ID code and other information stored in their memory. Tags can be either passive or active. Passive tags are activated by the electromagnetic field generated by the RFID antenna. Active tags, instead, are powered by an on-board battery (Finkenzeller, 2003).
Applications of RFID include inventory management, industry automation, ID badges and

access control, equipment and personnel tracking. Compared to conventional identification systems, such as barcodes, RFID tags offer several advantages, since they do not require

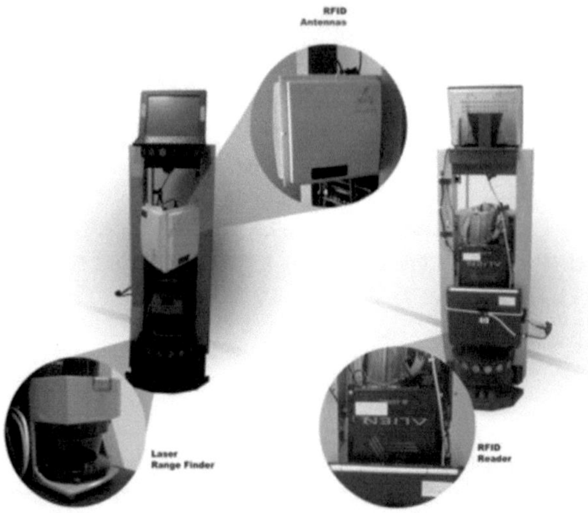

Fig. 1. The SMR equipped with RFID antennas and reader, and a laser rangefinder.

direct line-of-sight and multiple tags can be detected simultaneously. Owing to these properties, RFID has recently found its way in the mobile robotics field, promising to contribute new solutions to data association problems in basic navigation tasks, such as localization, mapping, goal reaching, and item detection (Kubitz et al., 1997; Tsukiyama, 2005; Schneegans et al., 2007; Milella et al. 2007). Nevertheless, in order for RFID sensors to be effectively used in mobile robotics applications, various issues have to be tackled. First, due to low cost and low power constraints, RFID devices are sensitive to interference and reflections from other objects. Therefore, RFID readings are generally affected by high uncertainty. Moreover, at least in the case of passive tags, an RFID reader can only determine whether a tag is present or not in its reading range, while it is not able to provide information about the position of the tag (Liu et al. 2006). These issues may be partially solved using active RFID (Kim & Chong, 2007; Chae & Han, 2005; Zhou et al., 2007). However, active transponders are more costly than passive ones and have a limited lifetime. Methods to localize passive tags, mostly based on probabilistic schemes, have been also developed by a few authors (Hähnel et al., 2004; Liu et al., 2006; Jia et al. 2006).
In this chapter, we present an alternative approach to passive RFID using fuzzy logic. Specifically, we present a fuzzy model of the RFID system, which describes the detection range of an antenna, as its distance and orientation relative to the tags vary. Based on this model, we develop two algorithms that allow a mobile robot equipped with RFID reader and antennas to localize passive tags deployed in the environment. The first one, named Fuzzy Tag Localization (FTL), aims at localizing accurately passive tags in the environment, generating what we refer to as an *RFID-augmented* map, i.e. a map of the environment enriched with RFID tags. Such a map can serve as a support for a variety of service robot tasks, like detecting items, obtaining information about the robot position, getting instructions to reach a given goal. The second method is called Fuzzy Tag Bearing

Estimation (FTBE). It allows one to estimate the bearing of a tag with respect to the mobile robot. The approach is suited when only the tag bearing relative to the robot is needed, like in some landmark-based self-localization algorithms (Stella & Distante, 1995; Milella et al., 2008), or when an approximate knowledge of the tag location is sufficient. The general use of both methods is in object identification and localization, map building, environment monitoring, and robot pose estimation.

Experimental verification of the proposed techniques has been performed in the ISSIA CNR Mobile Robotics Laboratory. The results suggest that both approaches are accurate in localizing RFID tags and can be effectively integrated with robotic mapping and surveillance systems.

The remainder of the chapter is organized as follows. After discussing related work in Section 2, the tag localization algorithms are detailed in Section 3. Section 4 provides a description of our SMR. Section 5 illustrates experimental results. Conclusions are drawn in Section 6.

2. Related Work

In the last decade, several worldwide projects have attempted to develop mobile surveillance platforms. A notable example is the Mobile Detection Assessment and Response System (MDARS) (Everett & Gage, 1999). The aim of this project was that of developing a multi-robot system able to inspect warehouses and storage sites, identifying anomalous situations, such as flooding and fire, detect intruders, and determine the status of inventoried objects using specialized RF transponders. In the RoboGuard project (Birk & Kenn, 2001), a semi-autonomous mobile security device uses a behaviour-oriented architecture for navigation, while sending video streams to human watch-guards. The Airport Night Surveillance Expert Robot (ANSER) (Capezio et al., 2005) consists of an Unmanned Ground Vehicle (UGV) using non-differential GPS unit for night patrols in civilian airports and similar wide areas, interacting with a fixed supervision station under control of a human operator. A Robotic Security Guard (Duckett et al., 2004) for remote surveillance of indoor environments has been also the focus of a research project at the AASS Learning Systems Laboratory, Örebro University, Sweden. The objective of this project was that of developing a mobile robot platform able to patrol a given environment, acquire and update maps, keep watch over valuable objects, recognise known persons, discriminate intruders from known persons, and provide remote human operators with a detailed sensory analysis. Another example of security robot is the one developed at the University of Waikato, Hamilton, New Zealand (Carnegie et al., 2004). It is named MARVIN (Mobile Autonomous Robotic Vehicle for Indoor Navigation) and has been designed to act as a security agent in indoor environments. In order for the robot to interact with humans, it is provided with speech recognition and speech synthesis software as well as with the ability to convey emotional states, verbally and non-verbally.

A current trend in the field of automated security is the growing use of RFID technology for object identification and tracking. Recently, several authors have shown RFID tags to be also effective in mobile robot navigation tasks, such as localization and mapping. For instance, the use of RFID as landmarks in topological maps for mobile robot navigation is proposed in (Tsukiyama, 2005). In (Kulyukin et al. 2004), passive RFID tags are manually attached to objects in an indoor environment to trigger local navigation behaviours of a mobile robot for

visually impaired assistance. Robot localization methods using particular tag arrangement on the floor are described in (Jing & Yang, 2007) and (Lim et al., 2006).
Although these methods are all effective in supporting mobile robot navigation, they either assume the location of the tags to be known a priori or require the tags to be installed to form specific patterns. This is reasonable in some industrial applications, while in office or home environments it is generally difficult to measure tag positions, and arrangement of multiple tags could turn into a difficult task. Moreover, tagged objects could be displaced, causing the necessity to recalculate their position. Hence, methods for localizing automatically RFID tags in the environment are generally desirable.
One of the first works dealing with the automatic localization of passive tags is the one in (Hähnel et al., 2004). The authors suggest a particle filtering method for localizing passive tags in a previously built map of the environment, using a mobile robot equipped with an RFID device and a laser rangefinder. Specifically, while the robot moves in the environment, the location of a tag is estimated starting from a set of particles, whose weights are updated at each successful detection of the tag, using the Bayes rule and a probabilistic model of the antenna. Bayesian solutions for tag localization are also adopted in (Liu et al., 2006; Jia et al. 2006; Alippi et al. 2006). In (Liu et al., 2006), two RFID tag-positioning algorithms are developed, namely an online approach and an offline approach. The offline method is equivalent to the one proposed in (Hähnel et al., 2004). The online algorithm is based on a simplified antenna model that defines a high probability region, instead of describing the probability at each location, in order to achieve computational efficiency. In (Jia et al. 2006), RFID tags are used for obstacle detection and avoidance. The Bayes rule is applied to estimate tag positions. Tags are also used as landmarks for robot localization based on visual input from a stereovision device. In (Alippi et al. 2006), the tag localization algorithm is formalized as a non-linear stochastic inversion problem. Several readers, equipped with rotating antennas, take observations. The reading units are connected in a local network with a server, which gathers the data and executes the localization task.
In this chapter, we introduce two novel methods for localizing passive tags. The algorithms are integrated in our surveillance mobile platform. As in (Hähnel et al., 2004), we refer to a model of the antenna reading range for estimating tag locations from a mobile robot. However, our approach is unique in that it uses fuzzy reasoning to both learn a model of the RFID system and localize the tags. Fuzzy logic has been widely recognized for its effectiveness and for the simplicity to define and understand the knowledge representation. It is especially useful when the process under analysis is complex, when the available source of information is inexact or uncertain, or for intelligent sensor integration and fusion (Mahajan et al., 2001). Our work shows that fuzzy logic is appropriate to deal with uncertainty in RFID systems.

3. Tag localization using fuzzy logic

Passive tags are not able to directly provide their location relative to the antenna or a distance measure. Only positive or negative responses whether a tag is present or not in the reading range are generated. Yet, positive readings can be used to estimate the tag position. As a matter of fact, a positive response reduces the potential locations of the tag to those that lie in the reading region of the device. Further improvement in tag position estimate can be achieved by considering that, whenever a tag is present in the reading range, the reader will detect it with a certain likelihood. Specifically, it has been shown that a tag closer to the centroid of the reading range is detected more frequently than a tag located at the boundary

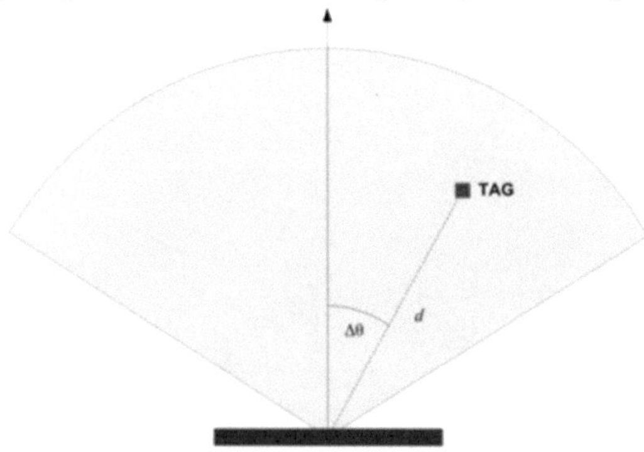

Fig. 2. Representation of bearing and range of a tag relative to the antenna.

(Liu et al., 2006). In summary, a successful detection provides a region that is likely to contain the tag and also allows the association of a detection rate to each point of the region. This region is usually referred to as the coverage map of the RFID device, and constitutes the observation or sensor model in probabilistic approaches.

Constructing an observation model for passive RFID systems is not trivial. RFID are sensitive to interference and reflections from the surroundings. The position of the tag relative to the receiver also influences the result of the detection process, since the absorbed energy varies accordingly and may become too low to power the chip inside the tag, causing the tag to not respond. These undesirable effects produce a number of false negative and false positive readings that lead to an incorrect idea about the tag location and, eventually, could compromise the overall performance of the system (Hähnel et al., 2004; Brusey et al., 2003). It is not feasible to explicitly account for all these factors, separately. Instead, a widely used approach to generate a model of the RFID device is that of manually mapping the probability of detecting a tag at different offsets from the reader by counting the detection frequencies over a 2D or a 3D grid of the environment. That leads to the construction of likelihood histograms, which are, then, typically, conservatively approximated with discrete models, consisting of two or, at most, three likelihood regions (Liu et al., 2006; Hähnel et al., 2004; Jia et al. 2006).

In this work, we propose a fuzzy logic solution to build a better approximation of the antenna detection field, though preserving computational simplicity. Then, based on this model, we develop two algorithms for estimating the position of passive tags using fuzzy reasoning: the Fuzzy Tag Localization (FTL) algorithm and the Fuzzy Tag Bearing Estimation (FTBE) algorithm.

In the rest of this section, first, we describe the fuzzy antenna model, then, we introduce the FTL and the FTBE methods.

3.1 Fuzzy Antenna Modeling

As a first step for RFID modelling, we generated a statistic histogram for our RFID system. Specifically, we rotated the robot in front of a tag, at different distances, several times, and we counted the number of successful detections for each pose in a discrete 2D grid. It was found that the coverage map of each antenna has approximately the shape of a sector with a radius of about 2.5*m* and an angular aperture of about 120°. Moreover, it was observed that detection rates tend to decrease smoothly at the boundaries of the coverage map.

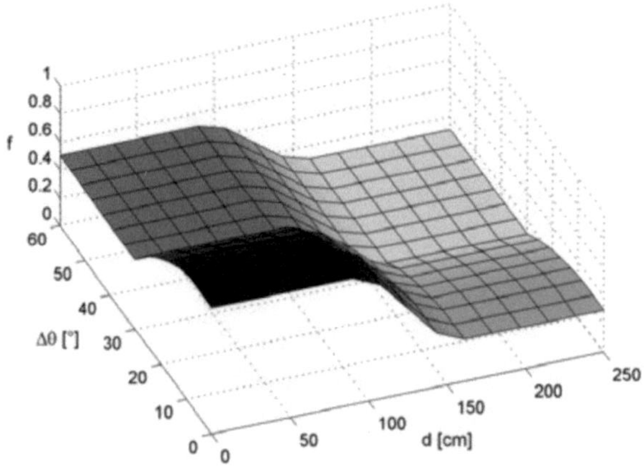

Fig. 3. Input-output surface of the fuzzy antenna model (see Table 1 for the fuzzy rules): darker grey denotes higher frequency of detection.

Such a result can be easily expressed using fuzzy logic. Specifically, we employ a zero-order Sugeno fuzzy inference system (Sugeno & Yasukawa, 1993) with two inputs and one output. With reference to the notation of Fig. 2, the inputs are the range d and the bearing $\Delta\theta$ of the tag relative to the antenna. The output f is an index defined in [0, 1] expressing the expected occurrence of detection, which we refer to as the frequency of detection of the tag. Two functions are defined for each input, labelled *Low* and *High*, respectively. The output, instead, consists of four constant values, labelled *Very Low* (VL), *Low* (L), *Medium* (M), and *High* (H). The parameters for these functions were tuned based on experimental data. The output f is given by the weighted average of all rule outputs. The *if-then* rules for fuzzy inference are reported in Table 1. They consist of heuristic rules, such as

IF Range (d) IS "Low" AND Bearing (Δθ) IS "Low" THEN Frequency (f) IS "High"

A representation of the input-output surface of the fuzzy logic system is shown in Fig. 3, with darker grey representing higher frequencies of detection.

3.2 Fuzzy Tag Localization (FTL)

In this section, we describe our approach to localize passive tags in the environment. The main idea underlying the proposed method is that of estimating the position of a tag as the

Rule #	Input 1: Tag Range (d)	Input 2: Tag Bearing (Δθ)	Output: Frequency of Det. (f)
1	High	High	Very Low
2	High	Low	Low
3	Low	High	Medium
4	Low	Low	High

Table 1. Fuzzy logic rules for modelling the RFID reading range.

most likely location among a set of potential locations. Specifically, as a tag is detected, a set of points P_j, for $j = 1, 2, \ldots, M$, is generated in a circular area around the current robot

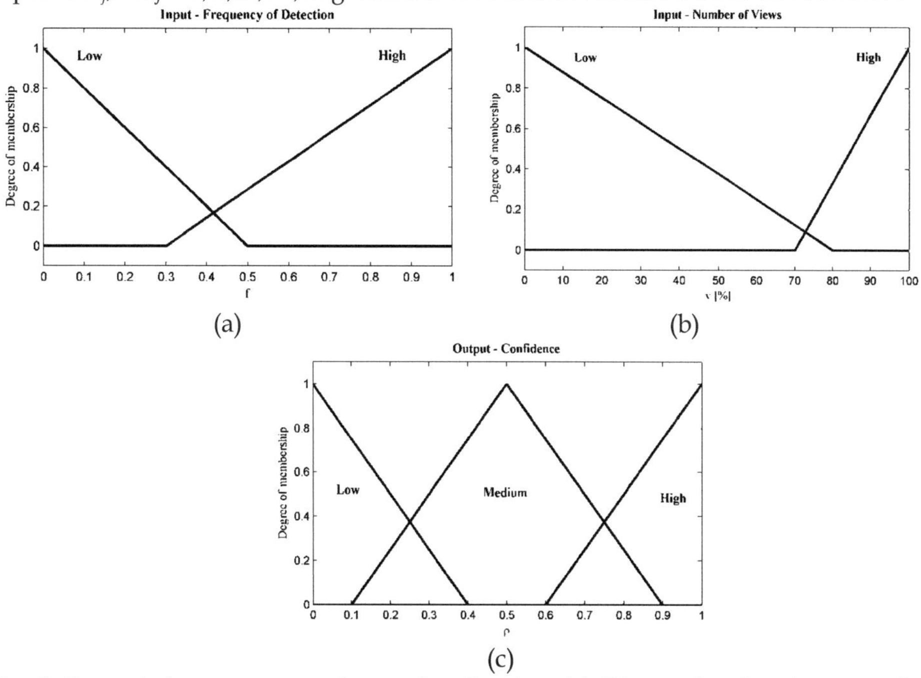

Fig. 4. Fuzzy inference system for tag localization: (a)-(b) membership functions for input variables, i.e. frequency of detection and number of views; (c) membership functions for the output variable, i.e. confidence.

position. The robot, then, moves around, performing multiple tag detections from different positions. It is assumed that the robot displacement from one position to another is known. At each new detection, a confidence value is assigned to every point P_j, expressing the likelihood that P_j corresponds to the actual tag location.

Our hypothesis in confidence estimation is that the higher is the detection frequency associated to a point according to the fuzzy antenna model, the higher is the possibility for that point of being the actual tag position. Furthermore, we assume that a point is more likely to correspond to the actual tag location if it belongs to the intersection region of the coverage maps drawn for the various robot poses during the localization procedure.

In order to express these hypotheses, we adopt fuzzy logic. The triangular membership functions used are shown in Fig. 4 (a)-(b) and Fig. 4 (c), for input and output variables,

Rule #	Input 1: Frequency (f)	Input 2: Num. of views (v)	Output: Confidence (ρ)
1	High	High	High
2	High	Low	Medium
3	Low	High	Medium
4	Low	Low	Low

Table 2. Fuzzy logic rules for tag localization.

respectively. The inputs are the detection frequency f_j^i associated to the point P_j at the *i-th* detection, which depends on the position of the point relative to the antenna, and the parameter v_j^i, expressing the number of times the point has been found to lie in the antenna detection area. The output is the confidence level ρ_j^i associated to the point P_j at the *i-th* iteration. The *if–then* rules for fuzzy inference are reported in Table 2.

To reduce the set of potential tag locations, each point P_j, is finally assigned an average confidence level. This is computed as the mean value of the confidence levels calculated for the same point in all the previous steps. Only the points whose average confidence value is greater than a threshold are retained. This process allows us to remove progressively, from the set of potential tag locations, those points which have low possibility of being the actual tag position, thus refining the estimate.

It is worth to note that if a map of the environment is available and the robot pose in the map is known from some global positioning system, then the described procedure allows us to localize the tags in the map. That leads to what we call an *RFID-augmented* map. Such a map may provide useful information about the environment in a simple form, since RFID tags can store data either to describe interest objects and regions, or to support mobile robot navigation tasks. In Section 5.1, we will show that such a map can improve the robot capability of performing environment monitoring tasks.

3.2 Fuzzy Tag Bearing Estimation (FTBE)

As a variation of the FTL method, we propose an algorithm to estimate only the bearing of a tag relative to the mobile robot, disregarding the range. This method is referred to as Fuzzy Tag Bearing Estimation (FTBE).

The strategy adopted is similar to the one used in the FTL approach. The main difference is that, since only the bearing of the tag has to be estimated, the points P_j representing the potential locations of the tag can be generated at a unique radial distance from the robot, arbitrarily chosen inside the antenna detection field, rather than in a predefined area around the vehicle. That leads to higher computational efficiency, making it more feasible an on line implementation of the approach. In addition, once the tag has been detected for the first time, the robot is not required to move around to perform multiple detections of the tag, but it has just to turn in place. Alternatively, a rotating antenna may be used.

With reference to the notation of Fig. 5, let us indicate with P_j, for $j = 1, 2,\ldots, M$, the points generated at the first detection of the tag, distributed at regular angular intervals and fixed radial distance r from the current robot position. Each point allows us to define a vector RP_j whose angle φ_j relative to the X_r- axis of a reference frame (R, X_r, Y_r) attached to the robot, represents a potential value for the tag bearing.

Once the point set has been generated, the robot starts to turn in place, while the reader continues to interrogate the tag. Every time a positive response is received, for each point P_j that falls in the antenna detection area, a frequency value is computed, based on the antenna model. If we approximate the centre of the antenna detection field with the centre of rotation of the vehicle, the inputs to the fuzzy antenna model for every point P_j at the *i-th* reading are

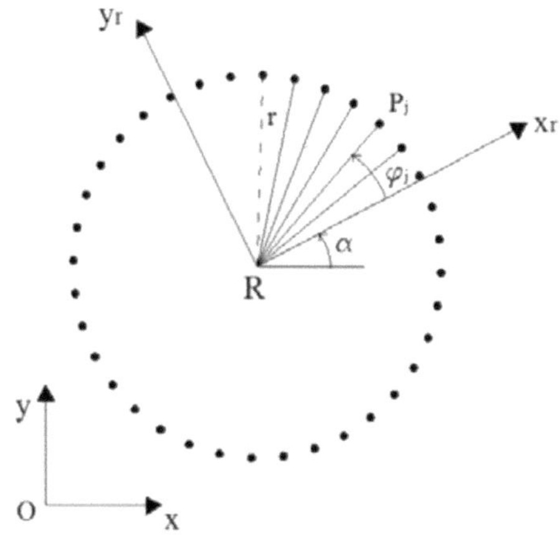

Fig. 5. Potential directions φ_j for estimating the bearing of a tag with respect to the robot reference frame (R, X_r, Y_r).

$$d_j^{\,i} = r \tag{1}$$

$$\Delta\theta_j^{\,i} = \varphi_j + \delta^i \tag{2}$$

where δ^i depends on the known orientation of the antenna relative to the vehicle (60° in our case) and on the robot rotation angle α. The output is the frequency of detection f_j^i.
Since we only have to manage a limited number of points (e.g. 180 points for a set of points generated at angular interval of 2°), we do not need to discard points at each novel reading. Instead, frequency values f_j^i are stored in a vector

$$[f_j^1 \quad f_j^2 \quad \dots \quad f_j^i \dots \quad f_j^{N_j}] \tag{3}$$

so that, at the end of the acquisition phase, for each point, we can calculate an average detection frequency value

$$\bar{f}_j = \frac{1}{N_j}\sum_{i=1}^{N_j} f_j^i \tag{4}$$

Furthermore, we can compute a parameter v_j

$$v_j = N_j / N \tag{5}$$

where we have denoted with N_j the dimension of the frequency vector for RP_j, which also represents the number of times the point P_j has fallen inside the antenna detection field, and with N the total number of detections. Then, similarly to what is done in the FTL module, fuzzy reasoning is used for confidence level computation. A two inputs-one output fuzzy inference system is employed. In order to eliminate the dependency of the frequency values on the chosen radius r, we normalize the average frequencies with respect to their maximum value f_j^{max}. Then, we compute the inputs to the fuzzy inference system for a point P_j which are $\bar{f}_j / \bar{f}_j^{\max}$ and v_j. The output is the confidence level ρ_j. The membership functions and the *if-then* rules are formally similar to those shown for the FTL module in Fig. 4 and Table 2, respectively. Only the bearing values with an associated confidence level higher than a threshold are retained. Finally, the median of these values is calculated and is assumed as the bearing measure.

It is worth to note that if the tag bearing estimate is performed from at least two different robot positions and the displacement of the robot from one position to the other is known, also the range of the tag relative to the robot can be computed by triangulation. Alternatively, if the tag position in the environment is known, the knowledge of the tag bearing relative to the robot may be exploited to get robot position information.

4. The mobile robot surveillance platform

To prove the feasibility of the proposed methods, we developed an RIFD managing and processing system integrated in our Surveillance Mobile Robot (SMR) (see Fig. 1).

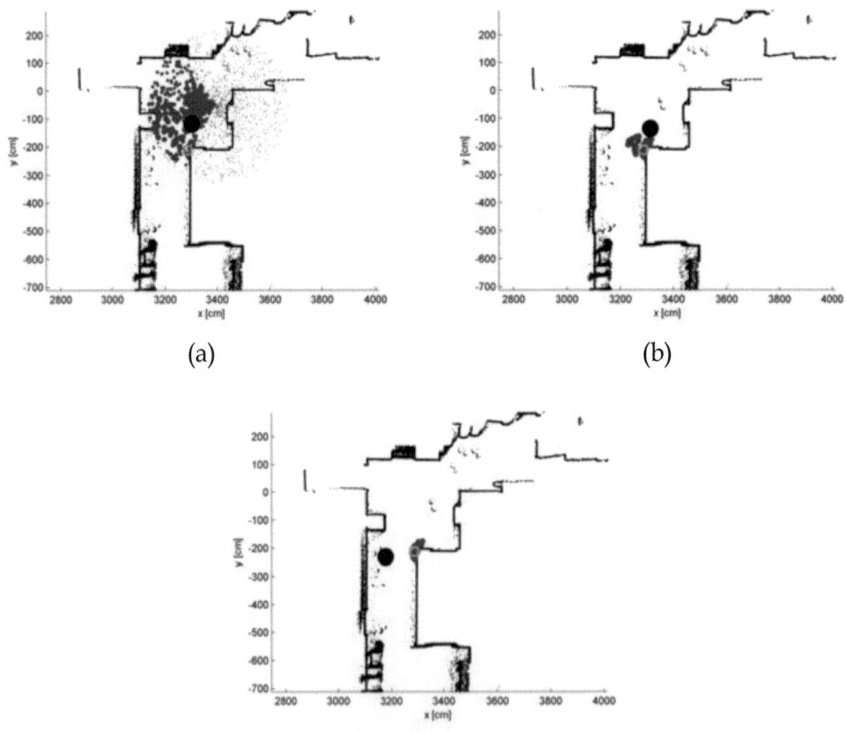

Fig. 6. Tag localization after (a) 5, (b) 15, and (c) 50 iterations using the FTL approach. In (a), (b), (c): a black circle represents the robot pose in the map; the actual tag position is indicated by a square; potential tag locations are shown as small circles with a radius proportional to the average confidence levels.

The latter consists of a commercial platform, i.e. PeopleBot by MobileRobots Inc., equipped with sonar and infrared sensors, a SICK LMS-200 laser range finder, an AVT Marlin IEEE 1394 FireWire monocular camera, and an Alien Technology's ALR-8780 reader with two external circularly polarized ALR 8610-C antennas. The RFID system works at 866MHz. Passive Alien's Class 1 128-bit NanoBlock tags are employed. They consist of rectangular targets with long side of about 10cm, containing, internally, an antenna for communication with the reader, and a microchip, which stores the ID code. Communication between the reader and the tags is performed through backscatter modulation.

The system has two processing units: the robot embedded PC and an additional laptop for RFID data acquisition and storage, vision processing, and application control. The software is based on an open source robotic development framework, named MARIE (Côté et al., 2006). The Java libraries provided by Alien Technology are employed for RFID data acquisition and storage.

The SMR is designed to perform active environment surveillance operations, taking advantage of an extremely modular software architecture that allows several sensor processing modules to be easily integrated. The robot is able to execute various missions. Each mission consists of a predetermined set of atomic tasks, which may be subject to precedence constraints (the end of one task may be a necessary prerequisite for the start of some other ones). Specifically, to control the robot and manage the surveillance tasks, we use a Matrix-based Discrete Event Control (M-DEC) approach (Tacconi & Lewis 1997). M-DEC is an integrated modelling and control framework based on Boolean matrices and vectors, which is intuitive (it can be interpreted using simple *if-then* rules), inherently modular (models of large systems can be obtained by assembling submodels of smaller size) and versatile. More details concerning the control system of the platform can be found in (Di Paola et al. 2009).

5. Experimental results

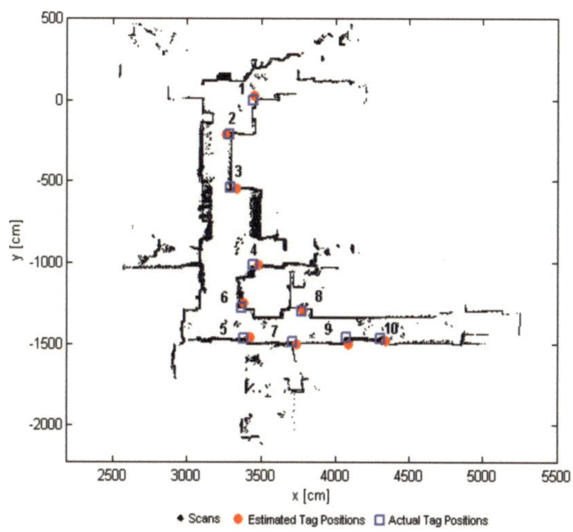

Fig. 7. Map of the environment with the tag locations estimated using the FTL module and those measured by a theodolite station.

Experiments were performed in the ISSIA CNR Mobile Robotics Laboratory of Bari, Italy, using the mobile platform described in Section 4. In the rest of this section, first, we show the results of tests concerning the FTL approach and the related RFID-augmented mapping, then, we present the tests carried out to verify the accuracy of the FTBE approach.

5.1 Fuzzy Tag Localization for RFID-Augmented Mapping

Ten tags were distributed in the environment, along an L-shaped corridor with a total length of about 40*m* and an average width of about 2*m*. Then, the robot was guided on a tour of the environment, acquiring laser and RFID data.

Both the geometric map of the environment and the robot trajectory were reconstructed using a laser-based SLAM routine.

Note that information concerning different tags can always be kept separate since a tag is univocally identified by its own code. Hence, at the end of the acquisition phase, for each tag, a set of robot poses is available for tag location estimate. Measurements of the tag positions were also performed with a theodolite station and were regarded as the ground truth.

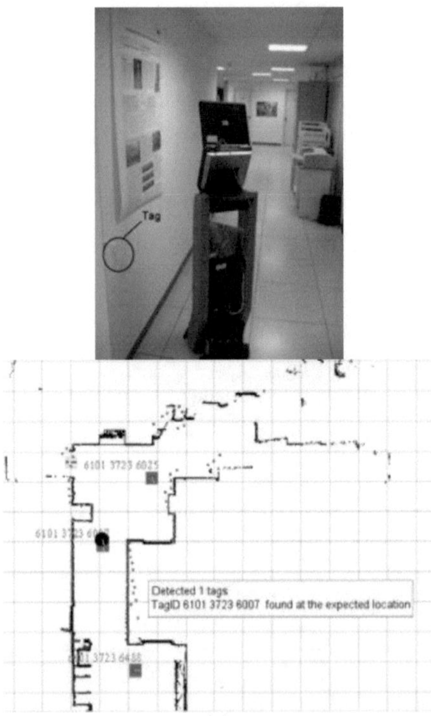

(a)

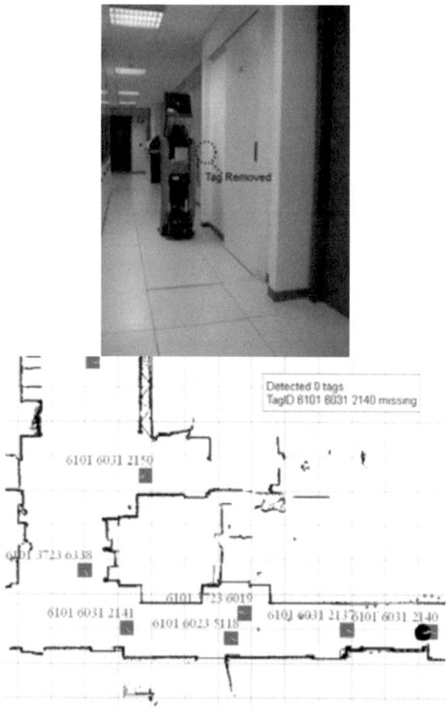

(b)

Fig. 8 RFID-based environment monitoring: (a) the robot detecting a tag; (b) the robot warning of a missing tag removed from its original location.

Fig. 6 shows the localization procedure using the FTL method for one of the tags. Whenever the tag is detected for the first time, a set of potential locations is generated in a circular area around the current robot position. As a new observation occurs, only those points whose confidence is greater than a threshold are retained. Then, at each step, the tag position is estimated as the weighted average of the residual points. In order to reduce the risk of eliminating valid points, at least ten robot poses are considered for the computation of average confidence levels before points are discarded for the first time. Fig. 6 (a) shows the sample set after 5 detections, while Fig. 6 (b) and Fig. 6 (c) display the distribution of the possible locations after 15 and 50 detections, respectively, showing how the estimate converges toward the actual tag position. In this test, after 200 detections, the error was of $10.0cm$. Similar results were obtained for all the tags.

Fig. 7 shows the map of the environment reconstructed by SLAM with overlaid the locations of the tag estimated by the FTL method and those measured using the theodolite station. We used, for each tag, 200 detections. The algorithm was initialized with 1500 samples. The average discrepancy between the tag positions estimated by the FTL algorithm and those measured using the theodolite was less than $35cm$ and in the worst-case measurement the discrepancy was less than $50cm$. These results demonstrate that the FTL method is accurate in localizing tags deployed at generic locations of an indoor environment, with the additional advantage of relying on simple fuzzy rules defined in the universe of discourse.

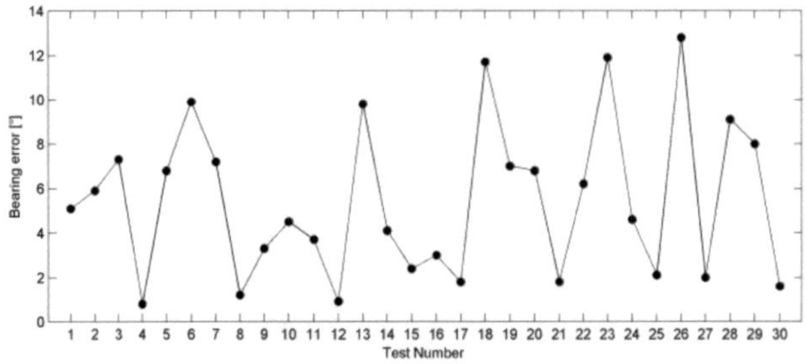

Fig. 9. Tag bearing error, estimated starting from 30 different robot poses around a tag, at distances ranging from 50*cm* to 170*cm* and orientations comprised between -90°÷90°.

The constructed RFID-augmented map can be used to support robot navigation tasks. For instance, based on RFID, a mobile robot can acquire information about its pose in the environment. Vice versa, knowing its pose from other sensors, it can use RFID information for environment monitoring.

As an example, Fig. 8 shows a navigation test, in which the robot was programmed to move in the laboratory using the constructed RFID-augmented map, in order to perform a typical surveillance task, based on the concept of goal points. A goal point is a location of the environment from which the robot observes the scene. Goal points were fixed in proximity of the tags. The robot was programmed to reach each goal and verify the presence or absence of the tag at the expected location. The pictures on the left portray the robot navigating in the environment. On the right, the output of the navigation module is shown with overlaid the output of the RFID system. Goal stations are denoted with small squares. ID codes of the tags located nearby each station are also reported.

5.2 Fuzzy Tag Bearing Estimation

In order to verify the accuracy of the FTBE approach, a first experimental session was carried out attaching a tag to a wall of the laboratory and running the algorithm from 30 different robot configurations around the tag. Tag range and bearing relative to the robot were comprised between 50*cm*÷170*cm* and -90°÷90°, respectively. In all tests, an angular interval of 2° was used for both data acquisition and point set generation.

The results were compared with ground-truth measures obtained using a theodolite station. Fig. 9 reports a graph of the bearing error e_b (angle β in Fig. 10) computed as the absolute difference between actual φ_a and estimated φ_e tag bearing, i.e.

$$e_b = |\varphi_a - \varphi_e| \tag{6}$$

It shows that the error has an average value of 5.4°, and is less than 13.0° in the worst case. Similar results were obtained in successive tests, performed using other tags wherever located in the environment, thus proving the effectiveness of the proposed approach.

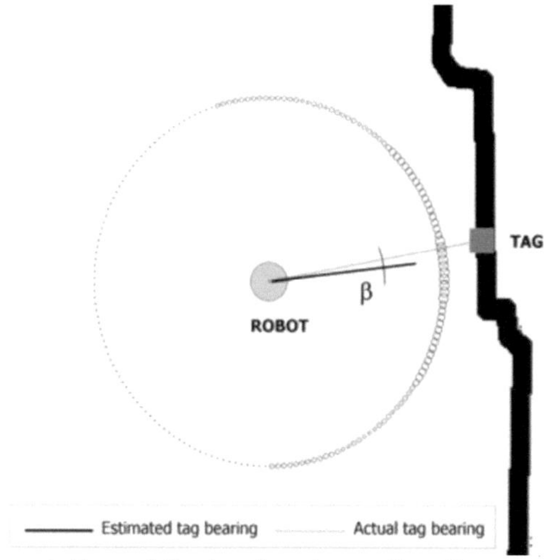

Fig. 10 Graphical representation of the tag bearing estimation process. The tag (light grey square) is attached on a wall of the environment (thick black line). Small circles located at $1m$ from the robot represent the points used for generating the potential tag bearings. The dimension of each circle is proportional to its confidence level.

6. Conclusion

We introduced two novel methods to estimate the location of passive RFID tags using a surveillance mobile robot equipped with RF reader and antennas, and a laser rangefinder. Both approaches are based on fuzzy logic and use a fuzzy model of the RFID system.
The first method, referred to as Fuzzy Tag Localization (FTL), allows us to accurately localize tags with respect to the robot. The positions of the tags in a map of the environment are also estimated, using a laser-based SLAM approach to globally localize the robot in the environment. That leads to what is called an RFID-augmented map. The second method, named Fuzzy Tag Bearing Estimation (FTBE), aims at estimating the bearing of a tag relative to the mobile robot. It can be used when an approximate knowledge about the tag location is needed, or only the bearing information is required, such as in some robot localization methods.
Experimental tests performed in a real indoor environment were presented to demonstrate the effectiveness of both methods. It was shown that fuzzy logic is appropriate in knowledge representation under uncertainty in RFID systems for mobile robotics applications like environment exploration and mapping.

7. References

Alippi, C., Cogliati, D., & Vanini, G. (2006). A statistical Approach to Localize Passive Tags, *Proceedings of IEEE International Symposium on Circuits and Systems (ISCAS)*, ISBN: 0-7803-9389-9, Island of Kos, Greece, May 2006.

Birk, A. & Kenn, H. (2001). An Industrial Application of Behavior-Oriented Robotics, *Proceedings of IEEE International Conference on Robotics and Automation*, Vol. 1, pp. 749 - 754, ISBN: 0-7803-6576-3, Seoul, Korea, May 2001.

Brusey, J., Floerkemeier, C., Harrison, M., & Fletcher, M. (2003). Reasoning about Uncertainty in Location Identification with RFID, *Proceedings of IJCAI-2003 Workshop on Reasoning with Uncertainty in Robotics,* ISBN: 0-127-05661-0, Acapulco, Mexico, August 2003, Elsevier Science Publishers.

Burgard, W., Moors, M., Fox, D., Reid, S., & Thrun, S. (2000). Collaborative multi-robot exploration, *Proceedings of IEEE International Conference on Robotics and Automation,* Vol. 1, pp. 476-481, ISBN: 0-7803-5886-4, San Francisco, CA, USA, April 2000.

Capezio, F., Sgorbissa, A., & Zaccaria, R., (2005). GPS-Based Localization for a Surveillance UGV in Outdoor Areas, *Proceedings of the Fifth International Workshop on Robot Motion and Control (RoMoCo'05),* pp. 157-162, ISBN: 83-7143-266-6, Poland, June 2005.

Carnegie, D. A., Prakash, A., Chitty, C., & Guy, B. (2004). A human-like semi-autonomous Mobile Security Robot, *2nd International Conference on Autonomous Robots and Agents,* ISBN: 0-476-00994-4, Palmerston North, New Zealand, December 2004.

Chae, H. & K. Han, K. (2005). Combination of RFID and Vision for Mobile Robot Localization, *Proceedings of the 2005 International Conference on Intelligent Sensors, Sensor Networks and Information,* pp. 75- 80, ISBN: 0-7803-9399-6, Melbourne, Australia, December 2005.

Côté, C., Brosseau, Y., Létourneau, D., Raïevsky, C., & Michaud, F. (2006). Robotic software integration using MARIE, *International Journal of Advanced Robotic Systems,* ISSN: 1729-8806, Vol. 3, No. 1, pp. 55-60, March 2006.

Duckett, T., Cielniak, G., Andreasson, H., Jun, L., Lilienthal, A., Biber, P., & Martínez, T. (2004). Robotic Security Guard - Autonomous Surveillance and Remote Perception, *Proceedings of IEEE International Workshop on Safety, Security, and Rescue Robotics,* Bonn, Germany, May 24-26, 2004.

Dehual, Z., Gang, X., Jinming, Z., & Li, L. (2007). Development of a mobile platform for security robot, *Proceedings of IEEE International Conference on Automation and Logistics,* pp. 1262–1267, ISBN: 978-1-4244-1531-1, August 2007.

Di Paola, D., Naso, D., Turchiano, B., Cicirelli, G., & Distante, A. (2009). Matrix-based Discrete Event Control for Surveillance Mobile Robotics, (to appear in) *Journal of Intelligent and Robotic Systems,* Springer, 2009.

Everett, H.R. & Gage, D.W., (1999). From Laboratory to Warehouse: Security Robots Meet the Real World, *International Journal of Robotics Research, Special Issue on Field and Service Robotics,* Vol. 18, No. 7, July 1999, pp 760-768, ISSN: 0278-3649.

Everett, H. (2003). Robotic security systems, *IEEE Instruments and Measurements Magazine,* ISSN: 1094-6969, Vol. 6, No. 4, pp. 30–34, December 2003.

Finkenzeller, K. (2003). *RFID Handbook: Fundamentals and Applications in Contactless Smart Cards and Identification,* Second Edition, Wiley, ISBN 13: 978-0-470-84402-1, England.

Hähnel, D., Burgard, W., Fox, D., Fishkin, K., & Philipose, M. (2004). Mapping and Localization with RFID Technology, *Proceedings of the IEEE International Conference on Robotics and Automation,* Vol. 1, pp. 1015-1020, ISBN: 0-7803-8232-3, New Orleans, LA, USA, April 2004.

Jia, S., Shang, E., Abe, T., & Takase, K. (2006). Localization of Mobile Robot with RFID Technology and Stereo Vision, *Proceedings of IEEE International Conference on Mechatronics and Automation,* pp. 508-513, ISBN: 1-4244-0465-7, Luoyang, China, June 2006.

Jing, L. & Yang, P. (2007). A Localization Algorithm for Mobile Robots in RFID System, *Proceedings of International Conference on Wireless Communications, Networking and Mobile Computing*, pp. 2109-2112, ISBN: 978-1-4244-1311-9, Shanghai, China, September 2007.

Kim, M. & Chong, N.Y. (2007). RFID-based mobile robot guidance to a stationary target, *Mechatronics*, Vol. 17, No. 4-5, May-June 2007, pp. 217-229, 2007, ISSN: 0957-4158.

Kubitz, O., Berger, M., Perlick, M., & Dumoulin, R. (1997). Application of Radio Frequency Identification Devices to Support Navigation of Autonomous Mobile Robots, *Proceedings of IEEE 47th Vehicular Technology Conference*, Vol. 1, pp. 126-130, ISBN: 0-7803-3659-3, Phoenix, AZ, USA, May 1997.

Kulyukin, V., Gharpure, C., Nicholson, J., & Pavithran, S. (2004). RFID in robot-assisted indoor navigation for the visually impaired, *Proceedings of IEEE/RSJ International Conference on Intelligent Robots and Systems*, Vol. 2, pp. 1979-1984, ISBN: 0-7803-8463-6, Sendai, Japan, September-October 2004.

Lim, H., Choi, B., & Lee, J. (2006). Efficient Localization Algorithm for Mobile Robots based on RFID Systems, *Proceedings of SICE-ICASE International Joint Conference*, pp. 5945-5950, ISBN: 89-950038-4-7, Bexco, Busan, Korea, October 2006.

Liu, X., Corner, M.D., & Shenoy, P. (2006). Ferret: RFID Localization for Pervasive Multimedia, *Proceedings of the 8th UbiComp Conference*, pp. 422-440, ISSN: 0302-9743, Orange County, California, September 2006, Springer Berlin.

Mahajan, A., Wang, K., & Ray, P.K. (2001). Multisensor Integration and Fusion Model that Uses a Fuzzy Inference System, *IEEE/ASME Transactions on Mechatronics*, Vol. 6, No. 2, June 2001, pp. 188-196, ISSN: 1083-4435.

Milella, A., Vanadia, P., Cicirelli, G., & Distante, A. (2007). RFID-based environment mapping for autonomous mobile robot applications, *Proceedings of IEEE/ASME International Conference on Advanced Intelligent Mechatronics*, pp. 1-6, ISBN: 978-1-4244-1264-8, September 2007.

Milella, A., Di Paola, D. Cicirelli, G., & D'Orazio, T. (2008). Using fuzzy RFID modelling and monocular vision for mobile robot global localization, *Proceedings of IEEE/RSJ Conference on Intelligent Robots and Systems*, pp. 4204-4204, ISBN: 978-1-4244-2057-5, September 2008.

Schneegans, S., Vorst, P., & Zell, A. (2007). Using RFID Snapshots for Mobile Robot Self-Localization, *Proceedings of the 3rd European Conference on Mobile Robots*, pp. 241-246, Freiburg, Germany, September 2007.

Stella, E. & Distante, A. (1995). Self-Location of a mobile robot by estimation of camera parameters, *Robotics and Autonomous Systems*, Vol. 15, No. 3, pp. 179-187, ISSN: 0921-8890, Elsevier, Amsterdam, Pays-Bas.

Sugeno, M. & Yasukawa, T. (1993). A fuzzy-logic-based approach to qualitative modeling, *IEEE Transactions on Fuzzy Systems*, Vol. 1, No. 1, 1993, pp. 7-31, ISSN: 1063-6706.

Tacconi, D. & Lewis, F. (1997). A new matrix model for discrete event systems: Application to simulation, *IEEE Control System Magazine*, Vol. 17, no. 5, 1997, pp. 6271.

Tsukiyama, T. (2005). World Map Based on RFID Tags for Indoor Mobile Robots, *Proceedings of the SPIE*, Vol. 6006, pp. 412-419, Boston, MA, USA, October 2005, David P. Casasent, Ernest L. Hall, Juha Röning, Editors.

Vig, L. & Adams, J.A. (2007). Coalition formation: From software agents to robots. *Journal of Intelligent and Robotic Systems*, Vol. 50, No. 1, pp. 85–118, ISSN: 0921-0296.

Zhou, Y., Liu, W., & Huang, P. (2007). Laser-activated RFID-based Indoor Localization System for Mobile Robots, *Proceedings of IEEE International Conference on Robotics and Automation*, pp. 4600 – 4605, ISBN: 1-4244-0601-3, Roma, Italy, April 2007.

3

Contact sensor for robotic application

Petr Krejci
Brno University of Technology
Czech Republic

Abstract
The chapter deals with design of contact force vector sensor. The information about interaction between robotic parts and surroundings is necessary for intelligent control of robot behavior. The simplest example of such interaction is mechanical contact between working part of robot and surroundings. Than the knowledge of contact characteristic is important for robot control. This mechanical contact could be described by vector of contact force which includes information about force magnitude as well as information about orientation and contact point. The information about contact force vector will allow to predict the geometry of object which is in the contact with robots parts and modify robots behaviour. This kind of sensor can be used for instance for control of robotic hand gripping force as well as for detection of collision between robot and surrounding.

1. Introduction

The design of contact force sensor was published by Schwarzinger, 1992. This design requires application of 24 strain gauges on active part of sensor. The quantity of strain gauges is sufficient for analytical determination of contact force vector.
Demand on small size of sensor for a lot of robotic applications (Grepl, R., Bezdicek, M., Chmelicek, J., Svehlak, M., 2004) disable application of a large number of strain gauges. Quantity of applied strain gauges and their size is limiting factor for using such design in our applications.
Our design of contact sensor supposes to use only three strain gauges on active part of sensor. However three strain gauges are not enough for the analytical expression of contact force vector. Due to this fact the neural network is used for force vector identification based on measured deformations of sensor body. The application of three strain gauges and new design will reduce size of sensor but requires a lot of numerical simulations for correct and accurate sensor behaviour.
The main advantage of using neural network is in low computational requirements for vector determination. It means fast response of sensor to contact load. The neural network is able to process measured data faster than nonlinear equations for force vector expression in analytical way.
The other advantage of our design is in reduced requirements for strain measurement by strain gauges. Generally, the Wheatstone bridge has to be used for strain measurement

(Hoffman, 1989) for each strain gauge. It means that our design significantly reduces complexity of electrical measuring unit. This reduction will allow us to build in control electronics into sensor body.

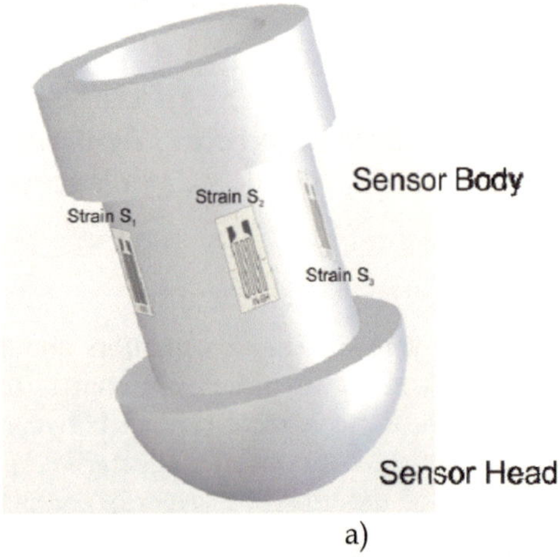

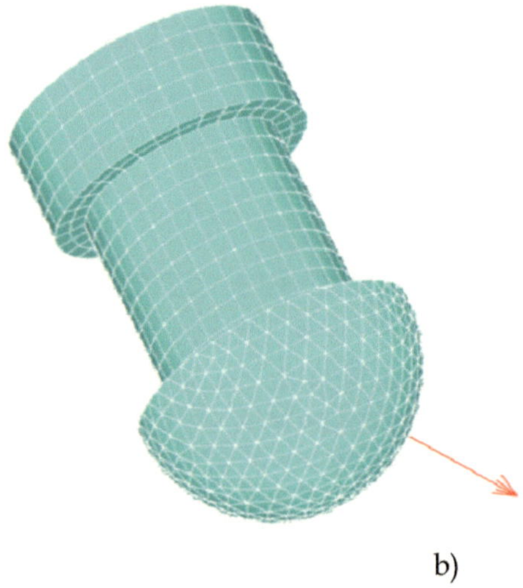

Fig. 1. Geometry of contact sensor and FE model of sensor

2. Sensor working principle

Basic principle of sensor is based on measuring of deformation of sensor body in three locations by three strain gauges on. Based on these deformations the contact force vector is identified by neural network. Huge matrix of training pairs is necessary for proper function of neural network. The training matrix contains pairs of deformations in three locations and force vectors corresponding to sensor body deformation. Finite element (FE) model of sensor was used for creating of training matrix. It is necessary to make a large number of numerical simulations with varied magnitude and position of contact force for properly work of neural network.

3. FE model of sensor

The FE model of sensor (see Fig. 1b) consists among others of sensor head, sensor body and sensor flange. The contact force is simulated as applied loads in selected nodes on sensor head. It is necessary to hold geometry and dimensions of model with real structure for correct work of sensor. Deformation of sensor during load is calculated in three positions. These positions correspond to positions of strain gauges on real sensor. For verification of FE model the force of 20N was applied on sensor head. Results of the verification are on Fig. 2 . The total strains in z-direction are S1= 2.9um/m, S2=12.6um/m and S3= -22.97um/m for this numerical simulation.

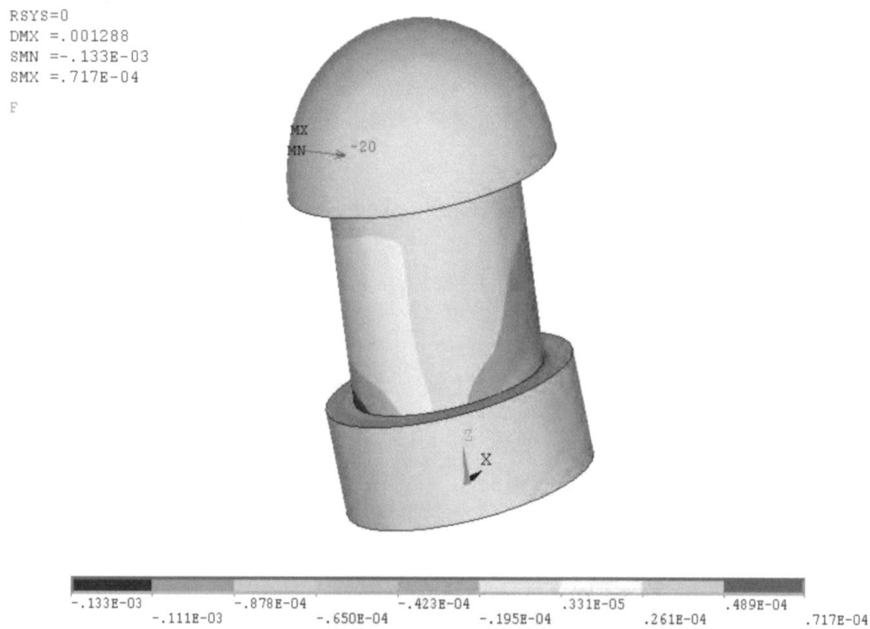

Fig. 2. Sensor deformation (applied load of 20 N) (unit of results are m/m)

4. Neural network

The architecture of artificial neural network (ANN) is shown on fig. 3. The input vector contains deformations of sensor body measured by strain gauges. The output vector contains information about contact force and position of contact force on sensor head.
The training matrix of 1000 training pairs was used for training of ANN in first step of this project. This amount of training pairs was used just for verification of functionality of suggested sensor. For better accuracy of ANN as well as sensor can by use much greater training matrix.

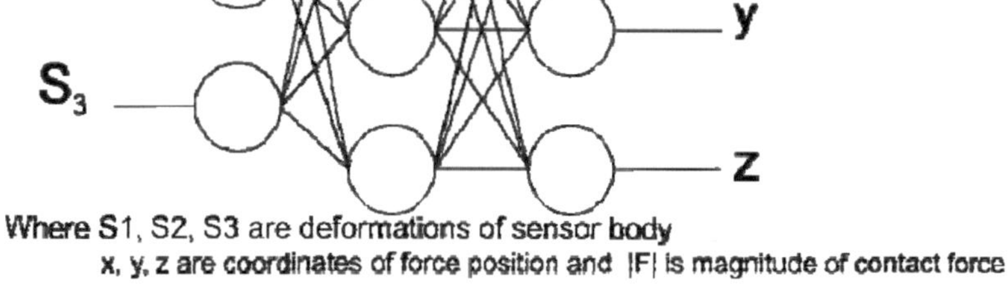

Fig. 3. Architecture of used neural network

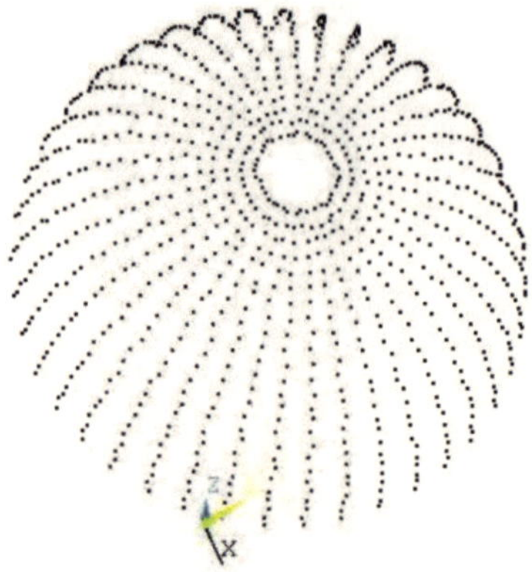

Fig. 4. Points of model load

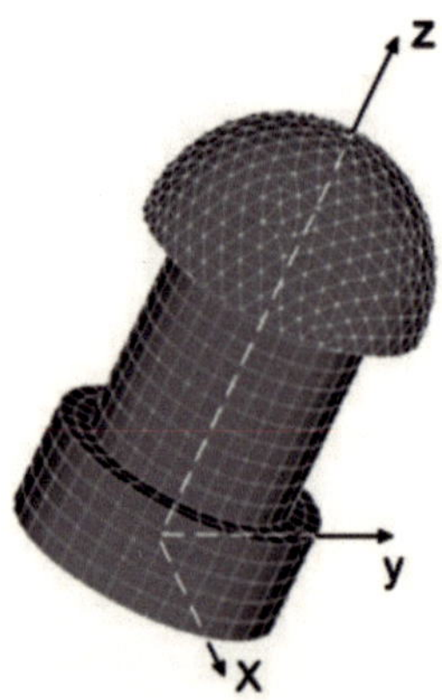

Fig. 5. Sensor coordinate system

5. Verification of ANN functionality

The force of 20N applied on sensor head was used for verification of ANN functionality. The position of the force was different than forces applied for training of ANN (points of model load used for training matrix creation are shown on Fig. 4). The x, y and z direction of applied force are 6.1mm, -3.31mm, 15.95mm respectively. The total strains in z-direction are S1=10.81um/m, S2=-23.57um/m and S3=10.04um/m. These strains were used as input vector of trained ANN. The result of contact force vector determination is in Table 1 (coordinate system of sensor is shown on .)

	Contact force coordinates		Accuracy [%]
	Point of FE model load	Position of contact force determined by Simulated by ANN	
x [mm]	6.10	6.03	98.85
y [mm]	-3.31	-3.29	99.39
z [mm]	15.95	15.97	99.87

Table 1. Result of verification

6. Experimental verification of sensor functionality

The sensor functionality was verified by experimental simulation in laboratory of Mechatronics. During experiment the loads of sensor was applied in several positions of sensor head. Gauging fixture (Fig. 6) was used for sensor positioning. Load was applied by materials testing machine Zwick Z 020-TND (Fig. 7, Fig. 8) where the real load force was measured. The deformation of sensor body was measured by strain gauges through HBM Spider 8 unit which is among other things designed for measuring of deformation by strain gauges.

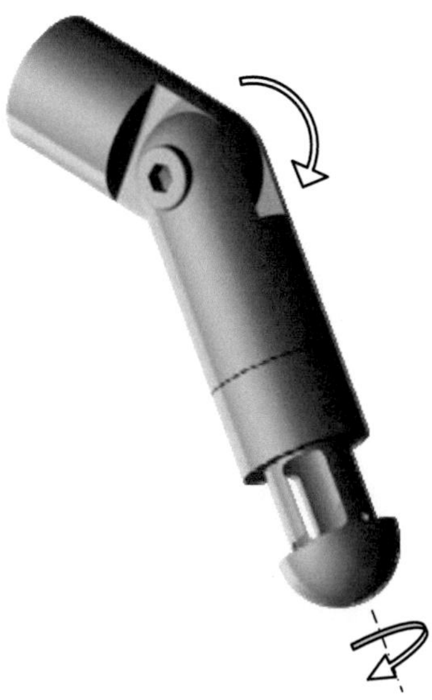

Fig. 6. Gauging fixture

Measured deformations was transferred to information about contact force position and magnitude by neural network implemented in Matlab software. The results of experimental verification for selected points are shown in Table 2 for four positions of load force and shows really good accuracy of designed sensor.

Load point	Direction [mm]	Contact force coordinates		Accuracy [%]
		Position of force during experiment	Simulated by ANN	
1	x	0.8	0.81	98.8
	y	-2.5	-2.75	91.0
	z	20.0	20.62	96.8
2	x	-1.0	-0.97	97.0
	y	-1.9	-2.03	93.6
	z	22.0	24.1	90.9
3	x	-1.0	-1.02	98.0
	y	0.1	0.11	90.9
	z	22.0	22.91	96.0
4	x	2.0	2.07	96.6
	y	-5.0	-4.52	90.4
	z	16.0	16.38	97.7

Table 2. Results of verification

Fig. 7. Testing machine Zwick Z 020-TND

Fig. 8. Loaded sensor during experimental verification of functionality

7. Optimization of sensor design

The low sensitivity of sensor was observed in axial direction during simulations and sensor testing. Therefore the topological optimization of sensor geometry was required for increasing of sensitivity for loads applied to sensor in axial direction.

The Finite element model of sensor in finite element software ANSYS was used for topological optimization procedure. Topological optimization (ref. ANSYS) is a form of "shape" optimization, sometimes referred to as "layout" optimization. The purpose of topological optimization is to find the best use of material for a body such that an objective criterion takes on a maximum/minimum value subject to given constraints (such as volume reduction).

The sensor body is the volume which was subjected to optimization process. Volumes located under supposed locations of strain gauges was excluded from process of optimization.

This optimization was done for two load steps - for load forces oriented in different directions. The first step of optimization procedure was done for axial load of sensor head while second step was done for radial load. The result of optimization is shown for volume reduction of 80% in Fig. 10. The figure shows distribution of pseudodensity in sensor body. The boundary conditions used during optimization procedure are shown in Fig. 9.

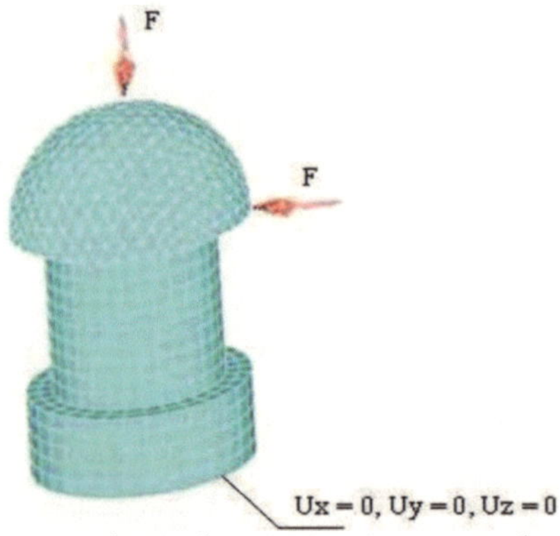

Fig. 9. Loads of sensor used in topological optimization procedure

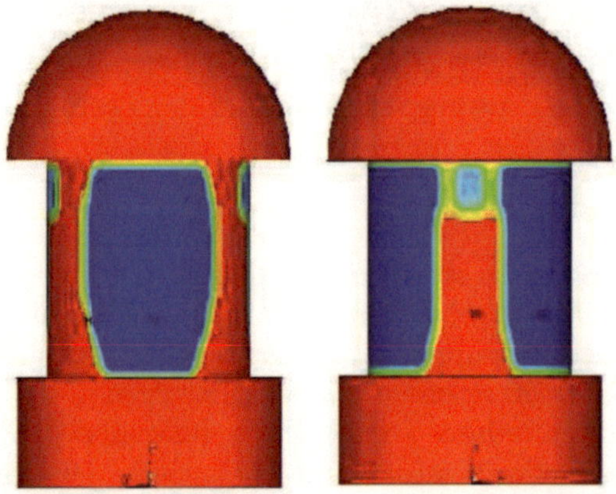

Fig. 10. Results of optimization (pseudodensity - red color means that volume will be included in final design, blue color mean that volume will be excluded from final design)

Optimized shape of sensor body need to by simplified by reason of good manufacturing. Due to this fact few shapes of cutting was designed with consideration of optimized shape (Fig. 10) and machining. Based on results of structural analysis rectangular shape of cutting with 1 mm hole (Fig. 11b)) produces the best results in terms of sensitivity. This shape is also suitable for simple machining. Fig. 12 shows prototype of optimized and non-optimized sensor which is made from aluminium alloy.

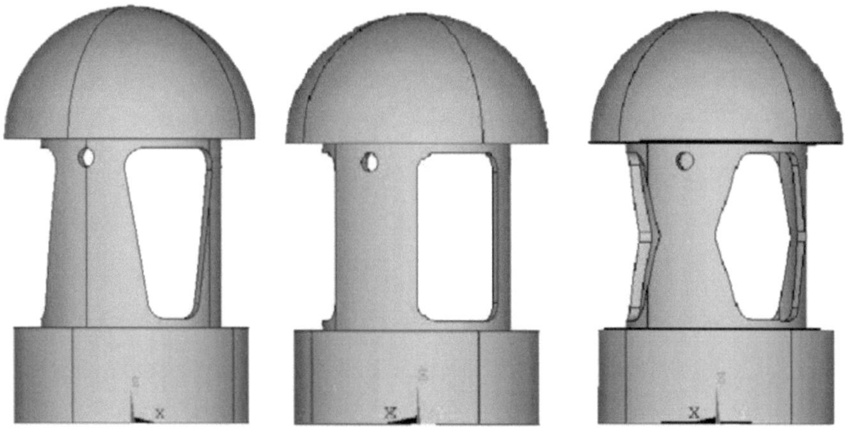

a) b) c)

Fig. 11. Optimized sensors with different shapes of cuttings

Fig. 12. Optimized and non-optimized sensor prototype

7.1. Structural analysis of sensor prototype

Structural analysis of optimized sensor was done in order to find out load limits where the linear behaviour of structure occurs. Results of this simulation are shown in Fig. 13 for load force of 140N. This value defines upper bound of sensor limits where plastic deformation of sensor body can occurs.

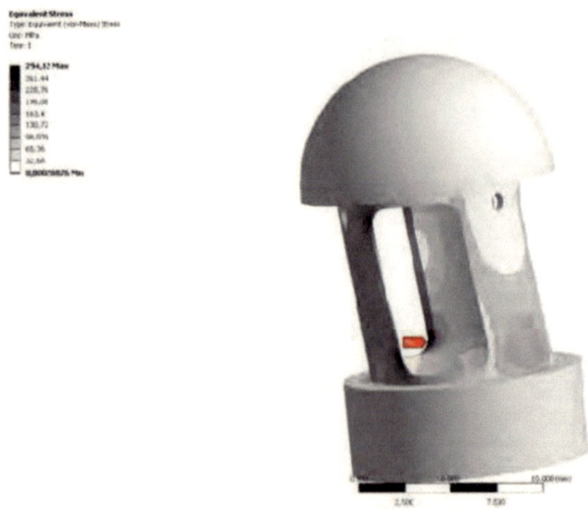

Fig. 13. Von-Misses stress (MPa) of sensor for load of 140 N applied in radial direction

8. Verification of optimized sensor functionality

Functionality of optimized sensor was also done by two methods. Finite elements model of sensor is used for calculation of body deformation caused by specified load in first method. The Second method using experimental verification of sensor subjected to real load. Deformations of sensor body observed by both methods are used as inputs of neural network which produces information about contact force magnitude and coordinates.

8.1. Verification of functionality by FEM simulation

Sensor functionality was proof by numerical simulation using FE model of sensor. Verification was done in same way as procedure described in section 4.

The maximal difference in load force position between force coordinates used for FEM model loading and simulated coordinates retrieved from ANN was up to 2%. This difference also shows error of trained neural network.

8.2. Experimental verification of functionality

The results of experimental verification that was done in same way as described in section 5 show that the maximum inaccuracy of sensor is up to 10%. This difference can be caused by inaccuracy in strain gauges application.

9. Conclusion

Presented chapter introduced new concept of contact sensor for robotic application that can be used to contact force vector determination. The problem of the sensor is low sensitivity for load in axial direction of sensor that was solved by topological optimization in ANSYS software. The reduction of 80% of sensor body volume was achieved and in this relation the sensitivity in axial direction increases. The functionality of sensor was proofed by numerical simulations and also by experimental verification using and simulating real load of sensor prototype. Verification was done for optimized and non-optimized prototype of sensor. Using only three strain gauges for deformation measurement of sensor body allow us to use SMD electronics parts and build up the unit to hollow sensor body. The sensor can be use for 10N to 140N load force range.

10. Acknowledgment

Published results were acquired using the subsidization of the Ministry of Education, Youth and Sports of the Czech Republic, research plan MSM 0021630518 "Simulation modeling of mechatronic systems" and GAČR project no. 101/08/0282.

11. References

Ansys, Inc., Theory Reference, Release 10, Southpointe, 275 Technology Drive, Canonsburg, PA 15317

Grepl, R., Bezdicek, M., Chmelicek, J., Svehlak, M.: Experimental quadruped walking robot: conception, design and control. Magazine - Elektronika, 8-9/2004, ISSN 0033-2089, Poland 2004.

Hoffman, K.: An Introduction to Measurements using Strain Gages, 1989, Hottinger Baldwin Messtechnik, GmbH, Darmstad

Krejci, P.: Contact sensor for robotic applications, Engineering Mechanics, Vol.12, (2005), No.A1, pp.257-261, ISSN 1210-2717

Krejci, P., Vlach, R., Grepl, R.: Contact sensor for robotic applications - Verification of Functionality, in conference proceedings: "Engineering Mechanics 2006", pp. 182-183, May 2006, Svratka, Czech Republic, ISBN 80-86246-27-2

Krejci, P., Vlach, R., Grepl, R.: Conception of Robotic Leg Design, in conference proceedings:"10th International seminar of Applied Mechanics", pp. 95-98, Wisla 2006, Poland, ISBN 83-60102-30-9

Schwarzinger Ch., Supper L. & Winsauer H. (1992) Strain gauges as sensors for controlling the manipulative robot hand OEDIPUS: RAM vol. 8, pp.17-22.

4

Develop a Multiple Interface Based Fire Fighting Robot

Ting L. Chien[1], Kuo Lan Su[2] and Sheng Ven Shiau[3]
[1]*Department of Electronic Engineering WuFeng Institute of Technology Ming-Hsiung, Chia 621, Taiwan cdl@mail.wfc.edu.tw*
[2] *Department of Electrical Engineering National Yunlin University of Science & Technology Douliou, Yunlin 640, Taiwan, sukl@yuntech.edu.tw*
[3]*Graduate school Engineering Science and technology National Yunlin University of Science & Technology Douliou, Yunlin 640, Taiwan g9610808@yuntech.edu.tw*

1. Abstract

The security of home, laboratory, office, factory and building is important to human life. We develop an intelligent multisensor based security system that contains a fire fighting robot in our daily life. The security system can detect abnormal and dangerous situation and notify us. First, we design a fire fighting robot with extinguisher for the intelligent building. The fire fighting robot is constructed using aluminum frame. The mobile robot has the shape of cylinder and its diameter, height and weight is 50 cm, 130 cm and 80kg. There are six systems in the fire fighting robot, including structure, avoidance obstacle and driver system, software development system, fire detection, remote supervise system and others. We design the fire detection system using two flame sensors in the fire fighting robot, and program the fire detection and fighting procedure using sensor based method. We design a low cost based obstacle detection module using IR sensors and ultrasonic sensors in the mobile robot. The man-machine interface of the fire fighting robot must be mobility and convenience. We use touch screen to display system status, and design a general user interface (GUI) on the touch screen for the mobile robot. Finally, we implement the fire detection system using fire fighting robot. If fire accident is true, the fire fighting robot can uses two flame sensors to find out fire source by the proposed method, and move to fire source to fight the fire using extinguisher.

Keywords：fire fighting robot、man-machine interface、touch screen

2. Introduction

Home can provide safety, convenience, and efficiency for people in the 21st century. An intelligent home system is integrated by many function and systems. One of the most important systems is the fire detection function system in an intelligent home [1,- 6]. The fire event may involve dangerous in life. In generally, the fire detection device is fixed on the wall or ceiling. But the method is not flexibility to detect fire event. It is not very convenience that uses many fire detection modules in the home. In the paper, we design a fire fighting robot to detect fire event, and use extinguish to fight the fire source, and can transmits the fire event to the remote supervised computer using RF interface and Internet, and transmit fire information to cell phone using GSM modern.

In the past literatures, many experts research in the service robot. Some research addressed in developing target-tracking system of service robot [11,12], such as Hisato Kobayashi et al. proposed a method to detect human being by an autonomous mobile guard robot [13]. Yoichi Shimosasa et al. developed Autonomous Guard Robot [10] witch integrate the security and service system to an Autonomous Guard Robot, the robot can guide visitors in daytime and patrol in the night. D. A. Ciccimaro developed the autonomous security robot – "ROBART III" which equipped with the non-lethal-response weapon [8,9]. Moreover, some research addressed in the robot has the capability of fire fighting [7]. There are some products that have been published for security robot. Such as SECON and SOC in Japanese, and International Robotics in USA, and Chung Cheng #1 in Taiwan [20]. Wang et al [14] develops a multisensor fire detection algorithm using neural network. One temperature and one smoke density sensor signal are fused for ship fire alarm system. Healey et al. [15] presents a real-time fire detection system using color video input. The spectral, spatial, and temporal properties of fire were used to derive the fire-detection algorithm. Neubauer [16] apply genetic algorithms to an automatic fire detection system. The on-line identification of stochastic signal models for measured fire signals was presented. Ruser and Magori [17] described the fire detection with a combination of ultrasonic and microwave Doppler sensor. Luo and Su [18,19] use two smoke sensors, two temperature sensors and two flame sensors to detect fire event, and diagnosis which sensor is failure using adaptive fusion method.

The chapter is organized as follows: Section II describes the system structure of the automatic fire fighting system. Section III presents the function of the fire fighting robot. Section IV explains the avoidance obstacle algorithm for the fire fighting robot, and the simulation and experimental result and scenario is implemented in section V. Section VI presents brief concluding comments.

3. System Architecture

The system architecture of the home security system is shown in Fig 1. The system contains fire fighting robot, television, remote supervise computer, GSM modern, security modules, appliance control modules and wireless RF controller. The robot and security device can receive the status of security module and appliance control module using wireless RF interface. In the security modules, they use one-way communication with the fire fighting robot. But the appliance control modules use two-way communication with the fire fighting robot. The fire fighting robot can communicate with GSM modern via RS232 interface, and can communicate with the remote supervise computer via wireless Internet. The display panel of the remote supervised computer is television, and users can control the fire fighting robot using wireless RF controller.

The fire fighting robot can get detection signals from security modules via wireless RF interface. The remote supervise computer can interact with the mobile robot via wireless Internet. Users can use the remote supervised computer to get security information of the home or building, and control the fire fighting robot to everywhere in the home. The mobile robot can transmit the reality status to the remote supervise computer via wireless Internet, and transmits the message to mobile phone using GSM modern.

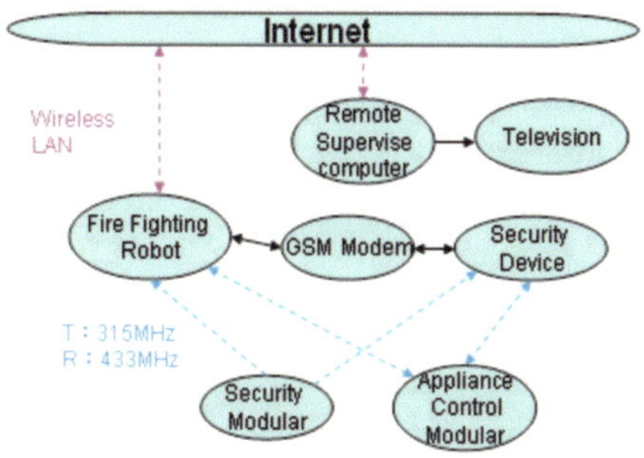

Fig. 1. The overview of the home security

The main controller of the fire fighting robot is industry personal computer (IPC). The hardware devices have display device (touch panel), extinguisher, CCD, two flame sensors and sensory circuits, driver system (DC servomotors, NI motion control card and MAXON drivers), GSM modem and wireless RF interface. The remote supervised computer is personal computer (PC) with a Pentium-IV 2.4G CPU. There are six systems in the fire fighting robot, including structure, avoidance obstacle and driver system, software development system, fire detection, remote supervise system and others. Fig. 2 is the structure of the fire fighting robot, and each system includes some subsystem. Each system contains some functions in the fire fighting robot.

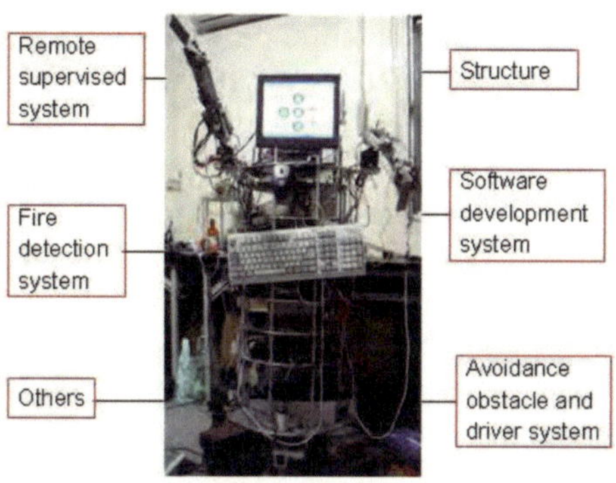

Fig. 2. The structure of the fire fighting robot.

4. Fire Fighting Robot

The block diagram of the fire fighting robot is shown in Fig. 3. In the drive system, there are fire fighting device, auto-charging device, and the robot can control appliance module via wireless RF interface. In the motion control function, the mobile robot can orders command to control two DC servomotors through motion control card and driver devices, and it can control two robot arms using the motion control card. In the avoidance obstacle function, the robot catches eight IR sensor signals by digital input terminal (in the motion control card), and measure distance of obstacle using eight ultrasonic sensors via series interface. We use microprocessor to drive eight ultrasonic sensors, and transmit distance range to main controller of the fire fighting robot via series interface. We design ultrasonic driver circuit and use multisensor fusion method to get exactly decision output. In the sensory system, there are fire detection, power detection and environment detection. The transmission interface between the remote supervise computer and the fire fighting robot is wireless Internet. There are some devices communicate with the main controller of the fire fighting robot. Such as touch screen, wireless image, wireless RF interface, alarm device and GSM modem.

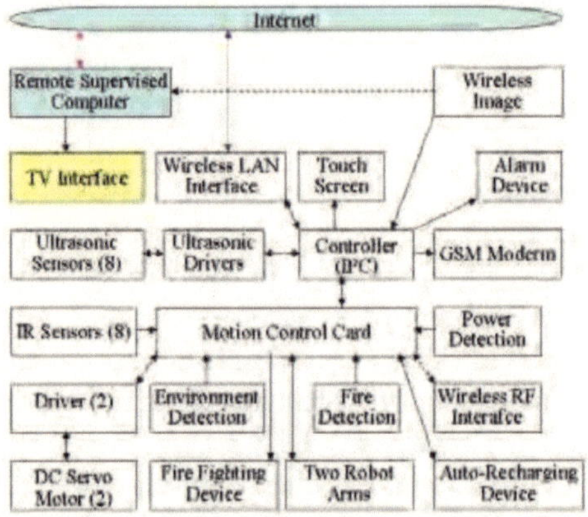

Fig. 3. The block diagram of the fire fighting robot.

In the hierarchy of the remote supervise system, it contains communication protocol, data base and user interface. The hardware configuration of the remote supervise system includes a fire fighting robot, supervised computer, GSM modem and wireless LAN, as show in Fig. 4.

The main controller of the fire fighting robot can get sensor signals to detect fire, intruder, gas, etc, and can interact with the remote supervise computer via wireless Internet. Users can control the fire fighting robot via wireless Internet, too. The remote user cans logo in the supervise computer to control the fire fighting robot, and can connect to the Web server or the robot server to get all information by computer or cell phone through the Internet.

The user interface of the fire fighting robot is shown in Fig. 5. The display panel contains five parts. In the security system, it can display detection results using variety sensors, and display the state of motor and battery, and display the detection results using sensors from the fire fighting robot. The program language of the user interface is designed by VB. The fire fighting robot communicates with mobile phone using GSM (Global System for Mobile) modular. The GSM modular (WMOD2) was made by Wavecom.. The modular is a seft-contained E-GSM900/GSM1800 (or E-GSM900/ GSM1900) dual band module.

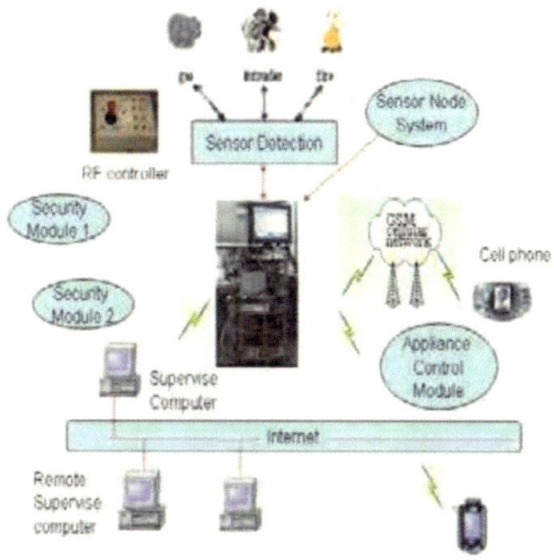

Fig. 4. The hardware configuration of the remote surveillance and control architecture.

Fig. 5. The user interface of the fire fighting robot

In the remote supervised system, the user interface is shown in Fig. 6. The panel can display the results using variety sensors from the fire fighting robot. The client-side program's function is to order command to the mobile robot and continuously update the sensory data, which receive from the remote supervise computer. The program of the client's user interface is designed by VB.

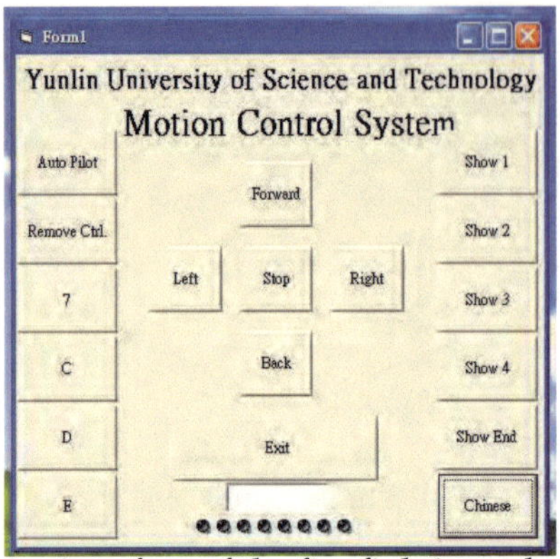

Fig. 6. The remote supervise interface of the fire fighting robot

The security module and appliance control module are designed by us. The controller of these modules is microprocessor (ATMEL89C2051). In the security module, it contains fire detection module, gas detection module, intruder detection module, etc. When the status is happen, these security modules can transmit RF signal to the mobile robot and the security device. The security module is shown in Fig 7(a). The appliance control module is shown in Fig 7(b). The appliance control modules can control appliance device using relay element, and feedback the action result to the mobile robot and the security device via wireless RF interface.

(a)

(b)

Fig. 7.The security module and appliance control module

5. Avoidance Obstacle System

In the avoidance obstacle and driver system, the fire fighting robot uses ultrasonic sensors and infrared sensors to detect obstacle and restructure the surrounding environment by using multisensor fusion and integration technique. The fire fighting robot uses eight pieces SMC-60R infrared reflection (IR) sensors to detect obstacle, and uses eight pieces Polaroid 6500 ultrasonic range sensors to measure the distance from obstacle. We fuse IR sensors and ultrasonic sensors to decide the map of environment. The detection module is shown in Fig. 8.

Fig. 8. The IR and ultrasonic detection module

There are two type objects are defined as "obstacle" and "free space", as shown in Fig. 9. Eight infrared sensors are used for obstacle detection in the uncertain environment, where each sensor is set on fixed location around the fire fighting robot. The angle is $\frac{\pi}{4}$ from X axis to the centre of infrared beam. The radius of the robot is R (R=25cm), and the distance between the robot and obstacle is Di at sensor modular i, and the safe distance (infrared reflection senor detection distance) between robot and obstacle is Ds. The Ds is a threshold to ensure that the fire fighting robot is not closes the obstacle. First, we define the relation function of infrared sensor is D_f

$$D_f = \begin{cases} 1 & \text{If } I_i \text{ has found obstacle} \\ 0 & \text{If } I_i \text{ has not found obstacle} \end{cases} \tag{1}$$

Then we define the relation function of ultrasonic sensor is D_u

$$D_u = \begin{cases} 1 & D_S \le D_{safe} \\ \dfrac{D_{safe}}{D_s} & D_s \ge D_{safe} \end{cases} \tag{2}$$

We can calculate the decision value of reliability of the obstacle detection at sensor module i to be defined D_i, and the decision value of reliability of the obstacle detection between the ultrasonic sensor module i and the ultrasonic sensor module i+1 is $D_{i,i+1}$

$$D_i = \frac{1}{2}\left(D_f + D_u\right) \tag{3}$$

$$D_{i,i+1} = \frac{1}{2}\left(D_i + D_{i+1}\right) \tag{4}$$

Finally, we can set a threshold value η, and $D_i \ge \eta$ or $D_{i,i+1} \ge \eta$. We can say "obstacle". Otherwise, $D_i \le \eta$ or $D_{i,i+1} \le \eta$, we can say "free space". The obstacle detection rule can written as Fig 10

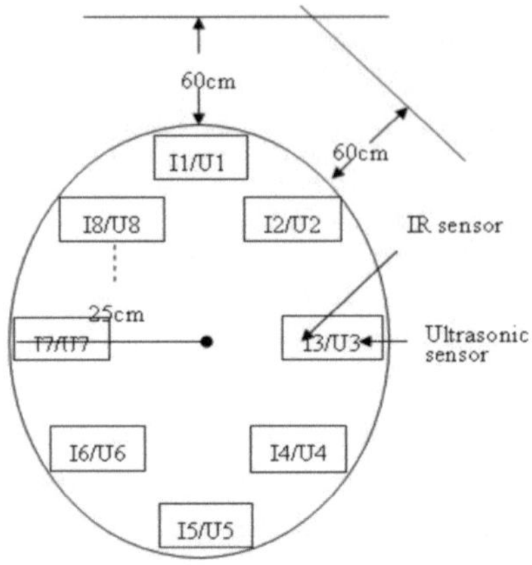

Fig. 9. The arrangement of IR and ultrasonic sensors

We use five IR sensors (I1, I2, I3, I7 and I8) and five ultrasonic sensors (U1, U2, U3, U7 and U3) to detect obstacle. The IR sensor can detects distance from obstacle to be 60 cm. The ultrasonic can detects distance range from 25cm to 10m. We fuse the advantages of these sensors to increase the precious for the obstacle detection. We use three IR sensors (I4, I5 and I7) to detect intruder and dynamic obstacle behind the fire fighting robot.

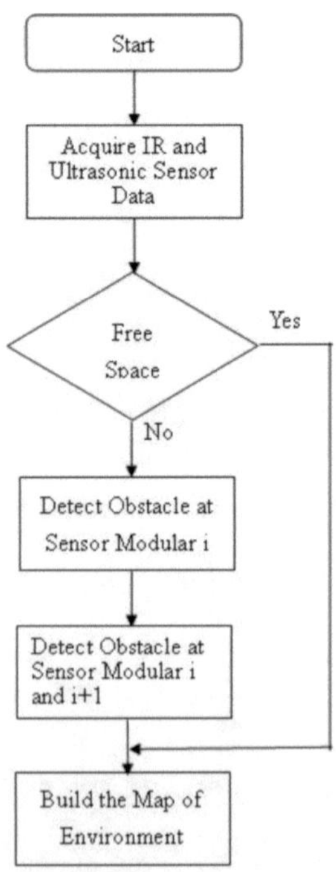

Fig.10. The obstacle detection rule of the fire fighting robot

6. Experimental Results

In the motion control experimental scenario of the fire fighting robot, we can select autonomous mode or wireless control mode. In the autonomous mode, the fire fighting robot can move according to environment state using IR sensors and ultrasonic sensors. In the wireless control mode, we can supervise the fire fighting robot for walking forward, walking backward, stop, rotation, turn right and turn left via multiple interface system (wireless RF interface or wireless RS232 interface). In the motion planning experiment, we program the fire fighting robot to have a maximum speed 40cm/sec, and a maximum rotation speed 100deg/sec for DC servomotor. Then we program the motion path is rectangle (see Fig 11). The experimental scenario of the fire fighting robot is shown in Fig. 12. First, the mobile robot start to move forward to the first goal (Fig 12(a)).if the robot move to the first goal and turn right, and it move .to the second goal. The experimental scenario is shown in Fig 12(b). Next it turns right and move to the third goal (Fig 12(c)). The robot moves to the third goal, and turn right to move start position. Finally, the fire fighting robot arrives at the start position, and stop. The experiment result is shown in Fig.12 (d).
Next, the fire fighting robot can uses IR sensors and ultrasonic sensors to construct environment. It can avoid state dynamic obstacle, and move in the free space. In the state avoiding, it uses five IR and ultrasonic sensor modules to detect obstacle on the front side of the mobile robot. The experimental result is shown in Fig. 13. In the Fig. 13 (a), it shows the mobile robot to detect the obstacle in right side. It can turn left to avoid obstacle, and move to the preprogramming path. The experimental scenario is shown in Fig. 13 (b).

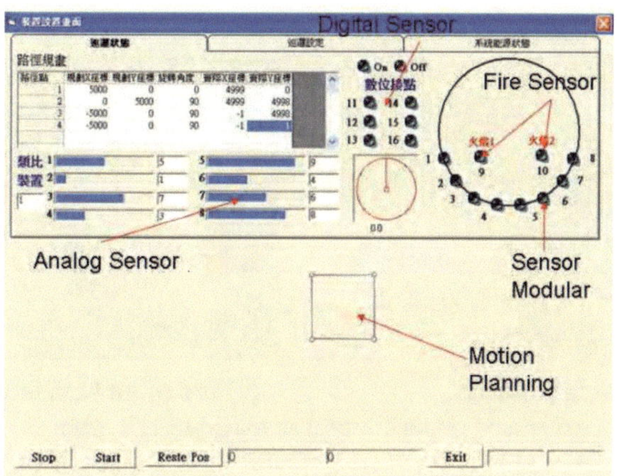

Fig. 11. The programming path is rectangle for the mobile robot

(a)The robot move to first goal (b)The robot turn right

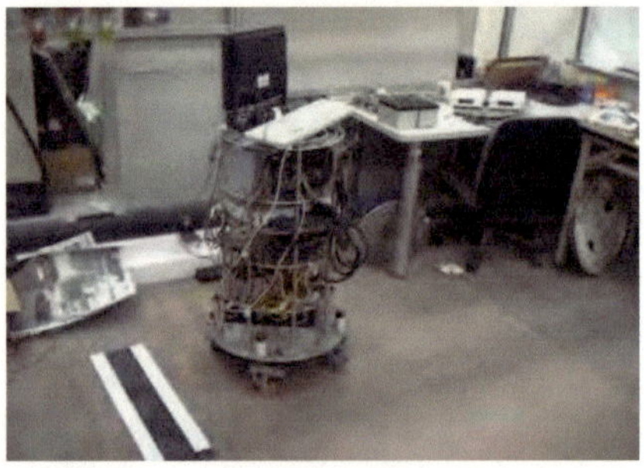

(C) The robot turn right to third goal (d)The robot move to start position

Fig. 12. The motion planning experimental scenario of the mobile robot

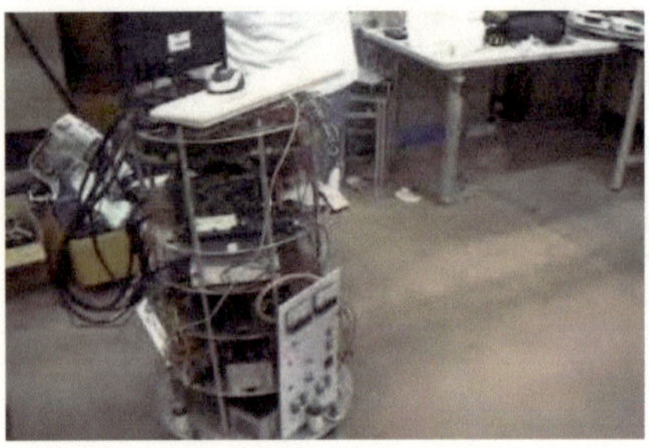

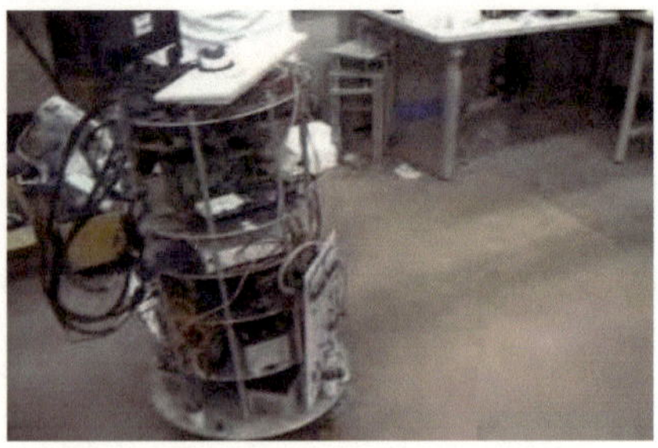

(a)The robot detect obstacle (b)The robot turn left

Fig. 13.The avoidance obstacle experimental scenario of the robot

In the fire detection experimental results, the fire fighting robot can move autonomous in the free space. The fire event may be detected using two flame sensors in the fire fighting robot. The flame sensor detects the fire event, and transmits the fire signal to the main controller (IPC) of the fire fighting robot using digital input of motion control card. The fire fighting robot moves to the fire location, and use two flame sensors to detect fire event again using multisensor rule. If the fire event is true, the fire fighting robot must fight the fire source using extinguisher. Otherwise, the flame sensors of the fire fighting robot detect the fire condition, and the fire fighting robot must be alarm quickly, and transmits the control signal to appliance control module (we use lamp instead of water, Fig 15(a)) to fight the fire source through wireless RF interface, and send the fire signal to the mobile phone using GSM modern (the experimental result is shown 15(b)), transmits the status to client computer via wireless Internet. In the intruder detection, the experimental results are the same as fire detection. The experimental result is shown in Fig. 14. The fire fighting robot can receives the wireless security signals from wireless security module, too.

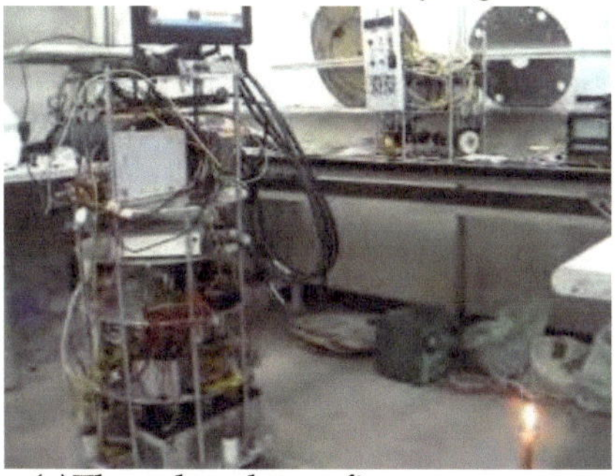

(a)The robot detect fire source

(b)The robot move to fire source

(c)The robot open extinguisher

(d)The robot fight the fire source

Fig. 14.The fire fighting experimental scenario of the mobile robot

(a) The lamp on

(b)Mobile phone

Fig. 15.The mobile executes fire detection

7. Conclusion

We have presented a multiple interface based real time monitoring system that is applied in home automation. The security system of the home and building contains fire fighting robot, security device, television, remote supervise computer, GSM modern, wireless RF controller, security modular and appliance control modular. The main controller of the fire fighting robot is industry personal computer (IPC). We order command to control the mobile robot to acquire sensor data, and program the remote supervised system using Visual Basic. The robot can receive security information from wireless RS232 interface, and design a general user interface on the control computer of the fire fighting robot. In the experimental results, the user controls the mobile robot through the wireless RF controller, supervised computer and remote supervised compute. The robot can avoid obstacle using IR sensor and ultrasonic sensor according to multisensor fusion method. It can use two flame sensors to find out the fire source, and fight the fire source using extinguisher. In the future, we want to design the obstacle detection modular using IR sensor and ultrasonic sensor using new fusion algorithm, and apply in the fire fighting robot. Then we want combine the laser range finder to get more exact and quickly environment map in the indoor and outdoor.

8. References

C. W. Wang and A. T. P. So, 1997, "Building Automation In The Century," in Proceedings of the 4-th International Conference on Advance on Advances in Power System Control, Operation Management, APCOM-97, Hong Kong, November,pp.819-824.

M. Azegami and H. Fujixoshi, 1993, "A Systematic Approach to Intelligent Building Design," IEEE Communications Magazine, October ,pp.46-48.

Kujuro and H. Yasuda, 1993, "Systems Evolution in Intelligent Building," IEEE Communication Magazine, October,pp.22-26.

M. R. Finley, J. A. Karakura and R. Nbogni, 1991, "Survey of Intelligent Building Concepts," IEEE Communication Magazine, April , pp.l8-20.

M. Fiax, "Intelligent Building," IEEE Communications Magazine April 1991, pp.24-27.

L. C. Fu and T. J. Shih, 2000,"Holonic Supervisory Control and Data Acquisition Kernel for 21st Century Intelligent Building System," IEEE International Conference on Robotics & Automation, Sam Francisco, CA, April, pp. 2641-2646

Bradshaw, , 1991 "The UK Security and Fire Fighting Advanced Robot project," IEE Colloquium on Advanced Robotic Initiatives in the UK, pp. 1/1-1/4.

Gilbreath, G.A., Ciccimaro, D.A., and H.R. Everett, 2000, "An Advanced Telereflexive Tactical Response Robot," Proceedings, Workshop 7: Vehicle Teleoperation Interfaces, IEEE International Conference on Robotics and Automation, ICRA2000, San Francisco, CA, 28 April.

Ciccimaro, D.A., H.R. Everett, M.H. Bruch, and C.B. Phillips, 1999, "A Supervised Autonomous Security Response Robot,", American Nuclear Society 8th International Topical Meeting on Robotics and Remote Systems (ANS'99), Pittsburgh, PA, 25-29 April.

Y. Shimosasa, J. Kanemoto, K. Hakamada, H. Horii, T. Ariki, Y. Sugawara, F. Kojio, A. Kimura, S. Yuta, 2000, "Some results of the test operation of a security service system with autonomous guard robot," The 26th Annual Conference of the IEEE on Industrial Electronics Society (IECON 2000), Vol.1, pp.405-409.

Sung-On Lee, Young-Jo Cho, Myung Hwang-Bo, Bum-Jae You, Sang-Rok Oh , 2000, "A stable target-tracking control for unicycle mobile robots," Proceedings of the IEEE/RSJ International Conference on Intelligent Robots and Systems, (IROS 2000) , Vol.3 , pp.1822-1827.

L. E. Parker, B. A. Emmons, 1997 ,"Cooperative multi-robot observation of multiple moving targets," Proceedings of the IEEE International Conference on Robotics and Automation,, vol.3, pp.2082-2089.

H. Kobayashi, M. Yanagida, 1995"Moving object detection by an autonomous guard robot," Proceedings of the 4th IEEE International Workshop on Robot and Human Communication, , TOKYO, pp.323-326.

W. Xihuai, X. Jianmei and B. Minzhong, 2000, "A ship fire alarm system based on fuzzy neural network,"in Proceedings of the 3rd World Congress on Intelligent Control and Automation, Vol. 3, pp. 1734 -1736.

Healey, G., Slater, D., Lin, T., Drda, B. Goedeke and A. D., 1993,"A system for real-time fire detection," in Proceedings of IEEE Computer Society Conference on Computer Vision and Pattern Recognition, pp. 605-606.

Neubauer A., "Genetic algorithms in automatic fire detection technology, 1997," Second International Conference On Genetic Algorithms in Engineering Systems: Innovations and Applications, pp. 180-185.

Ruser, H. and Magori, V., "Fire detection with a combined ultrasonic-microwave Doppler sensor," in Proceedings of IEEE Ultrasonics Symposium, Vol.1, 1998, pp. 489-492.

R. C. Luo, K. L. Su and K. H. Tsai, "Fire detection and Isolation for Intelligent Building System Using Adaptive Sensory Fusion Method," Proceedings of The IEEE International Conference on Robotics and Automation, pp.1777-1781.

R. C. Luo, K. L. Su and K. H. Tsai, 2002, "Intelligent Security Robot Fire Detection System Using Adaptive Sensory Fusion Method," The IEEE International Conference on Industrial Electronics Society (IECON 2002), pp.2663-2668.

5

Develop a Power Detection and Diagnosis Module for Mobile Robots

Kuo-Lan Su[1], Jr-Hung Guo[2] and Jheng-Shiann Jhuang[3]
[1]*Department of Electrical Engineering,National Yunlin University of Science & Technology,Douliou, Yunlin 640, Taiwan. sukl@yuntech.edu.tw*
[2]*Graduate school Engineering Science and technology National Yunlin University of Science & Technology,Douliou, Yunlin 640, Taiwan,g9710801@yuntech.edu.tw*
[3]*Department of Electrical Engineering,National Yunlin University of Science & Technology,Douliou, Yunlin 640, Taiwan. 9512710@yuntech.edu.tw*

1. Abstract

Autonomous mobile robot will be very flexibility to move in free space. But it is limited on power supply. The power of the mobile robot can provide a few hours of peak usage before the power is lack. The power detection system is an important issue in the autonomous mobile robot. In the chapter, we want to design a power detection and diagnosis module to measure the power condition of the mobile robot, and measure the voltage of the power system for mobile robots. We use multilevel multisensory fusion method to detect and diagnose current sensors and voltage signals of mobile robots. First, we use four current sensors to measure the power variety of the mobile robot. We use redundant management method and statistical predition method to detect and diagnosis current sensor status, and isolate faulty sensor to improve the power status to be exact. Then, we use computer simulation to implement the proposed method to be adequate. We design the power detection and diagnosis module using HOLTEK microchip. Users can select maximum and minimum current value and detection range of the power detection module. The power detection module can transmits the detection and diagnosis status to the main controller (Industry Personal Computer, IPC) of the mobile robot via series interface. Finally, we implement some experimental scenario using the module in the mobile robot, and can take some experimental results for some variety condition on sensor faulty.

Keywords - Autonomous mobile robot, redundant management method, statistical perdition method.

2. Introduction

With the robotic technologies development with each passing year, Mobile robots have been widely applied in many fields. Such as factory automation, dangerous environment detection, office automation, hospital, entertainment, space exploration, farm automation, military and security system. Recently more and more researchers take interest in the field especially intelligent service robot. There are some successful examples, ASIMO, KHR, QRIO and AIBO. In our laboratory, we have been designed a mobile robot (ISLR-I) to fight fire source. However the mobile robot has been working for a long time. The power of the mobile robot is lack, and it can not be controlled by the command, and some dangerous event may be happened. Thus, the mobile robot must quickly move to the recharging station. So we must detect power variety of the mobile robot all the time. Therefore, we must detect power variance of the mobile robot very carefully. We must calculate the residual power according to the power output of the mobile robot. The mobile robot has enough time to move to the recharging station autonomously.

We have designed a power detection system in the WFSR-I mobile robot. The contour of the robot is cylinder. The mobile robot has the shape of cylinder and its diameter, height and weight is 20cm, 30cm and 4kg. The robot is a four-wheeled platform equipped with a main controller (MCS-51 microprocessor). The power system of the mobile robot uses two rechargeable batteries [1,2,19]. We use laser line guard the mobile robot move to the recharging station. Next, we modify the power detection module applying in Chung-Cheng I security robot using microprocessor (MCS51), too. The Chung-Cheng I security robot has the shape of cylinder and its diameter, height and weight is 50cm, 150cm and 80kg. The module can calculate the exact current variety of the Chung-Cheng I security robot, and use image guard the security robot move to the recharging station. The experimental results are very successful [3,5]. Now we design the power detection module applying in the ISLR-I mobile robot using HOLTEK microchip. The new module wants to reduce the cost of the power detection module, and extend more and more functions for mobile robots. The module can transmit the power detection results to the main controller of the mobile robot via series interface.

In the past literature, many researches have been proposed current detection methods. A. J. Melia and G.F. Nelson postulate that monitoring of the power supply current could aid in the testing of digital integrated circuits [6,7]. Levi was one of the first to comment upon the characteristics of CMOS technology which make it special amenable to IDD Testing [8]. Malaiy and Su use IDD testing and estimating the effects of increased integration on measurement resolution [9,10]. Frenzel proposed the likelihood ration test method applying on power-supply current diagnosis of VLSI circuits [11]. Horming and Hawkins reported on numerous experiments where current measurements have forecast reliability problems in devices which had previously passed conventional test procedures[12,13].Then, many researches dedicated to improving the accuracy of measuring current [14,15]. Maly et al proposed a build-in current sensor which provides a pass/fail signal when the current exceeds a set threshold [16,17].

The chapter is organized as follows: Section II describes the system structure of the power detection system for the ISLR-I mobile robot. Section III presents the hardware structure of power detection system for the mobile robot. The detection and diagnosis algorithm is explained in section IV. Section V explains the user interface of the power detection system for the mobile robot. Section VI presents the experimental results for power detection and isolation scenario of mobile robot. Section V presents brief concluding remarks.

3. System Architecture

The mobile robot is constructed using aluminium frame. The mobile robot has the shape of cylinder and its diameter, height and weight is 50 cm, 110cm and 40 kg. Figure 1 (a) shows the hardware configuration of the mobile robot (ISLR.-I). The main controller of the mobile robot is industry personal computer (IPC). The hardware devices have GSM modern, batteries, NI motion control card, wireless LAN, fire fighting device and sensory circuits, touch screen, distributed control module, power detection and diagnosis module, driver system, DC servomotors, color CCD and some hardware devices [18].

There are six systems in the mobile robot, including structure, avoidance obstacle and driver system, software development system, detection system, remote supervised system and others. Figure 1 (b) is the hierarchy structure of the mobile robot, and each system includes some subsystem. For example, the detection system contains power detection system, fire fighting device, fire detection rule and fire detection hardware… etc.

Manuscript must contain clear answers to following questions: What is the problem / What has been done by other researchers and where you can contribute / What have you done / Which method or tools you used / What are your results / What is new and good, what is not good / Future research.

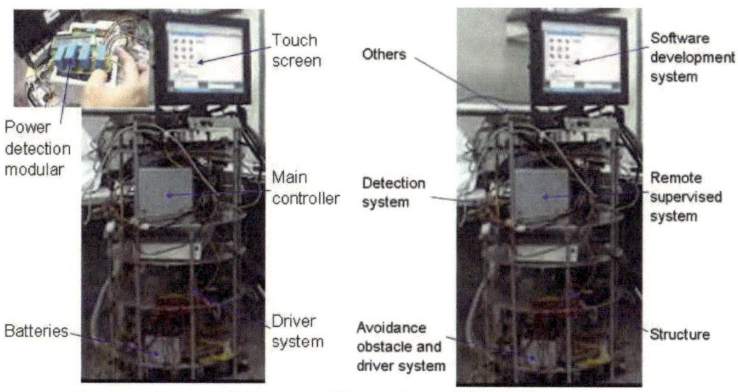

Fig. 1. The contour and structure of the mobile robot (ISLR-I).

4. Power Detection System

The power detection system of the mobile robot is shown in Figure 2. We proposed a power detection and diagnosis system using four current measured values and four voltage measured values, and use a multilevel multisensor fusion method to decide the exact power output of mobile robot. The power detection system contains six parts (see Figure 2). They are main computer, auto-switch, A/D and I/O card, the power detection and isolation module, batteries and three detection algorithms. The main computer implements the statistical signal prediction method and polynomial regression algorithm, and control the A/D and I/O card. The A/D and I/O card can control the auto-switch to cut off the power of the mobile robot. The main controller of the mobile robot can calculate power value according the current and voltage measured values. The redundant management method is implemented in the power detection and isolation module.

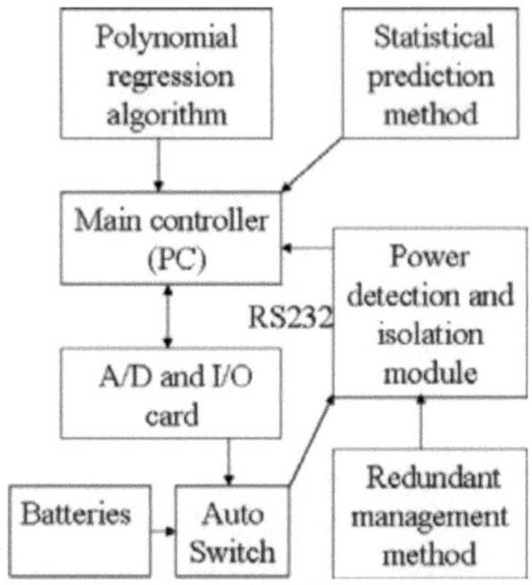

Fig. 2. The power detection and prediction system of the mobile robot

The power detection system of the mobile robot contains four DC type current sensors, a HOLTECK microchip (controller), display and alarm device, some hardware devices and a series interface. The hardware block diagram of the power detection and isolation module is shown in Figure 3. The controller is a HOLTEK microchip (HT46R25), and detects the power variance using four DC type current sensors and voltage measured values. The input signal has scale selection switch and mode selection switch. The output signal contains safety switch, series interface, display and alarm device. The safety switch may be used to turn on or off the power of the mobile robot according to the real status. The power detection module can measure maximum current up to about 50A. The prototype of the power detection, diagnosis and isolation module is shown in Figure 4.

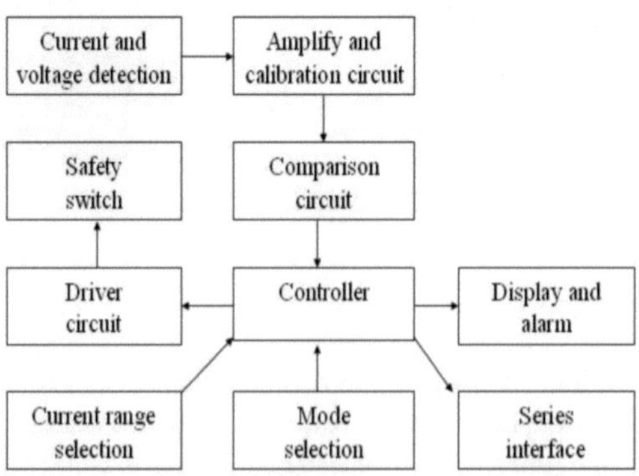

Fig. 3. The hardware block diagram of the power detection module

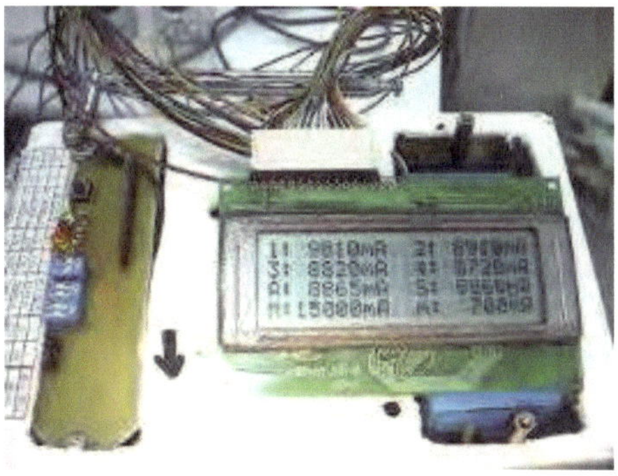

Fig. 4. The prototype of the power detection module

5. Detection Algorithm

In the power detection, diagnosis and isolation module, we use redundant management method and statistical perdition method to detect and diagnose sensory status and isolate faulty sensors. The redundant measurements of a process variable are defined as [4].

$$M = HX + E \tag{1}$$

$$|\varepsilon_i| \le b_i \tag{2}$$

Where:

M = the measure vector (l x 1) is generated from sensors.
H = the measurement matrix (l x n).
X = the n-dimensional measured true values.
E = the measurement error.
b_i = The specified error bound of the measurement m_i

The fault detection and isolation procedure presented in the chapter is applicable to scalar measurements only. The measurement matrix in equation (1) can be chosen as $H = [1,1,...1]^T$, without loss of generality. A pair of scalar measurements m_i and m_j. The magnitude of $(m_i - m_j)$ is compared with the sum (b_i + b_j) of the respective error bounds for a consistency check. Any two scalar measurements m_i and m_j at the sample time k are defined to be consisted if

$$|m_i(k) - m_j(k)| \le (b_i(k) + b_j(k)) \tag{3}$$

Otherwise, we can say the two scalar measurements m_i and m_j to be inconsistency. In this condition, the consistency index of a measurement m_i is defined at a given sample time as

$$I_i = \sum_{j=1}^{l} f\left[\left|m_i - m_j\right| \le (b_i + b_l)\right] \quad i = 1...l \tag{4}$$

Where l sensor numbers and the indicator function are $f[*]$ to be defined as

$$f[*]=\begin{cases}1, \text{if } * \text{ is true}\\ 0, \text{if } * \text{ is false}\end{cases} \tag{5}$$

For each sensory measurement m_i, the degree of inconsistency I_i provides l distinct range from 0 to l, If m_i is more fault than m_j at the given sample time. The I_i value of m_i is smaller than I_j. Then the estimate value $\hat{x}$ of the measured parameter is obtained by a weighted average of the remaining measurements at the sample time

$$\hat{x} = \frac{\sum_{i=1}^{l} m_i I_i}{\sum_{i=1}^{l} I_i} \tag{6}$$

A flowchart of the redundancy sensory management method is shown in Figure 5. We use the proposed method to detect which measurement value to be faulty. We set the parameter initial values and these initial parameter values contain maximum and minimum value, threshold values,...etc. Then the microchip acquires sensory signals using analogy input interface. Next the microchip calculate the *Ii* values using equation (4) ,and calculate indicator function values for each sensor using equation (5) . Finally, we can calculate the estimated value using equation (6) for the power detection module.

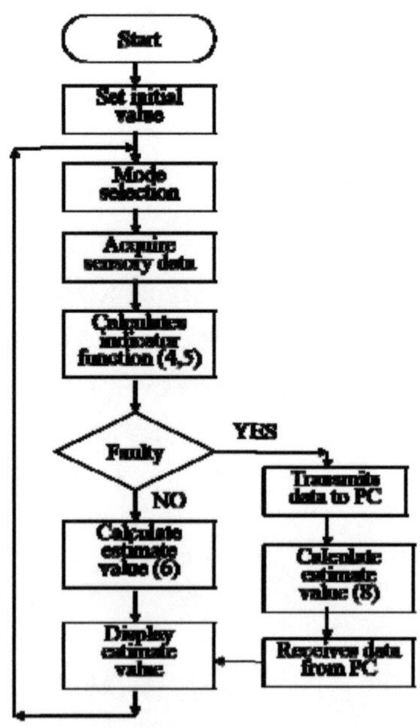

Fig. 5. Flowchart for redundancy sensory management method.

We use redundant sensor management method to detect and diagnose sensory status. When the method is faulty on the $N+1$ measurement value of the power detection module, we can predict the measurement value using P estimation value as before. In the chapter, we select P is 100.

In the level 2, the fusion method is statistical signal perdition method. The fusion decision output of level 1 transmits to main controller (industry personal computer) via series interface (RS232). We model the observed system as the sum of three signal components, to be shown in equation (7).

$$M = X + E \tag{7}$$

If the signal is deterministic and the noise E is Gaussian with zero mean, then we can calculate the mean value from P estimated values. The mean value $\bar{x}$ and standard deviation S_i is [20]:

$$\bar{x} = \frac{\sum_{k=N-P}^{N} \hat{x}_i(k)}{P} \tag{8}$$

$$S_i = \sqrt{\frac{1}{P-1}\sum_{k=1}^{P}\left(m_i(k) - \hat{x}(k)\right)^2} \qquad i = 1,\ldots l \tag{9}$$

Then we use the same b_i as threshold value, and compute the error between the sensors measured value $m_i(N+1)$ and mean value $\bar{x}$. The error is over the threshold, and we can say the sensor measured value $m_i(N+1)$ to be faulty. Otherwise, we can say the sensor measured value is exact. That is

$$w_i(N+1) = \begin{cases} 0 \ , & \left|\dfrac{m_i(N+1) - \bar{x}}{\bar{x}}\right| \geq 0.05 \\ 1 \ , & \left|\dfrac{m_i(N+1) - \bar{x}}{\bar{x}}\right| < 0.05 \end{cases} \qquad i = 1,2,\ldots l \tag{10}$$

$$\hat{x} = \frac{\sum_{i=1}^{l} w_i(N+1)\, m_i(N+1)}{\sum_{i=1}^{l} w_i(N+1)} \tag{11}$$

Finally, we can compute the estimated value of these exact measurement values using equation (11). In the condition, we can say the measurement value of the sensor is faulty. We can not say the sensor to be faulty. That is to say, the proposed method can detect current variety of the mobile robot power system, and isolate the faulty measurement value. But it can not decide the sensor to be broken. We can compute the mean and standard deviation of the each sensor, and compare the standard deviation to decide the broken sensor.

In the redundant management method and statistical signal method, we can get an exact power value for power detection, and isolate faulty signal from current sensor and voltage signal. Then we want to predict the residual power of the mobile robot. First we must fit the curve from the power detection value of the mobile robot. Next the user can set the critical value of the power. The main controller of the mobile robot can calculate the extrapolation value from the critical value, and can calculate the residual working time for the mobile robot.

We fit a second-order polynomial regression

$$y = a_0 + a_1 x + a_2 x^2 + e \tag{12}$$

The sum of the squares of the error is

$$S_r = \sum_{i=1}^{n} (y_i - a_0 - a_1 x_i - a_2 x_i^2)^2 \tag{13}$$

To generate the least squares fit, we take the derivative of Equation (13) with respect to each of the unknown coefficient of the polynomial, and we can get

$$\begin{aligned} na_0 + (\sum x_i)a_1 + (\sum x_i^2)a_2 &= \sum y_i \\ (\sum x_i)a_0 + (\sum x_i^2)a_1 + (\sum x_i^3)a_2 &= \sum x_i y_i \\ (\sum x_i^2)a_0 + (\sum x_i^3)a_1 + (\sum x_i^4)a_2 &= \sum x_i^2 y_i \end{aligned} \tag{14}$$

Finally we can calculate $a_0, a_1,$ and a_2 from Equation (14). Then we set the power critical value to be P_S and

$$a_2 x^2 + a_1 x + a_0 = P_S \tag{15}$$

We can calculate the x value (the unit is second) from Equation (15). The sample time of the power system is 1 second.

6. User Interface

The main controller of the mobile robot is industry personal computer (IPC). The main controller can receive power status of the mobile robot using power detection module. Then the power detection module can transmits four current measured and four voltage measured values, maximum and minimum current values, detection range, detection status, average value and estimate value to the main controller via series interface (RS232). The power detection interface of the mobile robot is shown in Figure 6. The supervised computer can receives measured values from the mobile robot via wireless internet, and display power status of the mobile robot is shown in Figure 7.

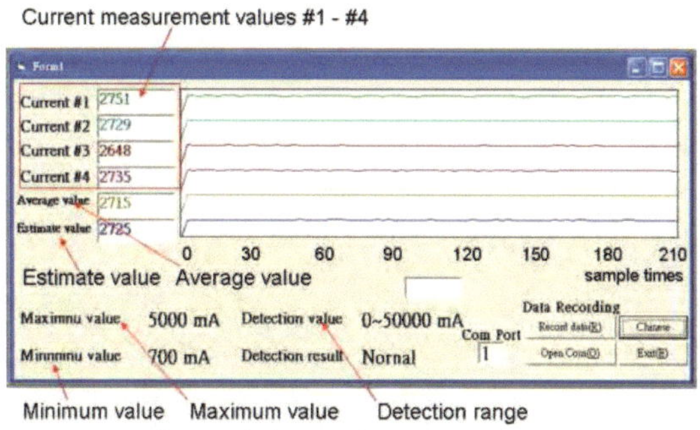

Fig. 6. The power detection interface of the mobile robot

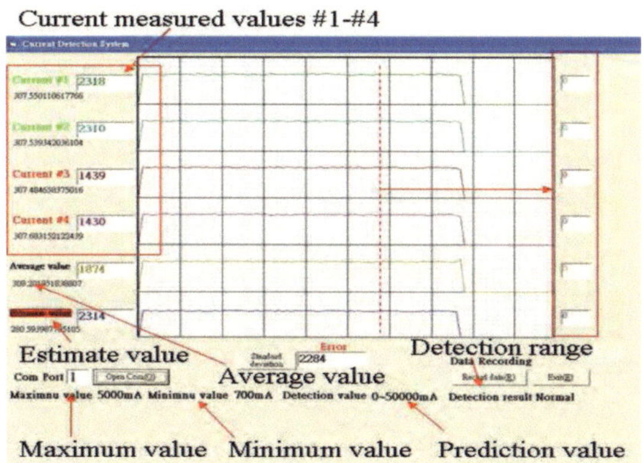

Fig. 7. The power detection interface of the supervised computer

In the monitor, it can display four current, average current and estimation current values on real-time, and plot the curves for these measured values. Users can select any sample time point of these curves using mouse, and display these measured values on the left side of the monitor. It can display the maximum and minimum values, detection range and the standard deviation on the bottom of the monitor.

Another display interface is shown in Figure 8. It can display the standard deviation values on real-time for four current measured values., average measured value and estimated value. It can plot the standard deviation curve for these measurement values. The sample time is one second. The residual power prediction interface of the mobile robot is shown in Figure 9. The upper of the monitor display four current and four voltage measured values, current and voltage average values, and current and voltage estimated values. We plot the curve of power measured value on real-time, and use the proposed method to fit the polynomial curve by the previous one hundred data. Then we set the power critical value to calculate the residual time. It can display on the bottom of the monitor.

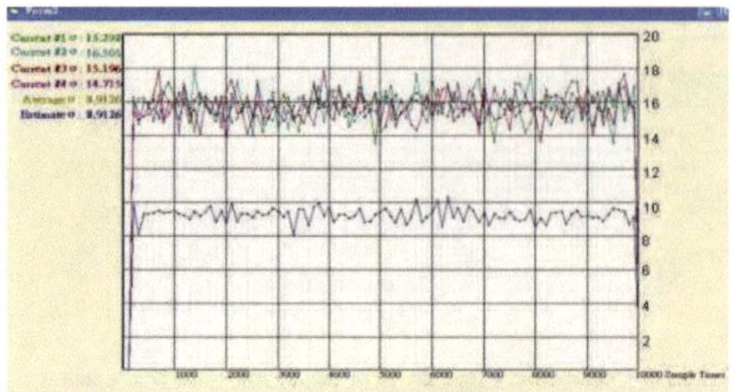

Fig. 8. The standard deviation values for current measurement

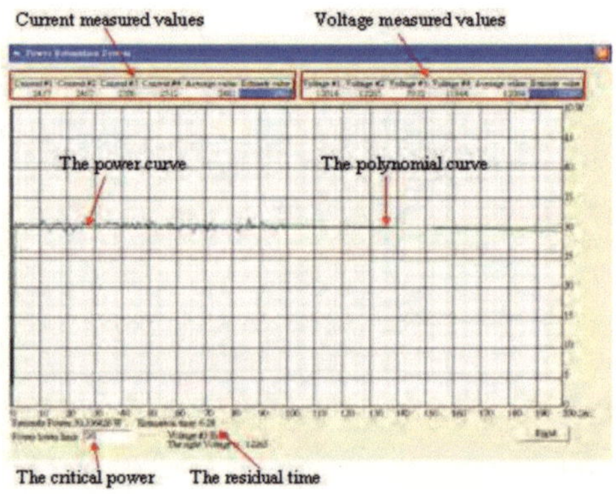

Fig. 9. The residual power prediction

7. Experimental Results

In the power detection system, we use four DC type current sensors to detect the current variety of the mobile robot. The power detection and diagnosis module is equipped in the mobile robot. We test the functions of the power detection module. We can see the module display the current measurement value to be about 9010 mA, and the Amp meter of the mobile robot is 9A. The module can measure the exact current value to be used in the mobile robot, and the experimental result is shown in Figure 10. In the module, we can select the detection mode using switch input to be shown in Figure 11(a). We can select the maximum current is 15A. That is to say, the module must cut off the power output from the power system of the mobile robot, and the power output is over 15A. We can select the minimum current 500mA in the module, too. The power output is small than the minimum current. The module must transmit the signal to mobile robot to find out the recharging station. That is to say, the power of the mobile robot is lack. The experimental is shown is shown in Figure 11(b).

Fig. 10. The power output is about 9 A in the mobile robot

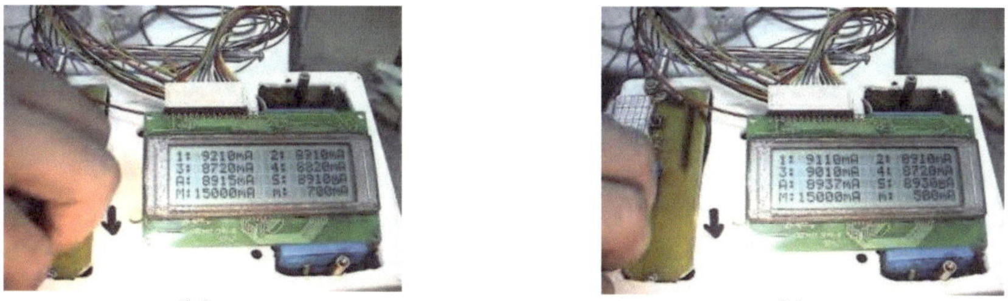

(a) (b)

Fig. 11. The maximum and minimum current selection mode

Then we use four experiments to implement the diagnosis function of the power detection and diagnosis module, and set the threshold value is 5% of the reference value. In the Figure 12, we pick up the current sensor #1 from the power detection module; we can see the LCD panel display 0mA on the current measurement value. The average value is (0mA+ 1470mA+ 1760mA+1470mA)/4=1750mA. The current value is wrong. The exact (estimate) current is (1470mA+ 1470mA) /2=1470mA. The detection value of current sensor #1 is wrong. We must isolate the detection value, and the differential value (1470mA-0mA) is bigger than threshold. The current value (1760mA) of current sensor #3 is wrong, too. We must isolate the detection value (1760mA) that the differential value (1760mA-1470mA) is bigger than threshold.

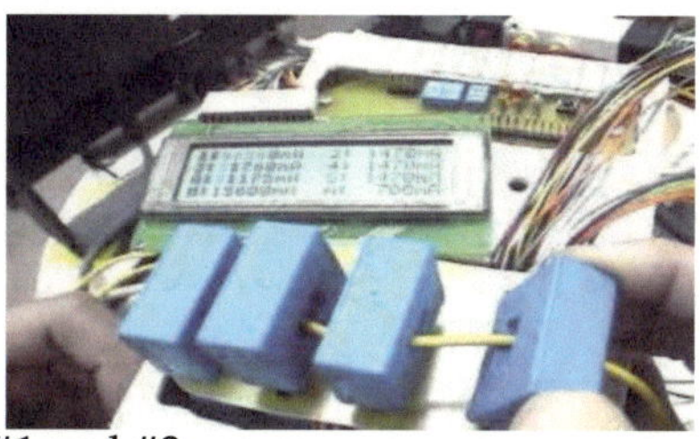

Fig. 12. The current sensor #1 and #3 are wrong

In the Figure 13, We pick up the current sensor #2, and the measurement value of the sensor #4 is wrong. We can see the LCD panel display 90mA on the current measured value. The average value is (2150mA+90mA+2250mA+1860mA)/4 =1587mA. The current value is wrong. The exact (estimate) current is (2150mA+ 2250mA) /2=2100mA. The detection value of current sensor #2 is wrong. We must isolate the detection value for current sensor #2 and #4, and the differential values (2150mA-90mA) and (2150mA-1860mA) are bigger than threshold. We pick up the current sensor #3, the measurement value is error to be shown in the Figure 14. We can see the LCD panel display 0mA on the power detection module. The average value is (1760mA+ 1760mA+ 0mA+ 1660mA)/4=1295mA. The current value is wrong. The exact (estimate) current is (1760mA+ 1760mA+ 1660mA) /3=1726mA. The detection value of current sensor #3 is wrong. We must isolate the detection value for current sensor #3, and the differential value (1760mA-0mA) is bigger than threshold.

Fig. 13. The current sensor #2 and #4 are wrong

Fig. 14. The current sensor #3 is wrong

In the Figure 15, we pick up the current sensor #4, and the measure value of the sensor #3 is wrong. We can see the LCD panel display 90mA. The average value is (2050mA+ 2150mA+ 3620mA+90mA)/4=1977mA. The current value is wrong. The exact (estimate) current is (2050mA+ 2150mA) /2=2100mA. The detection value of current sensor #4 is wrong. We must isolate the detection value, and the differential value (2050mA- 90mA) is bigger than threshold. The current value (1760mA) of current sensor #3 is wrong, too. We must isolate the detection value (3620mA) that the differential value (3620mA-2050mA) is bigger than threshold.

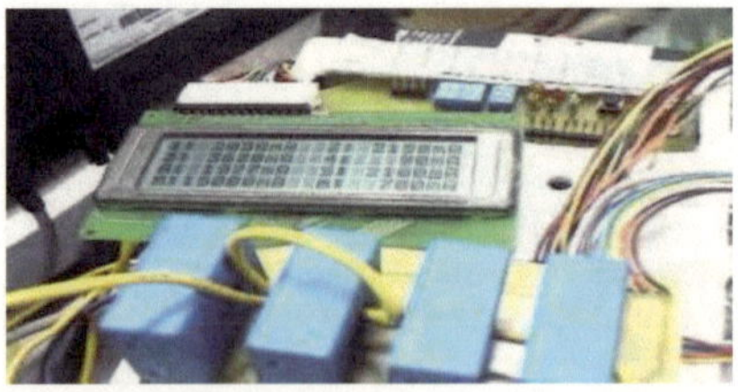

Fig. 15. The current sensor #3 and #4 are wrong

We implement the human machine interface function in the power detection of the mobile robot. The module can transmits the power status and four current measured values to the controller of mobile robot. The experimental scenario is shown in Figure 16. The touch panel can display power status of the mobile robot, and receives four current measured, detection range and maximum and minimum current set values.

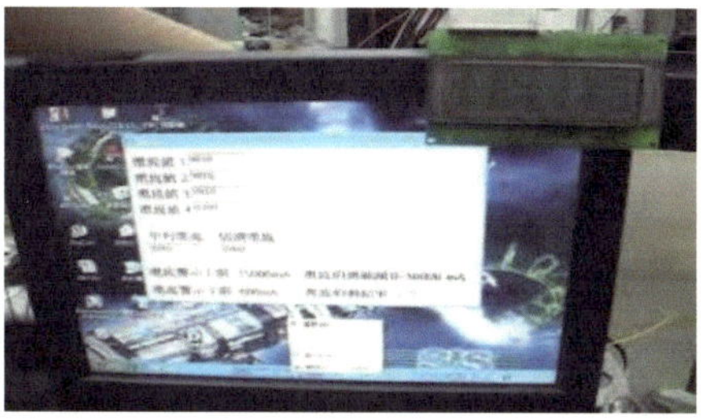

Fig. 16. The interface of mobile robot display the power status

In the current detection and isolation module of the mobile robot, we use four DC type current sensors to detect the current variety for the mobile robot. In the chapter, we assume that a priori of all sensors are the same, and we use computer simulation for four cases. Case I is all sensors are consistent. Case II is one faulty current sensor m_4. There are three sensors m_1, m_2 and m_3 to be consistent. Case III is two consistent pairs, (m_1, m_2) and (m_3, m_4), that are mutually inconsistent. Case IV presents m_1, m_2, m_3 and m_4 to be mutually inconsistent. The simulation experimental results are shown in [5]. Then we use four experiments to implement the diagnosis function of the current detection and isolation module.

CASE I : If all measurements are consistent, i. e., $I_i = 0, i = 1,2,3,4$, the estimated value is the same as average value, and the estimation value is 2640mA, and we can use the equation (16). The experimental results are shown in Fig. 17. The interface of the mobile robot displays the power status in Figure 17 (a). The interface of the supervised computer displays the power status in Figure 17 (b).

$$\begin{aligned}\hat{x} &= \frac{m_1 w_1 + m_2 w_2 + m_3 w_3 + m_4 w_4}{w_1 + w_2 + w_3 + w_4} \\ &= \frac{2647 + 2639 + 2640 + 2641}{4} \approx 2640\end{aligned} \tag{16}$$

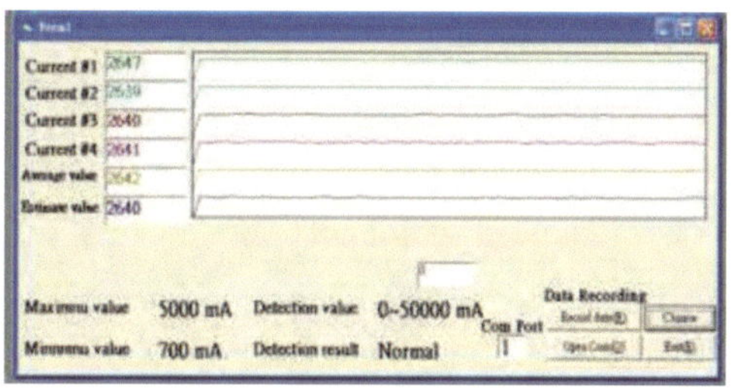

(a) The display status in the mobile robot

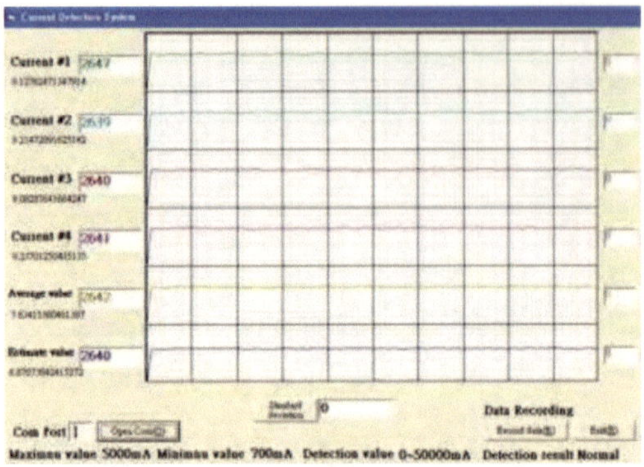

(b) The display status in the supervised computer

Fig. 17. The display status for case I

Case II: The current detection and isolation module has one current sensor to be faulty. We can calculate the Ii and indicator function as follow

For sensor #1(true) and I_1

$$\left|\frac{2569\ mA - 2560\ mA}{2569\ mA}\right| \le 0.05 \qquad f[*]=0 \tag{17}$$

$$\left|\frac{2569mA - 2441mA}{2569mA}\right| = 0.049 \le 0.05 \qquad f[*]=0 \tag{18}$$

$$\left|\frac{2569\ mA - 22429\ mA}{2569\ mA}\right| \ge 0.05 \qquad f[*]=1 \tag{19}$$

$$I_1 = 0+0+1=1 \tag{20}$$

For sensor #4(faulty) and I_4

$$\left|\frac{2429\ mA - 2569\ mA}{2420\ mA}\right| \ge 0.05 \qquad f[*]=1 \tag{21}$$

$$\left|\frac{2429mA - 2560mA}{2429mA}\right| = 0.051 \ge 0.05 \qquad f[*]=1 \tag{22}$$

$$\left|\frac{2429\ mA - 2441\ mA}{2429\ mA}\right| \le 0.05 \qquad f[*]=0 \tag{23}$$

$$I_1 = 1 + 1 + 0 = 2 \tag{24}$$

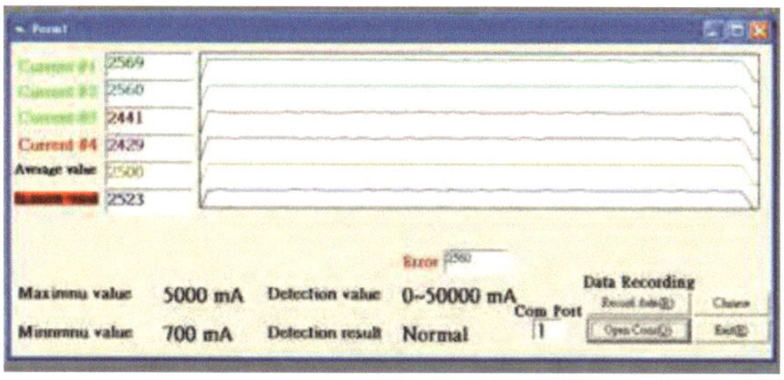

(a) The display status in the mobile robot

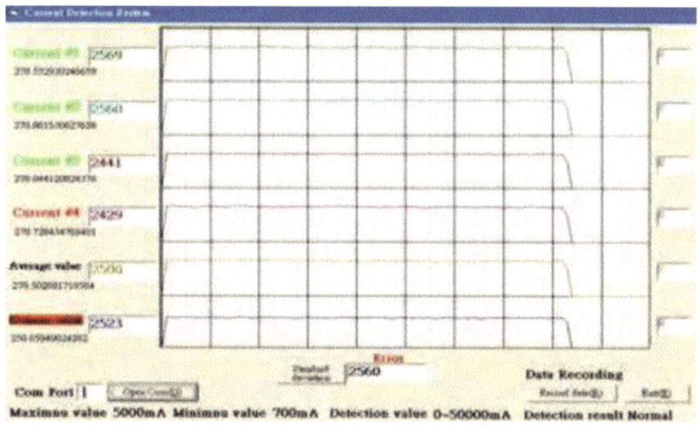

(b) The display status in the supervised computer

Fig. 18. The display status for case II

We can compute $I_1 = I_2 = I_3 = 1$, and $I_4 = 2$. We compute the average value (2569mA+ 2560mA+ 2441mA+ 2429mA)/ 4 =2500mA to be wrong. The current value is wrong. The exact (estimate) current is (2569mA+ 2560mA+ 2441mA)/3=2523mA. The detection value of current sensor #4 is wrong. We must isolate the measured value #4. The experimental result is shown in Fig. 18. The estimation value is

$$\hat{x} = \frac{m_1 w_1 + m_2 w_2 + m_3 w_3}{w_1 + w_2 + w_3} \tag{25}$$

Case III: if two measurements m_3, and m_4 faulty simultaneously and identically, they are two consistent pairs, (m_1, m_2) and (m_3, m_4), that are mutually inconsistent, and we can fine $I_1 = I_2 = I_3 = I_4 = 2$, the result is that no estimate value can be obtained because there is a possible common-mode faulty. That is to say, we can't decide (m_1, m_2) or (m_3, m_4), which pair will be right in the current detection module, and transmits measured values of current sensors to IPC. It can find out faulty sensor using statistical prediction method, and decide an exact estimate value. The experimental result is shown in Fig. 19. The average value is (2318mA+ 2310mA+1439mA+1430mA)/4=1874mA. The current value is wrong. The current detection and isolation module can not calculate exact (estimate) current. The redundant sensor management method is wrong for the case. We must use statistical signal detection method to calculate the exact current value for the case. We can compute the mean value is 2284mA, and

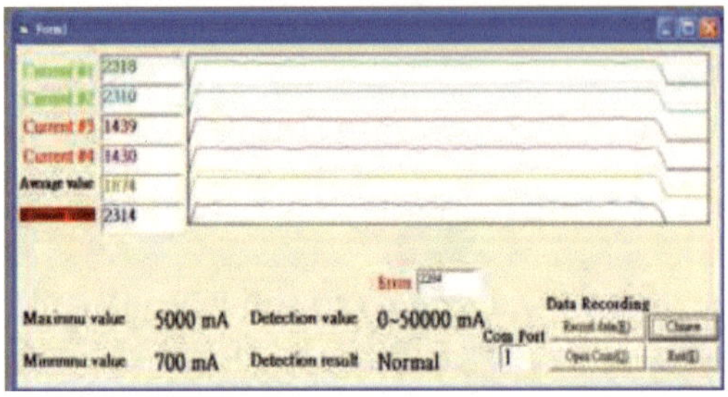

(a) The display status in the mobile robot

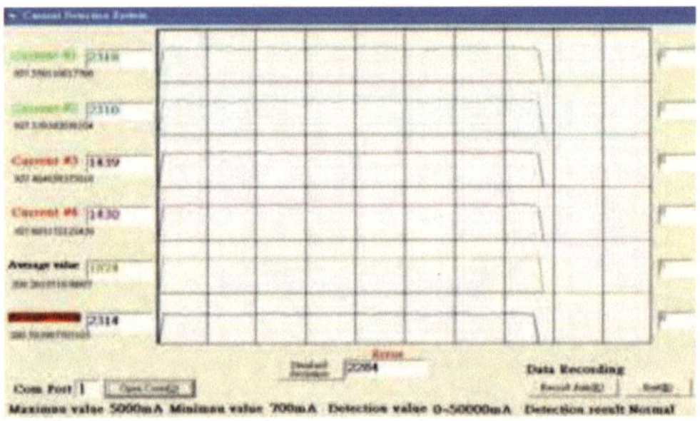

(b) The display status in the supervised computer

Fig. 19. The display status for case III

$$\left|\frac{2284\ mA - 1439\ mA}{2284\ mA}\right| = 0.37 \geq 0,05 \tag{26}$$

$$\left|\frac{2284\ mA - 1430\ mA}{2284\ mA}\right| = 0.37 \geq 0.05 \tag{27}$$

$$\left|\frac{2284\ mA - 2318\ mA}{2284\ mA}\right| = 0.015 \leq 0.05 \tag{28}$$

$$\left|\frac{2284\ mA - 2310\ mA}{2284\ mA}\right| = 0.013 \leq 0.05 \tag{29}$$

We can say the exact current detection is (2318mA+2310mA)/2=2314mA.

Case IV: if m1, m2, m3 and m4 are mutually inconsistent, and we can fine I1=I2=I3=I4=3, no estimate value can be obtained because all measurements are inconsistent. That is to say, we can't decide which sensor will be right in the current detection module, and transmits measured values to IPC. It can find out faulty sensor, and decide an exact estimate value. The experimental result is shown in Fig 20. The average value is (2489mA+ 2219mA+2029mA+2649mA)/4=2347mA. The current value is wrong. The current detection and isolation module can not calculate exact (estimation) current. The redundant sensor management method is wrong for the case. We must use statistical signal detection method to calculate the exact current value for the case. We can compute the mean value is 2198mA, and

$$\left|\frac{2489\ mA - 2198\ mA}{2198\ mA}\right| = 0.013 \geq 0.05 \tag{30}$$

$$\left|\frac{2219\ mA - 2198\ mA}{2198\ mA}\right| = 0.009 \leq 0.05 \tag{31}$$

$$\left|\frac{2029\ mA - 2198\ mA}{2198\ mA}\right| = 0.077 \geq 0.05 \tag{32}$$

$$\left|\frac{2649\ mA - 2198\ mA}{2198\ mA}\right| = 0.21 \geq 0.05 \tag{33}$$

We can say the exact current detection is 2219mA.

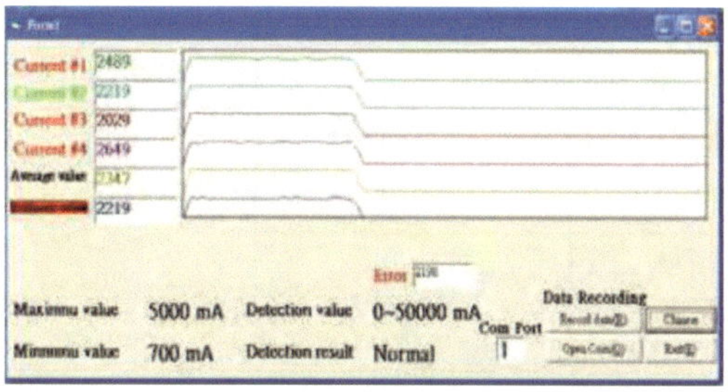

(a)The display status in the mobile robot

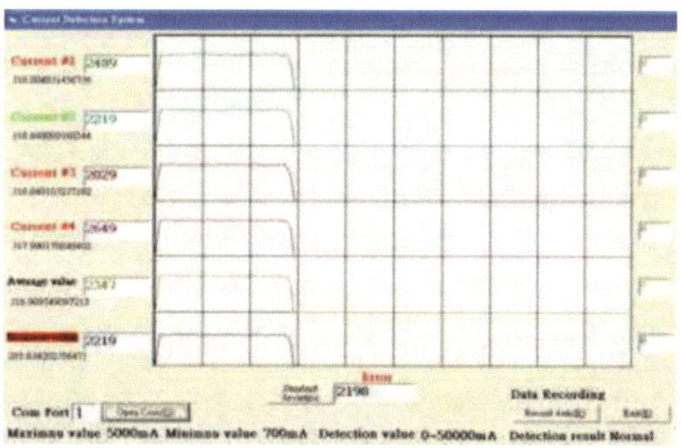

(b) The display status in the supervised computer

Fig. 20. The display status for case IV

In the experimental results of the current sensor diagnosis, we compute the standard deviation of each sensor. If the standard deviation is bigger than threshold, we can say the current sensor to be broken. We must isolate the measured value of the current sensor, and replace it with other current sensor. For case I, all sensors are consistent, and standard deviation varieties of four current sensors are shown in Fig 21. We can see the value is almost 10, and these values are not over threshold.

In the case II, the measured value of current sensor #4 is broken. We can see the standard deviation (145.9) is bigger than threshold. We can diagnose the current sensor #4 to be broken. The experimental result is shown in Fig. 22.

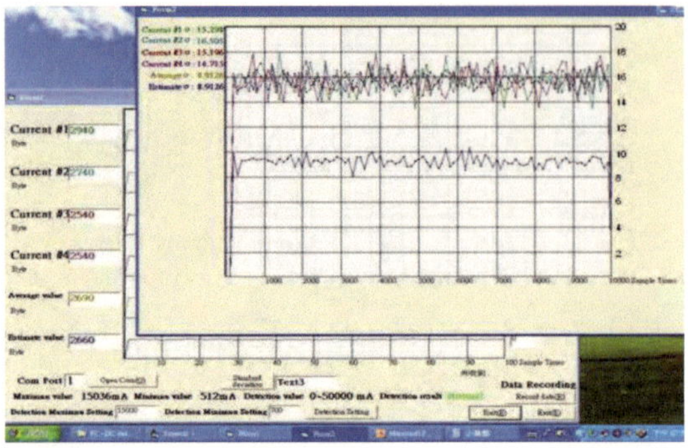

Fig. 21. The standard deviation variety of case I

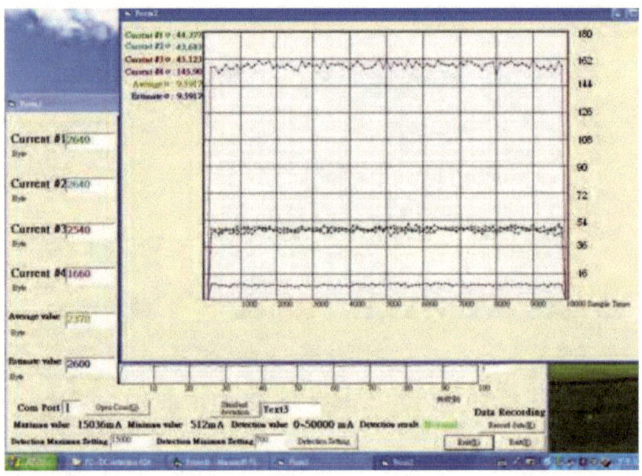

Fig. 22. The standard deviation variety of case II

In the residual power prediction experiment, the user can set the critical power. The proposed method can calculate the residual time, and the power of the mobile robot down to the critical value. In the Fig. 23, the user set the critical power to be 26 W. first, the mobile robot can fit the second-order curve using polynomial regression method. Then it can compute the residual time, display on the bottom of the monitor.

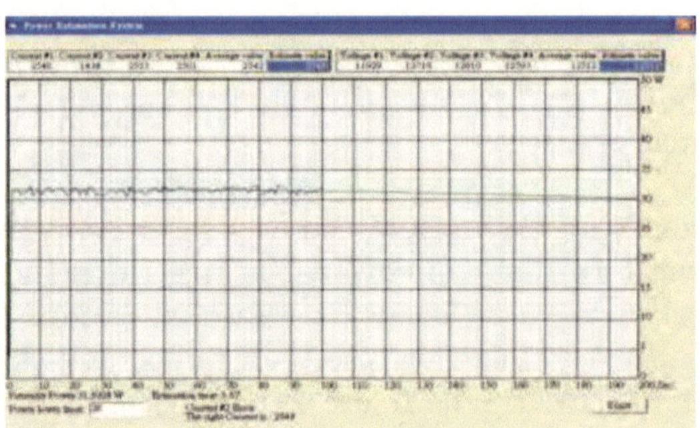

Fig. 23. The residual power prediction for 26W

8. Conclusion

We successful designed a power detection and isolation module that has been integrated in the ISLR-I mobile robot, and calculate the residual power on real-time for the mobile robot. The controller of the power detection and faulty isolation module is HOLTEK microchip. The module can measure maximum current is 50 A, and users can select the current detection range and the detection mode. The detection, isolation and diagnosis algorithm use multilevel multisensor fusion method. There is redundant management method and statistical signal detection method. It can isolate faulty sensor, and estimates an exact power detection value for mobile robot. The module can transmits really current and voltage values, maximum and minimum values, detection range, and detection results to main controller (IPC) of the mobile robot via series interface (RS232). The IPC can transmit the power status to the supervised computer via wireless Internet. The main controller of the mobile robot can fit a second-order polynomial curve using auto-regression method. Then the user can select the critical power value to prediction the residual time on real-time for the mobile robot moving in free space.

9. References

Kuo L. Su , Ting L. Chien and Jr H. Guo, (2004), "Design An Multiagent Based Supervise System Through Internet for Security Robot," The 1nd International Conference on New Technological Innovation for Position, pp.133-138, June 9-11, Congress Center Act city Hamamatsu, Japan.

Ting L. Chien , Kuo L. Su and Jr H. Guo, (2004), "Develop a Multi Interface Based Detection Module for Home Automation," The 1nd International Conference on New Technological Innovation for Position, pp.289-294.

Ren C. Lui, Kuo L. Su and Chi W Deng, (2003), "Power Supply Diagnosis System Using Redundant Sensor for Intelligent Security Robot," IEEE International Conference on Industrial Electronic, Control, and Instrumentation, pp.2500-2506.

H. P. Polenta, A. Ray and J. A. Bernard, (1998), "Microcomputer-based Fault Detection Using Redundant Sensors," IEEE Transactions on Industry Application, Vol.24, No. 5, September/October, pp.905-912.

Ren C. Luo, Kuo L. Su and Chi W. Deng, (2003), "Multisensor Based Power Supply Diagnosis System for Intelligent Security Robot," IEEE International Conference on Industrial Electronic, Control, and Instrumentation, pp.2500-2506.

A.J. Melia, (1978), "supply-current analysis (SCAN) as a screen for bipolar integrated circuits," Electronics Letters, Vol.14, No. 14 , pp.434-436.

G. F. Nelson, W. F. Boggs, (1975), "parametric tests meet the challenge of high_density ICs," Electronics, Dec. PP108-111.

M. W. Levi, (1981), "CMOS is most testable," Proceedings of International Test Conference. pp.217-220.

Y. K. Malaiya, (1984), "Testing stuck-on faults in CMOS integrated circuits", Proceedings of International Conference on Computer-Aided Design, pp.248-250.

Y. K. Malaiya, S. Y. H. Su, (1982), "A new fault model and testing technique for CMOS devices," Proceedings of International Test Conference, pp.25-34

J. F. Frenzel, (1994), "Power-Supply Current Diagnosis of VLSI Circuits," IEEE Transaction on reliability Vol. 43, No.1, pp.30-38.

L. K. Horning et al, (1987), "Measurements of quiescent power supply current for CMOS ICs in production testing", Proceedings of International Test Conference, pp.300-309.

M. sodden and C. F. Hawkins, (1986), "Test considerations for gate oxide shorts in CMOS ICs", IEEE DEsesign & Test, pp.56-64.

C. Crapuchettes, (1987), "Testing CMOS IDD on large devices," Proceedings of International Test Conference, pp.310-315.

M. Keating and D. Meyer, (1987), "A new approach to dynamic IDD testing," Proceedings of International Test Conference, pp.316-321.

L. R. Carley and W. Maly, (1988), "A circuit breaker for redundant IC systems, Proceedings of Custom Integrated Circuits Conference, pp.27.6.1-27.6.6.

W. Maly and P. Nigh, (1988), "Build-in current testing- Feasibility study," Proceedings of International Conference on Computer-Aided Design, pp.340-343.

Kuo L. Su, (2006), "Automatic Fire Detection System Using Adaptive Fusion Algorithm for Fire Fighting Robot," IEEE International Conference on System, Man and Cybernetics, Grand Hotel, Taipei, Taiwan, October 2006, pp.966-971.

Kuo L. Su, Ting L. Chien and Jr H. Guo, (2004), "Design a Low Cost Security Robot Applying in Family," International Conference on Autonomous Robots and Agents, December 13-15, Palmerston North, NZ, pp.367-372.

Kuo L. Su, Ting L. Chien and Jr H. Guo, (2005), "Decelop a Seft-diagnosis Function Auto-recharging Device for Mobile Robot," IEEE International Workshop on Safety, Security, and Rescue Robot, pp.1-6.

6

Design and Implementation of Intelligent Space: a Component Based Approach

Takeshi Sasaki and Hideki Hashimoto
Institute of Industrial Science, The University of Tokyo
Japan

1. Introduction

In recent years, the research field on smart environments, which are spaces with multiple embedded and networked sensors and actuators, has been expanding (Cook & Das, 2004). The smart environments observe the spaces using distributed sensors, extract useful information from the obtained data and provide various services to users. Such an environment is also referred to as smart space, intelligent environment, etc., and many researchers have developed smart environments for providing informative services to the users (e.g. support during meeting (Johanson et al., 2002), health care (Nishida et al., 2000), support of the elderly (Mynatt et al., 2004), information display using a pan-tilt projector (Mori et al., 2004)). On the other hand, smart environments are also used for support of mobile robots that work in complicated human living environments. In this type of smart environments, mobile robots inside the space get necessary information from multiple distributed sensors and various functions such as localization, path planning and human-robot interaction are performed with the support of the system (Mizoguchi et al., 1999), (Sgorbissa & Zaccaria, 2004), (Koide et al., 2004).

Aiming to provide both informative and physical services to the users, we have also been developing a smart environment, called Intelligent Space (iSpace), since 1996 (Lee & Hashimoto, 2002). Fig. 1 shows the concept of iSpace. In iSpace, not only sensor devices but also sensor nodes are distributed in the space because it is necessary to reduce the network load in the large-scale network and it can be realized by processing the raw data in each sensor node before collecting information. We call the sensor node devices distributed in the space DINDs (Distributed Intelligent Network Device). A DIND consists of three basic components: sensors, processors and communication devices. The processors deal with the sensed data and extract useful information about objects (type of object, three dimensional position, etc.), users (identification, posture, activity, etc.) and the environment (geometrical shape, temperature, emergency, etc.). The network of DINDs can realize the observation and understanding of the events in the whole space. Based on the extracted and fused information, actuators such as displays or projectors embedded in the space provide informative services to users. In iSpace, mobile robots are also used as actuators to provide physical services to the users and for them we use the name mobile agents. The mobile agent can utilize the intelligence of iSpace. By using distributed sensors and computers, the mobile agent can operate without restrictions due to the capability of on-board sensors and computers. Moreover, it can understand the request from people and offer appropriate service to them.

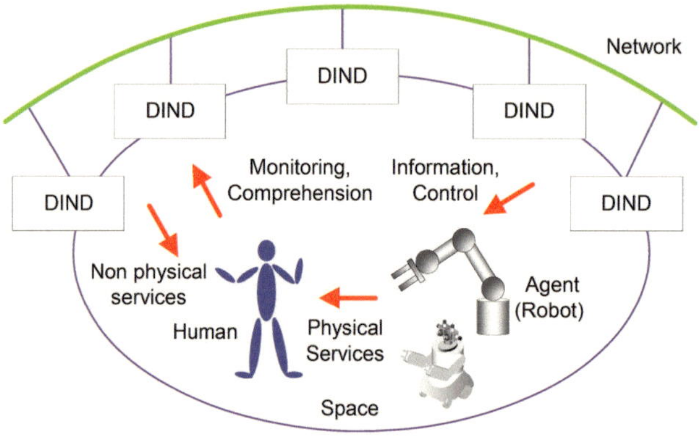

Fig. 1. Concept of Intelligent Space (iSpace)

Smart environments should have flexibility and scalability so that we can easily change the arrangement of embedded devices and switch applications depending on the size of the space, technological advances, etc. Therefore, the system integration becomes an important issue. In this chapter, the problem of implementation of iSpace is addressed. To be more precise, component based system integration of iSpace is described. In order to implement the system efficiently, a development support platform of robot systems is utilized as a RT (Robot Technology) Middleware.

Next section gives a selection of a development support platform. Component design is discussed in section 3. Section 4 describes the implementation of robot technology components. In section 5, experimental results of mobile robot navigation is presented. Conclusion and future work are given in section 6.

2. Development Support Platform of Robot System

2.1 Review of existing platforms

Until now, various types of development support platform of robot systems have been developed. We review some existing platforms in this subsection.

Player/Stage/Gazebo (Player Project, http://playerstage.sourceforge.net/) (Gerkey et al., 2003) is one of the most famous platforms for mobile robot control. Player is a network server for mobile robot control. Player provides interfaces to obtain information from various types of sensors and control actuators. Stage and Gazebo are a 2D and a 3D simulator, respectively. Since the simulators have same interface as that of actual robots, algorithms tested on the simulators can be applied to the actual systems without making major changes in the source codes. Another development support platforms for mobile robot applications is Miro (Neuroinformatik: Robotics, http://www.informatik.uni-ulm.de/neuro/index.php?id=301&L=1) (Utz et al., 2002) that is a middleware for mobile robots. Since Miro uses distributed object technology CORBA (Common Object Request Broker Architecture), the platform can ensure the connectivity of programs that operate on different

operating systems. In addition, Miro provides some basic functions used for mobile robot control including self-localization and map building as class of objects.
Some platforms are also developed for other specific applications. ORiN (Open Resource interface for the Network / Open Robot interface for the Network) (Mizukawa et al., 2004) is a communication interface to provide unified access method to the devices. ORiN also uses DCOM (Distributed Component Object Model) or CORBA as a distributed object middleware. The main target application of ORiN is FA systems. PEIS Middleware (Broxvall, 2007) is a middleware for ubiquitous applications developed in PEIS Ecology project (PEIS Ecology Homepage, http://www.aass.oru.se/~peis/). PEIS Middleware realizes cooperation of distributed devices by using a tuple space which is a kind of a shared memory space. This project also develops Tiny PEIS kernel for tiny networked embedded devices which do not have enough memory.
The other platforms are designed for multiple purposes so that various kinds of robot elements can be developed. In ORCA project (orca-robotics project, http://orca-robotics.sourceforge.net/) (Brooks et al., 2007), a structure of component is defined and the system is developed based on these components. ORCA uses Ice (Internet Communication Engine) as a distributed object middleware. This project has component repository and various components can be downloaded from the website. OpenRTM-aist (RT-Middleware: OpenRTM-aist Official Website, http://www.is.aist.go.jp/rt/OpenRTM-aist/) (Ando et al., 2005) also supports component based system development. OpenRTM-aist utilizes CORBA for ensuring the connectivity between components on the network. The members of this project are working on standardization of Robotic Technology Component in OMG (Object Managing Group) (OMG, 2008) and the latest version of OpenRTM-aist complies with the specification adopted by OMG. OpenRTM-aist also promotes improvement of the development environment and offers a template code generator which makes a source code of a component from the specification of the component (e.g. number of I/O ports, etc.) and a system design tool which provides graphical user interface to change the connection of components and start/stop the system. In addition, a lightweight middleware RTC-Lite is developed for embedded systems that have insufficient resources to operate OpenRTM-aist.
Microsoft Robotics Studio (Microsoft Robotics, http://msdn. microsoft.com/en-us/robotics/default.aspx) (Jackson, 2007) is a development environment for robot systems which adopts service oriented architecture. Microsoft Robotics Studio provides various kinds of tools for implementing robot systems efficiently, for example, a library for asynchronous programming, a visual programming language, a 3D simulator and so on.

2.2 Selection of platform for development of iSpace

In order to determine a platform that is used in this research, we consider following criteria.

1. *Modularity:* In module or component based systems, independent elements (modules or components) of functions of the systems are first developed and the systems are then built by combining the modules. The modularization increases maintainability and reusability of the elements. Moreover, flexible and scalable system can be realized since the system is reconfigured by adding or replacing only related components.
2. *Standardization:* In order to receive the benefit of modularization, it is necessary to ensure the connectivity between components. This means that the interface between components should be standardized. Considering the cooperation of components that are developed by various manufactures, international standardization is desirable.

3. *Suitability for iSpace application:* As mentioned above, some platforms are developed for specific applications, for example, mobile robots and industrial robots. The platforms which are aimed at network sensing or ubiquitous computing are appeared to be suitable for iSpace. However, as also shown in the network robot (Hagita, 2006) and the ubiquitous robot (Kim et al., 2007) concept, iSpace pays attention to cooperation of various types of robots including mobile robots and software robots. Therefore, development support platforms should be applied to various applications.

Table 1 shows a comparison of the existing platforms mentioned in the previous subsection from the view point of modularity, standardization and suitability for iSpace application. We can find that only OpenRTM-aist (AIST, Japan) meets all of our requirements described above. So we decided to use OpenRTM-aist in this research.

Platform	Style of system development	Network	Standardization activity	Target application
Player/Stage/Gazebo	Object oriented	Socket	-	Mobile robot
Miro	Object oriented	CORBA	-	Mobile robot
ORiN (ORiN Ver.2)	Object oriented	DCOM, CORBA	ISO	FA system
PEIS Middleware	Component based	P2P / tuple space	-	Ubiquitous device
ORCA (ORCA2)	Component based	Ice	-	Versatile
OpenRTM-aist	Component based	CORBA	OMG	Versatile
Microsoft Robotics Studio	Service oriented	SOAP	-	Versatile

Table 1. Comparison of development support platforms of robot systems

3. Design of Components

From the point of view of component based system development, the design of functional unit (granularity) of components and interface between components (input and output) is important. In this section, we consider that the observation function of iSpace consists of the information acquisition part and the information integration part. In the following subsections the component design for these two sub-functions and mobile robot navigation function is discussed.

3.1 Design of the information acquisition part

Sensors which can get the same sort of data should be replaced by changing the components without any other modification. Moreover, it is desirable that processing methods can easily be changed depending on the purpose. So we decided to make sensor components and information processing components in the information acquisition part. The outputs of the sensor components are raw sensor data. In addition, the sensor components have interface which provides hardware dependent information used for information processing. The information processing components receive data from the sensor components, process the data and extract required information. Since the outputs of the information processing components are the inputs of the information integration part, the information processing components send reliability information, which is used for information fusion, as well as the extracted information. Fig. 2 shows the configuration of components in the information acquisition part.

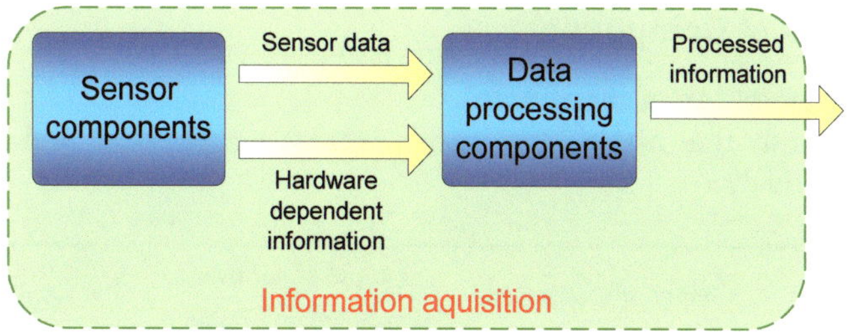

Fig. 2. Component design of information acquisition function

3.2 Design of the information integration part

The inputs and outputs of the information integration part use same data structure as those of the outputs of the information acquisition part so that, in a small system, applications can also receive information from the information acquisition components directly. The information integration part should support various usages of the obtained information, for example, utilizing all obtained information, extracting patterns from the long-term observation information, selecting information which has specific properties and so on. Therefore information from the information acquisition part is fused by the information fusion components and the fused information is stored in the database components. Fig. 3 shows the configuration of components in the information integration part.

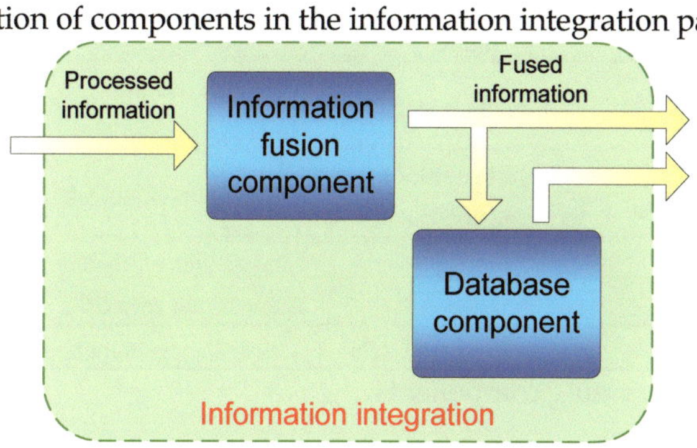

Fig. 3. Component design of information integration function

3.3 Design of mobile robot navigation function

In our previous research (Lee & Hashimoto, 2003), in order to reduce the cost of the system, mobile robots do not have external sensors and iSpace controls the mobile robots completely. However, considering the recent progress of information and robot technology, it is desirable to give autonomy to mobile robots and realize services by cooperation between iSpace and the robots. Therefore, sensors are attached to mobile robots and the mobile robots get information from iSpace according to their request or situations.

Mobile robot navigation can be realized by localization and mapping, path planning and obstacle avoidance, etc. so it is natural to make these functional units as components. We note that we can reuse some sensor and processing components of iSpace as those of mobile robots.

4. Implementation of Components

4.1 Information acquisition components

Table 2 and 3 show the major developed sensor components and data processing components, respectively.

Sensor	Output of sensor data	Output of hardware dependent information	Available devices
CCD camera	RGB image	Image properties (width, height, etc.)	OpenCV compatible cameras
Range imaging camera	Depth image	Image properties (width, height, etc.)	SwissRanger SR-3000 (MESA Imaging)
Laser range finder	Distance to objects, scan area	Observable area information	URG series (Hokuyo automatic)
Microphone	Sound data	Sampling parameters	Any
3D ultrasonic positioning system	3D positions of ultrasound tags	-	ZPS-3D (Furukawa)

Table 2. Sensor components

Sensor	Processing	Input	Output	Method
Camera	Object segmentaion	Camera image	Object positions in image coordinates	Background subtraction, color histogram matching
	3D reconstruction	Object positions in image coordinates	3D positions of objects	Epipolar geometry
Laser range finder	Tracking	Scan data	2D positions of objects Background scan data	Background subtraction
	Mapping	Background scan data	Occupancy grid map	Ray casting

Table 3. Information processing components

The sensors that are distributed in iSpace and mounted on mobile robots - CCD cameras, range imaging cameras, laser range finders, a 3D ultrasonic positioning system and so on were modularized as robot technology components.

To develop information processing components, we consider acquisition of positions of objects since position information is one of the most basic information for iSpace.

For laser range finder, we made components for measuring 2D positions of objects based on background subtraction and clustering. Fig. 4 shows the tracking process. Background subtraction is processes of extracting moving objects. The static parts of scan (background) are subtracted from the scan data in order for determining which parts of the scan are due to moving objects (foreground). The scan points in the foreground are clustered based on the Euclidian distance between them using a nearest neighbor classifier. This divides the foreground to a number of clusters, each belonging to one of the tracked object. Clusters with a small number of scan points are discarded as measurement noise. The details of our laser range finder based tracking method and evaluation of the method are described in (Brscic & Hashimoto, 2006). Fig. 5 shows examples of tracking of a mobile robot and multiple people using laser range finders. In the experiments, three laser range finders were used and the tracked positions obtained in each laser range finder were sent to a position server mentioned below. The figure shows the fused position information in the server.

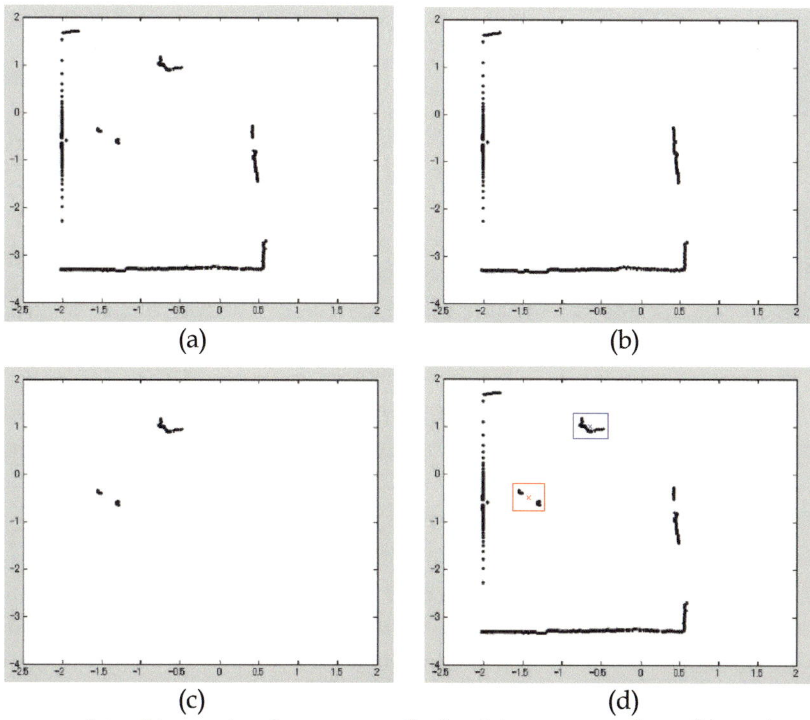

Fig. 4. Process of tracking using laser range finder (a) raw scan data, (b) static part of the scan (background), (c) result of background subtraction (foreground), (d) result of clustering and centers of the clusters. The units of x and y are in meters.

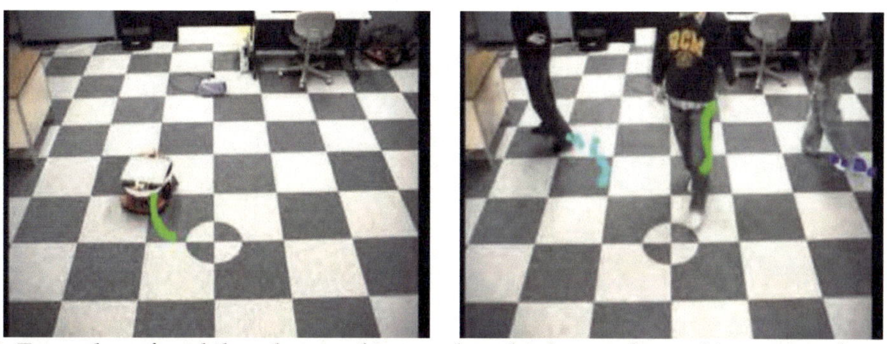

Fig. 5. Examples of mobile robot tracking and multiple people tracking using three laser range finders. The tracking results obtained in a position server mentioned below are overlaid on video images.

For CCD camera, similar algorithm to the one using laser range finder was implemented for detecting moving objects in iSpace. Fig. 6 shows an example of object segmentation based on background subtraction and color histogram matching using a CCD camera. In addition, a component of 3D reconstruction based on epipolar geometry was developed in order to measure 3D positions of the objects.

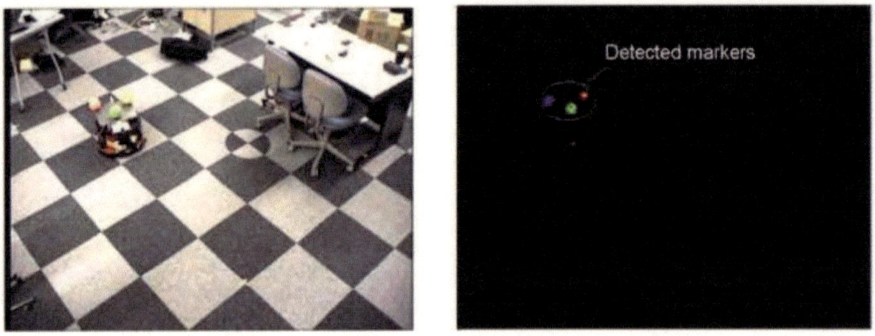

Fig. 6. Detection of color markers on a mobile robot using a camera

Furthermore, geometrical structures of environments are also important since it can be used for obstacle avoidance of mobile robots. Therefore static parts of the space extracted from scan data of laser range finders (backround data) are output as an occupancy grid map that represents whether the part of the environment is occupied by the obstacles or not.

4.2 Design of the information integration part

Information fusion components are implemented for information obtained by the information acquisition parts. So we developed a position server component and a map server component to fuse information. Since it is difficult to determine how many sensor nodes are connected to the server components, DynamicPorts which realize dynamic addition and deletion of data ports were also implemented. Fig. 7 shows an integration of occupancy grid maps obtained by two laser range finders. Fig. 7 (a) and (b) are maps from laser range finder 1 and 2, respectively, and Fig. 7 (c) is an integrated map.

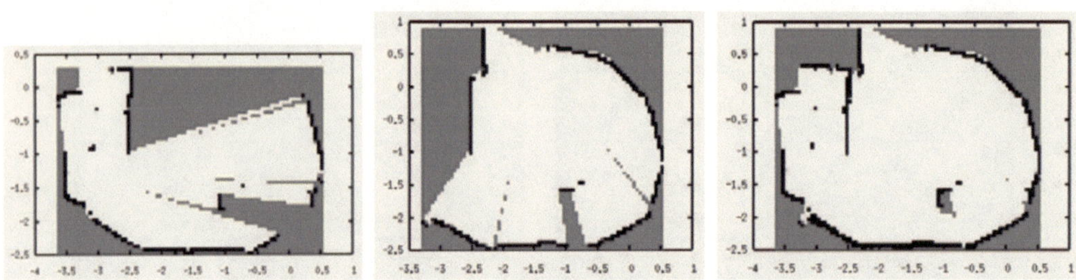

(a) (b) (c)

Fig. 7 Integration of occupancy grid maps in the map server. (a) (b) map from each laser range finder, (c) integrated map. The while, black and gray cells denote free (no obstacles), obstacle and unknown regions, respectively. The units of x and y are in meters.

Moreover, database components for position and map information were developed by using PostgreSQL.

4.3 Utility components

We also developed some components which were considered to be useful for component based system integration. These components are listed in Table 4.

Component	Input	Output	Operation
Console input	-	Data from console	Get data from console
Console output	Data to console	-	Output data to console
Data viewer	Data to be plotted	-	Visualize data
Coordinate transformation	Data to be transformed, transformation parameters	Transformed data	Perform coordinate transformation
Data conversion	Data to be converted	Converted data	Convert data type, e.g. from int to double, etc.
Calibration	Set of corresponding points	Calibration parameters	Calibration support (camera, laser range finder)

Table 4. Utility components

Some of them are implemented for efficient operation of the system. These components include coordinate transformation components, data conversion components, input/output components and so on. Fig. 8 shows results of visualization of sensor data, scan data from a laser range finder component and a depth image obtained from a range imaging camera component, by using the developed viewer component. Such visualized information of outputs of a component can be used for, for example, monitoring of the space and failure detection of the component.

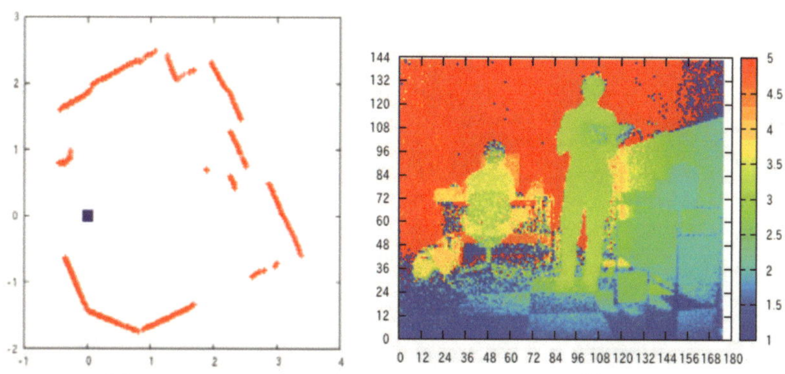

Fig. 8. Examples of visualization of sensor data (a) scan data from a laser range finder URG-04LX. The blue square shows the position of the laser range finder. The units of x and y are in meters. (b) depth map from a 3D camera SwissRanger SR-3000. The bar represents the distance from the sensor to the object in meters. The units of x and y are in pixels.

These tools also contain the components that focus on the development support of iSpace. The representative example of them is calibration support components. Although calibration is needed for proper calculation from the local coordinate system to the world coordinate system, it takes a great deal of time and effort to calibrate many sensors distributed in iSpace. The developed camera and laser range finder calibration support components provide various types of calibration methods including the conventional

manual calibration and the automated calibration based on object tracking (Sasaki & Hashimoto, 2006), (Sasaki & Hashimoto, 2009).
Fig. 9 shows manual calibration of a laser range finder. In the case of manual calibration, in order to estimate the pose of a laser range finder placed in the space (Fig. 9 (a)), a calibration object (an object which can be well detected by a laser range finder) is placed in turn on several points with known global coordinates (Fig. 9 (b) (c)). The calibration parameters are then calculated based on the positions of the calibration object in the global and local (laser range finder's) coordinate system (Fig. 9 (d)).
Fig. 10 shows an example of automated calibration of two laser range finders based on human tracking. In this experiment, the coordinate system of laser range finder 1 was considered as the reference coordinates (Fig. 10 (a)). The result of object tracking in each laser range finder is overlaid on the video image. Before calibration (Fig. 10 (b)), the positions of the person in each sensor's coordinates were stored as corresponding points. At that time, the tracked positions of two laser range finders were not coincident since the pose of laser range finder 2 was unknown. The calibration process was then performed based on the set of corresponding points (Fig. 10 (c)). As a result, almost the same tracked positions were obtained in laser range finder 1 and 2. By fusing these measurements, more reliable estimations could be acquired (Fig. 10 (d)).

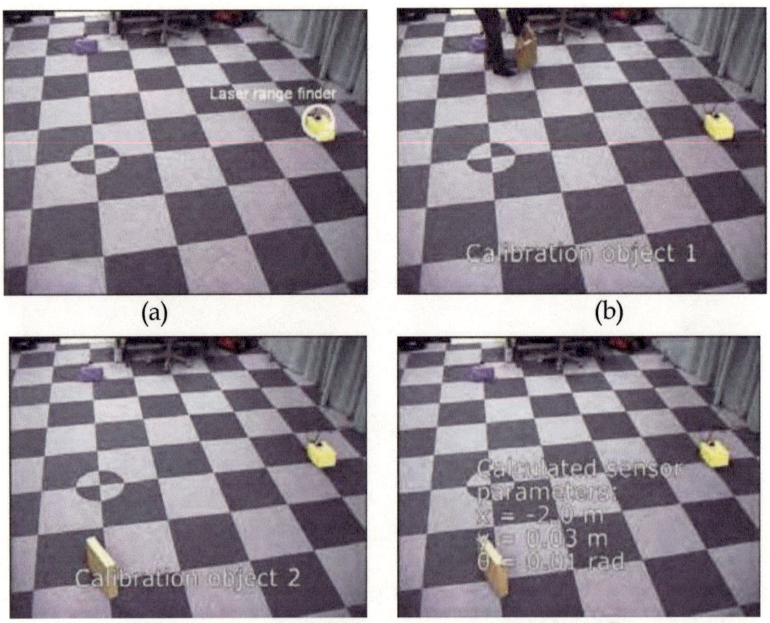

(a) (b) (c) (d)

Fig. 9. Manual calibration of laser range finder using a calibration object (a) arrangement of a laser range finder, (b) (c) placement of a calibration object, (d) calculated calibration parameters

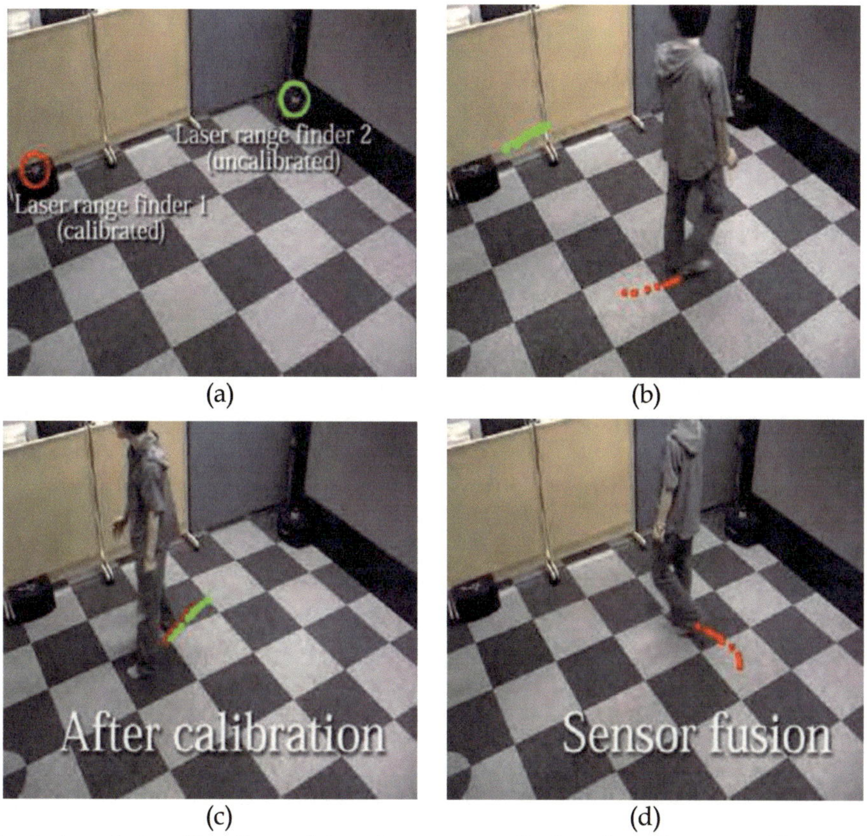

Fig. 10. Automated calibration of two laser range finders based on a calibration object (a) arrangement of laser range finders, (b) tracked positions of a person before calibration, (c) tracked positions after calibration, (d) result of sensor fusion.

4.4 Mobile robot navigation components

In addition to a mobile robot component, localization, mapping and path planning and obstacle avoidance components were developed as mobile robot navigation components. In our mobile robot localization method, the position of the mobile robot measured by distributed sensors in iSpace and the mobile robot on-board sensor data (wheel encoder) are fused using EKF (Extended Kalman Filter) to minimize the position error. To find the best way for the robot to move through the space towards a goal (path planning), we use the Field D* method (Ferguson & Stentz, 2005). Moreover, in order for the robot to follow the calculated path and at the same time avoid bumping into obstacles, the Dynamic Window Approach (DWA) (Fox et al., 1997) is used as a local control algorithm. Table 5 shows the developed mobile robot navigation components.

Component	Input	Output	Method
Localization	On-board sensor data, position information from iSpace	Robot pose	Extended Kalman filter
Path planning and obstacle avoidance	Current pose, occupancy grid map, goal position	Control input	Dynamic Window approach, Field D*
Mobile robot platform	Control input	Encoder data	-

Table 5. Mobile robot navigation components

5. Experiments

5.1 Component based system integration

Fig. 11 shows the procedure of system integration based on RT components. Building a map of the environment by using laser range finders is illustrated in the figure. First, the components needed for the designed system are selected from the component repository and the data (input and output) ports are connected. Here a laser range finder component, a mapping component and a map server component are picked up and connected (Fig. 11 (a) (b)). These components are then activated (Fig.11 (c)) and the tasks are executed: each component receives data from the input ports, processes the data and sends the obtained results to the output ports. In this case, the laser range finder component gets scan data, the mapping component extracts a map of the space from the scan data and the map is stored in the map server component (Fig.11 (d)). As discussed above, one of the major advantages of component based integration is scalability and flexibility of the developed system. By adding second laser range finder and mapping components, the observation area can readily be expanded (Fig. 11 (e) (f)). Moreover, the system can be modified by changing the connection of the components. For example, the output of a laser range finder component can be monitored by connecting it to a viewer component (Fig. 11 (g) (h)).

5.2 Development of mobile robot navigation system

We also present a mobile robot navigation system based on the developed components as an application. The configurations of components for mobile robot navigation in iSpace and on a mobile robot are shown in Fig. 12 and Fig. 13, respectively. The information obtained by distributed devices is sent to the corresponding information servers. The information servers fuse the information and store it to the database. A mobile robot moves to a goal point that is sent from an application or a user while getting necessary information from the databases and onboard sensors.

Fig. 14 shows the experimental environment. In this experiment, a mobile robot moved from the point (-1.5, 1), and made the rounds of three points (-1, 0), (1, -1) and (1.5, 1) and returned to the point (-1, 0). In the room, two laser range finders (LRF 1 and 2) were placed at around (0.5, -2) and (2.5, 0), respectively, and sent the observed map information to the map server. These laser range finders were calibrated by using calibration components in advance. Moreover, a 3D ultrasonic positioning system was installed in the environment. This system involves 96 ultrasonic receivers installed on the ceiling.

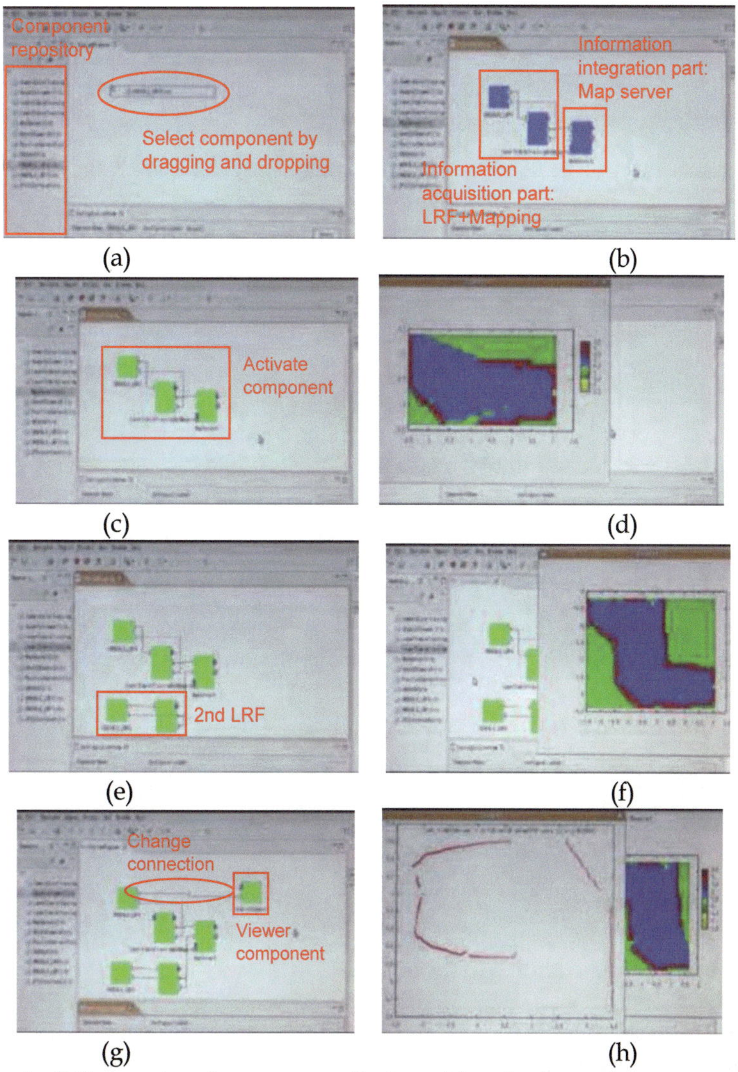

Fig. 11. Map building using laser range finders (a) selection of components from the repository, (b) connection of data ports, (c) activation of components, (d) built map of the space, (e) addition of a second laser range finder, (f) map from two laser range finders, (g) modification of the system , (h) raw sensor data

This system can measure the three dimensional position of an ultrasonic transmitter attached on the top of the mobile robot to an accuracy of 20-80 millimeters using triangulation. So the mobile robot could get the position and map information from the corresponding servers. The mobile robot also had a laser range finder and could detect obstacles around the robot. This means

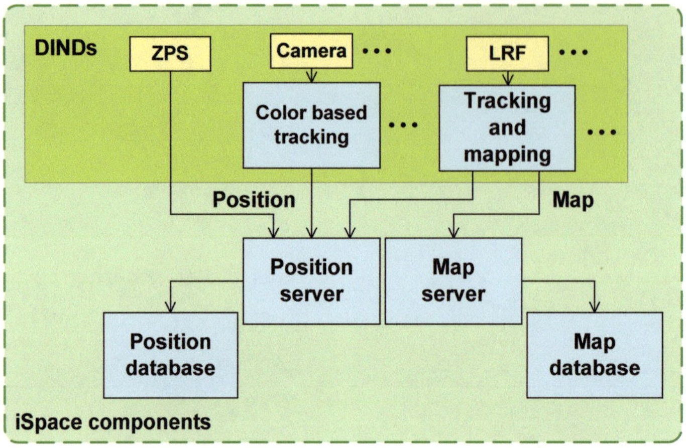

Fig. 12. Component configuration of iSpace

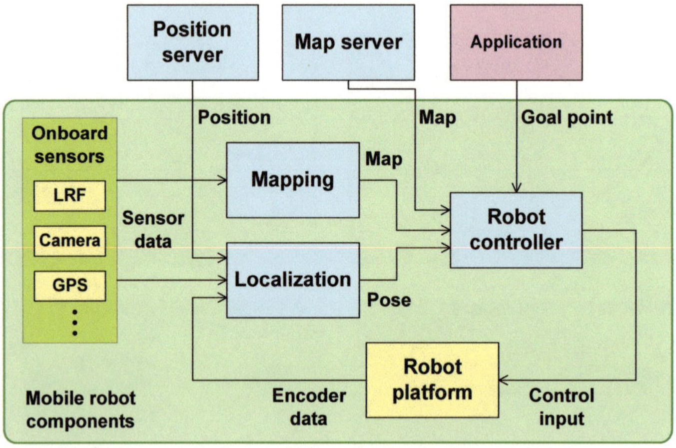

Fig. 13. Component configuration of a mobile robot

that, in this experiment, the 3D ultrasonic positioning system and laser range finders were used as sensors - for localization of mobile robot and map building, respectively.

On the other hand, the mobile robot has a laser range finder to detect obstacles around the robot. For estimating the position of the robot by the 3D ultrasonic positioning system, an ultrasound transmitter is installed on the top of the mobile robot. It is also equipped with a wireless network device to communicate with iSpace. Fig. 15 shows occupancy grid maps of the environment obtained by distributed laser range finders. Since the observable areas of the distributed laser range finders cannot cover the whole space, for example, an obstacle placed around (-0.3, 0.7) is not observed. Therefore, the mobile robot has to detect obstacles in the region by using onboard sensors.

Fig. 16 and 17 show the result and the snapshots of mobile robot navigation. We can find that the mobile robot planed the path by using the map from iSpace and passed through first three subgoals (-1, 0), (1, -1), (1.5, 1). Moreover, on the way from (1.5, 1) to (-1, 0) where the global map was not given, the mobile robot built the local map based on data from the onboard laser range finder and reached the goal point successfully.

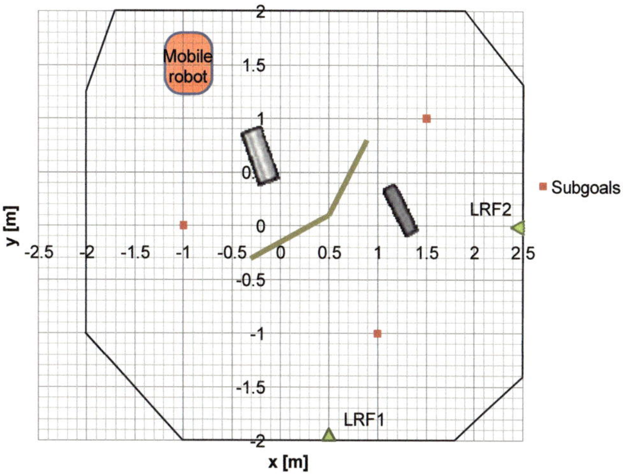

Fig. 14. Experiment environment

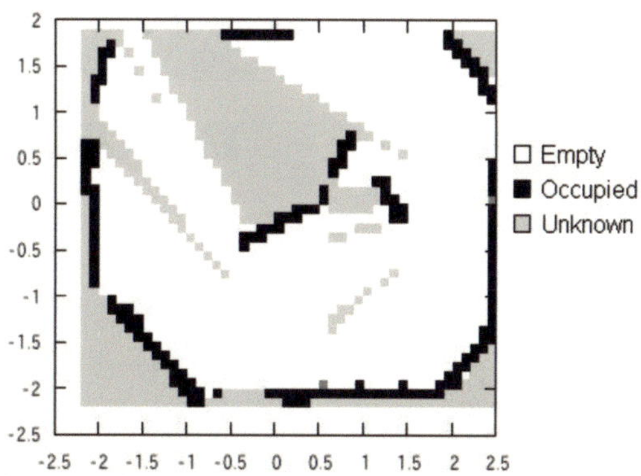

Fig. 15. Occupancy grid maps obtained by distributed laser range finders. The units of x and y are in meters.

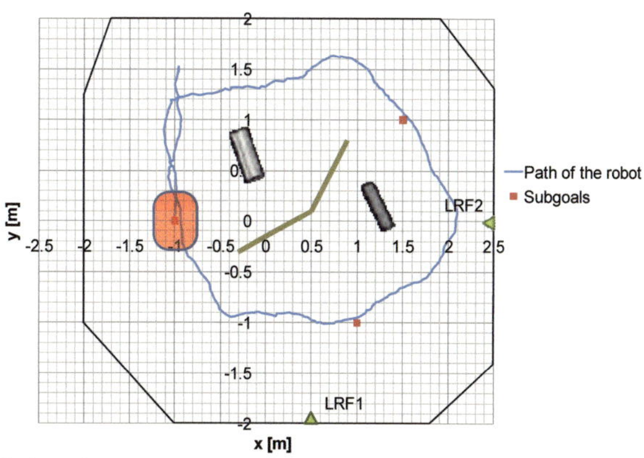

Fig. 16. Result of mobile robot navigation

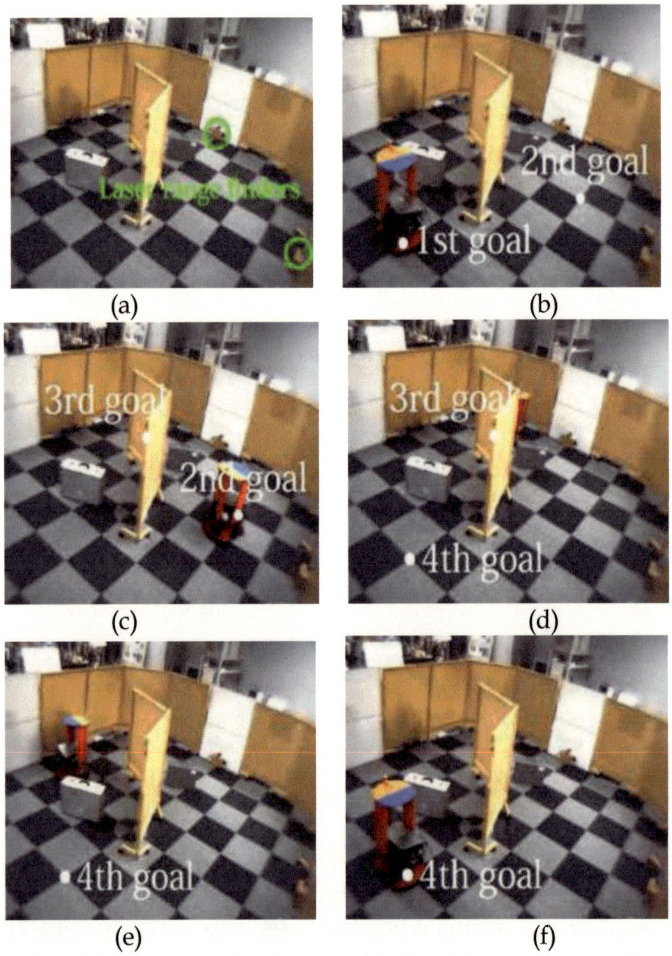

Fig. 17 Snapshot of mobile robot navigation experiment (a) placement of laser range finders, (b)-(d) passage of the first, second and third subgoals, (e) avoidance of a previously undetected obstacle, (f) passage of the fourth subgoal

6. Conclusion

Intelligent robot systems are developed by integration of mechatronics and software technologies. However, the systems are getting more complicated since the cooperation of various types of robots is necessary to realize advanced services for users. Therefore, the system integration becomes an important issue. In order to realize a flexible and scalable system, Intelligent Space (iSpace) is implemented using RT (robot technology) middleware. First we discussed the component design of the information acquisition function and the information integration function. The information acquisition part consists of sensor components and data processing components whereas the information integration part is composed of fusion components and database components. The developed components were then introduced and the operations of them were demonstrated. As an application, a mobile robot navigation system which can utilize information obtained from both distributed and onboard sensors is developed. For future work we will develop sensor and processing components as well as application and actuator components to provide iSpace platform that can realize various types of services to users.

7. References

Ando, N.; Suehiro, T.; Kitagaki, K.; Kotoku, T. & Yoon, W.-K. (2005). RT-Middleware: distributed component middleware for RT (Robot Technology), *Proceedings of the 2005 IEEE/RSJ International Conference on Intelligent Robots and Systems,* pp.3555-3560, ISBN 0-7803-8912-3, Alberta Canada, Aug. 2005.

Brooks, A.; Kaupp, T.; Makarenko, A.; Williams, S. & Oreback, A. (2007). Orca: a component model and repository, In: *Software Engineering for Experimental Robotics (Springer Tracts in Advanced Robotics 30),* Brugali, D. (Ed.), pp.231-251, Springer-Verlag Berlin and Heidelberg GmbH & Co. KG, ISBN 978-3-540-68949-2.

Broxvall, M. (2007). A middleware for ecologies of robotic devices, *Proceedings of the First International Conference on Robot Communication and Coordination,* 30 (1)-(8), ISBN 978-963-9799-08-0, Athens, Greece, Oct. 2007.

Brscic, D. & Hashimoto, H. (2006). Tracking of objects in Intelligent Space using laser range finders, *Proceedings of the 2006 IEEE International Conference on Industrial Technology,* pp.1723-1728, ISBN 1-4244-0726-5, Mumbai, India, Dec. 2006.

Cook, D. J. & Das, S. K. (2004). *Smart Environments: Technologies, Protocols, and Applications (Wiley Series on Parallel and Distributed Computing),* Wiley-Interscience, ISBN 0-471-54448-7, USA.

Ferguson, D. & Stentz, A. (2005). The Field D* algorithm for improved path planning and replanning in uniform and non-uniform cost environments, *Technical Report CMU-RI-TR-05-19,* Robotics Institute, Jun. 2005.

Fox, D.; Burgard, W. & Thrun, S. (1997). The dynamic window approach to collision avoidance, *IEEE Robotics and Automation Magazine,* Vol.4, No.1, (Mar. 1997) pp.23-33, ISSN 1070-9932.

Gerkey, B.; Vaughan, R. T. & Howard, A. (2003). The Player/Stage project: tools for multi-robot and distributed sensor systems, *Proceedings of the 11th International Conference on Advanced Robotics,* pp.317-323, Coimbra, Portugal, Jul. 2003.

Hagita, N. (2006). Communication robots in the network robot framework, *Proceedings of the 9th International Conference on Control, Automation, Robotics and Vision,* pp.1-6, ISBN 1-4244-0341-3, Singapore, Dec. 2006.

Jackson, J. (2007). Microsoft Robotics Studio: a technical introduction, *IEEE Robotics and Automation Magazine,* Vol.14, No.4, (Dec. 2007) pp.82-87, ISSN 1070-9932.

Johanson, B.; Fox, A. & Winograd, T. (2002). The Interactive Workspaces project: experiences with ubiquitous computing rooms, *IEEE Pervasive Computing,* Vol.1, No.2, (Apr.-Jun. 2002) pp.67-74, ISSN 1536-1268.

Kim, J.-H.; Lee, K.-H.; Kim, Y.-D.; Kuppuswamy, N. S. & Jo, J. (2007) Ubiquitous robot: a new paradigm for integrated services, *Proceedings of the 2007 IEEE International Conference on Robotics and Automation,* pp.2853-2858, ISBN 1-4244-0601-3, Rome, Italy, Apr. 2007.

Koide, Y.; Kanda, T.; Sumi, Y.; Kogure, K. & Ishiguro, H. (2004). An approach to integrating an interactive guide robot with ubiquitous sensors, *Proceedings of the 2004 IEEE/RSJ International Conference on Intelligent Robots and Systems,* Vol.3, pp.2500–2505, ISBN 0-7803-8463-6, Sendai, Japan, Sep.-Oct. 2004.

Lee, J-H. & Hashimoto, H. (2002). Intelligent Space - concept and contents, *Advanced Robotics,* Vol.16, No.3, (Apr. 2002) pp.265-280, ISSN 0169-1864.

Lee, J.-H. & Hashimoto, H. (2003). Controlling mobile robots in distributed intelligent sensor network, *IEEE Transaction on Industrial Electronics,* Vol.50, No.5, (Oct. 2003) pp.890-902, ISSN 0278-0046.

Mizoguchi, F.; Ohwada, H.; Nishiyama, H. & Hiraishi, H. (1999). Smart office robot collaboration based on multi-agent programming, *Artificial Intelligence,* Vol.114, No.1-2, (Oct. 1999) pp.57-94, ISSN 0004-3702.

Mizukawa, M.; Sakakibara, S. & Otera, N. (2004). Implementation and applications of open data network interface 'ORiN', *Proceedings of the SICE 2004 Annual Conference,* Vol.2, pp.1340-1343, ISBN 4-907764-22-7, Sapporo, Japan, Aug. 2004.

Mori, T.; Hayama, N.; Noguchi, H. & Sato, T. (2004). Informational support in distributed sensor environment sensing room, *Proceedings of 13th IEEE International Workshop on Robot and Human Interactive Communication,* pp.353-358, ISBN 0-7803-8570-5, Kurashiki, Japan, Sep. 2004.

Mynatt, E. D.; Melenhorst, A.-S.; Fisk, A.-D. & Rogers, W. A. (2004). Aware technologies for aging in place: understanding user needs and attitudes, *IEEE Pervasive Computing,* Vol.3, No.2, (Apr.-Jun. 2004) pp.36-41, ISSN 1536-1268.

Nishida, Y.; Hori, T.; Suehiro, T. & Hirai, S. (2000). Sensorized environment for self-communication based on observation of daily human behavior, *Proceedings of the 2000 IEEE/RSJ International Conference on Intelligent Robots and Systems,* Vol.2, pp.1364-1372, ISBN 0-7803-6348-5, Takamatsu, Japan, Nov. 2000.

OMG (2008). Robotic Technology Component Specification Version 1.0, OMG Document Number: formal/2008-04-04, Apr. 2008.

Sasaki, T. & Hashimoto, H. (2006). Camera calibration using mobile robot in Intelligent Space, *Proceedings of SICE-ICASE International Joint Conference 2006,* pp.2657-2662, ISBN 89-950038-4-7, Busan, Korea, Oct. 2006.

Sasaki, T. & Hashimoto, H. (2009). Calibration of laser range finders based on moving object tracking in Intelligent Space, *Proceedings of the 2009 IEEE International Conference on Networking, Sensing and Control,* pp.620-625, ISBN 978-1-4244-3491-6, Okayama, Japan, Mar. 2009.

Sgorbissa, A. & Zaccaria, R. (2004). The artificial ecosystem: a distributed approach to service robotics, *Proceedings of the 2004 IEEE International Conference on Robotics and Automation,* Vol.4, pp.3531-3536, ISBN 0-7803-8232-3, New Orleans, LA, USA, Apr.-May 2004.

Utz, H.; Sablatnog, S.; Enderle, S. & Kraetzschmar, G. (2002). Miro - middleware for mobile robot applications, *IEEE Transaction on Robotics and Automation,* Vol.18, No.4, (Aug. 2002) pp.493-497, ISSN 1042-296X.

Application of robotic and mechatronic systems to neurorehabilitation

Stefano Mazzoleni[1], Paolo Dario[1], Maria Chiara Carrozza[1] and Eugenio Guglielmelli[2]

[1] *ARTS Lab, Scuola Superiore Sant'Anna, Pisa, Italy*

[2] *Laboratory of Biomedical Robotics & EMC, Campus Bio-Medico University, Rome, Italy*

1. Introduction

During the last decades, the potentiality of robotics as a tool for neuroscientific investigations has been demonstrated, thus contributing to increase knowledge on biological systems. On the other hand, a detailed analysis of the potentialities of these systems (Dario et al., 2003) based on recent neuroscientific achievements, in particular about the mechanisms of neurogenesis and cerebral plasticity underlying the motor learning and the functional recovery after cerebral injury, highlights the advisability of using the robotic technologies, as systems able to contribute to a breakthrough in the clinical procedures of neurorehabilitative treatments. Several examples of robotic machines applied to both neuroscience and neurorehabilitation can be found in the literature (Krebs et al., 1998; Colombo et al, 2000).

One of the main scientific and technological challenges is represented by the design and development of innovative robotic and mechatronic systems able to i) simplify interaction modalities during assisted motor exercises, ii) enhance adaptability of the machines to the actual patient performance and residual abilities, iii) provide a comprehensive picture of the psycho-physiological status of the patient for assessment purposes, through the integrated use of brain imaging techniques.

The basic assumption of this work relies on a human-centred approach applied to the design of robotic and mechatronic devices aimed at carrying out neuroscientific investigations on human sensorimotor behaviour, delivering innovative neurorehabilitation therapies and assessing the functional recovery of disabled patients. Special attention is paid to the issues related to human-machine interaction modalities inspired to human motor mechanisms and the design of machines for the analysis of human motor behaviour and the quantitative assessment of motor performance.

2. Background

In industrialized societies, several factors contribute to a growing need for rehabilitative services, as complement and support to surgical and pharmacological treatments. The main

of them are the increasing longevity of the population, the trend towards reducing the duration of hospitalization, the use of therapies that can treat highly progressive debilitating diseases, the increased incidence of severe and moderate disabilities resulting from the activities at risk of injury and trauma, the use of advanced techniques of resuscitation.
The need for appropriate rehabilitative therapies has an increasing importance in many motor disorders of neurological origin: in this case we speak more specifically of neurorehabilitation.
Millions of people worldwide suffer from motor disorders associated with neurological problems such as stroke, brain injuries, spinal cord injuries, multiple sclerosis, Parkinson's disease.
A brief outlook to the Italian situation can help to understand the impact: each year in Italy about 196,000 strokes occur[1] with approximately 20% of affected people who die within the first month following the acute event and 30% of survivors are severely disabled. Of these 196,000, 80% are first episodes, whereas 20% are relapses.
Stroke represents the third leading cause of death in industrialized countries, after cardiovascular diseases and cancer and the leading cause of disability with a significant impact at individual, family and social level (Feygin et al., 2003; Murray et al., 1997; Marini et al., 2004).
The incidence of stroke progressively increases with age: it reaches the maximum value in people over 85 years old (24.2%) with a male predominance (28.2%) than females (21.8%). The prevalence of stroke in the Italian elderly population (age 65-84 years) is equivalent to 6.5% and is slightly higher in men (7.4%) than in women (5.9%). Stroke affects, although to a lesser extent, young people: every year about 27,000 people in productive age (<65 years) are affected (SPREAD, 2007).
In the U.S., the estimated cost of hospitalization due to stroke in 1998 is $68.9 billion (Heart Disease and Stroke Statistics 2009 Update) .

The traditional therapy methods present some limits, which is important to focus on. In many of the above mentioned cases, the traditional motor rehabilitative approaches involve manipulation of the paretic upper limb by the therapist. Usually the treatment is planned by assessing *ex ante* the residual abilities of the subject and can last several hours a day: it can be often a long and exhausting exercise for both the patient and the therapist.
The therapeutic treatments can be extended for several months after hospitalization, during which patients must travel daily to the clinical facilities and face hard discomforts for themselves and their family.
Moreover, for many motor disorders is not yet sufficiently clear what are the therapeutic approaches and clinical protocols that are objectively more effective for a better recovery of motor function; it partly derives from the fact that the residual abilities of the patient are often assessed by using largely subjective methods of measure, and that makes difficult an adequate evaluation of rehabilitation treatment's effects on the patient.
The nature of these treatments, which have to be administered by therapists on a patient at a time, and the lack of methodologies and tools able to compare the different rehabilitative therapies and their effectiveness make the costs associated to rehabilitation services typically high; thus, the ratio between the number of qualified human resources to be used for the rehabilitative services and the number of patients is often higher than one. It is also difficult

[1] Data extrapolated from the population in 2001.

to define methods for assessing and improving the cost/effectiveness ratio related to specific rehabilitation programs.

The use of robotic machines for neurorehabilitation is inspired by the neurophysiological evidence showing that, starting from the cellular level, synaptic connections undergo continuous changes, in response to physiological events, environmental stimuli (processes of learning and memory) and damages to the Central Nervous System (CNS)[2].

The topology of the motor and sensory cortex is not fixed, but flexible and adaptable to learning and experience (Donoghue et al., 1996). This characteristic of the motor cortex has important implications for rehabilitation: a) rapid changes in cortical activity can occur, b) the intensive training of cortical area may occur at the surrounding areas' expense and c) cortical areas can adapt their functions to those changes.

Thanks to the brain imaging techniques, such as functional Magnetic Resonance Imaging (fMRI), Positron Emission Tomography (PET), Transcranial Magnetic Stimulation (TMS) associated to Motor Evoked Potentials (MEP) and electrical stimulation, changes in CNS's excitability and topology can be shown. Through the use of such techniques, it is possible to identify regions that have suffered a damage and apply a specific therapy.

The sensorimotor learning is influenced by physical (sensory feedback such as vision, hearing and proprioception), psychological (pleasure/pain, motivation, emotional impulses and desire) and cognitive (decision making, planning, reasoning, concentration and attention, language and understanding, previous experiences) factors.

It can be facilitated by the repetition of movements directed to specific targets (goal-oriented movements), the strengthening of muscles and the increase of the range of motion (ROM), the modulation of spasticity, an increased demand of focusing attention on the movement and the increase of sensory stimuli.

In recent years different research groups have studied and developed innovative robotic and mechatronic systems able to let the patient perform repetitive and goal-oriented movements. These systems can provide a safe and intensive training[3] that can be carried out in association with other types of treatment, appropriate to the different residual motor abilities, potentially able to significantly improve the rehabilitation outcomes, to perform an objective assessment and to improve the planning and use of healthcare resources.

In the rehabilitation assisted by a robot, the patient's role is undoubtedly central: the machine supports, and, if necessary, completes, the movement performed by the patient according to his/her residual motor abilities ("assisted as needed" control strategy).

---

[2] The terms *neuroplasticity* or *neural plasticity* are used to point out the sequence of changes in chemical (interaction between neurotransmitter and receptor), electrical (long-term depression and long-term potentiation) and molecular (activation of transcription factors and protein synthesis) responses, which lead to a reorganization of connections in the cerebral areas and, consequently, to cognitive changes and stable behaviours.

[3] During each training session using robotic systems, a high number of movements can be performed: the repetition of motor actions is a factor which can promote the recovery of motor functions.

People suffering from motor disorders can perform the rehabilitation therapy with the support of a "rehabilitation machine"[4].

The patient, through the interaction with these systems, receives different sensorimotor and cognitive inputs, such as proprioceptive and visual *stimuli*, motivational incentives[5]: by using appropriate sensors, the machine is capable of measuring dynamic variables of clinical interest during the performance of active and passive movements by the patient. Thus, a quantitative assessment of specific physiological mechanisms, of motor recovery and functional skills can be carried out. This type of assessment is much more accurate than those using traditional methods. In addition, the machine may enable the therapist to plan the treatment and let the patient execute a wide sequence of movements, which can be useful for the limb rehabilitation.

The application of machines to rehabilitation is sometimes limited by technical and functional factors; their real advantage in clinical applications has been only partially proved. However, there are solid arguments that encourage researchers to design and develop innovative systems for rehabilitation, which derive a direct benefit from the scientific and technological progress in the field of bioengineering, particularly in biomedical robotics and mechatronics.

The clinical potential of these machines, however, is clearly significant as they can, on the one hand, assist the therapist in the administration of a patient-specific physical therapy, with the accuracy and repeatability, typical of the robotic systems, and on the other hand, to acquire quantitative information on the patient's movements.

Such information may be useful for the evaluation of both the patient's motor function and the mechanisms of motor recovery. These machines can also enable the patient to perform rehabilitative sessions in a semi-autonomous modality, and, in principle, even at his/her own home, thus reducing the need for the therapist's continuous commitment[6].

The technological innovation in robotics and mechatronics has contributed to achieve encouraging results in the knowledge of motor recovery mechanisms and to a real progress in the rehabilitation field, with a potential high impact.

---

[4] A "rehabilitation machine" is a mechatronic or robotic system able to support the therapist during the administration of programmable and customized rehabilitation programs. It is composed by a mechanical structure where the following modules are present: 1) actuators, 2) energy supply, 3) proprioceptive and exteroceptive sensors, providing information on the machine status and the interaction between the machine and the environment, respectively, 4) a microcontroller, dedicated to the processing of data from sensors and generation of motor control commands and 5) a human-machine interface (graphical user interface), dedicated to user inputs, data recording and feedback output.

[5] The patient feels often rewarded by the use of high-tech systems for rehabilitation. Besides, motivational incentives are strongly stimulated by the use of graphical interfaces which provide a feedback on the performed movements, which are linked to the recovery of essential functionalities for his/her daily life.

[6] This therapy known as "tele-rehabilitation" is based on the integration of high-tech systems (i.e., a robotic system for rehabilitation) and telecommunication infrastructures (i.e., cable connections, optical fibres, wireless networks and satellite systems): it is aimed at enabling the execution of rehabilitation treatments at own home or rehabilitation centre, through the direct remote supervision and monitoring by physicians and therapists.

In the wide range of technological applications developed in the context of biomedical robotics, undoubtedly a class of particular importance is represented by the systems for the rehabilitation of patients who have a reduced mobility, following an injury or disease.
In the next paragraphs, robotic systems for upper and lower extremities rehabilitation and mechatronic systems for the functional assessment and the movement analysis for this type of patients will be described.

3. Robotic systems for upper limb rehabilitation

The World Health Organization estimates that each year 15 million people worldwide are affected by a stroke and 5 million of those are living with a permanent disability (WHO). The majority of post-stroke patients is able to recover an independent walking, but many of them fail to obtain a functional use of upper limbs, even after a prolonged rehabilitation treatment: these functional limitations are responsible for a significant reduction quality of life (Nichols-Larsen et al., 2005).
One year after the acute event, patients are usually considered chronic and rehabilitative therapies are often suspended, but several studies have shown that improvements in motor abilities induced by rehabilitative therapies may also occur in patients with chronic damage from 6 to 12 months after the acute event (Duncan et al., 1992; Hendricks et al., 2002).
Recent approaches that involve a repetitive training of upper limbs with activities aimed at task-oriented movements have provided evidence of improvements in hemiparetic patients more than a year after the stroke. In particular, the Constraint-Induced Movement Therapy (CIMT), based on an intense practice functionally oriented to hemiparetic upper limbs tasks obtained by a restriction of the unimpaired upper limb seems to be effective in reducing long-term disability (Miltner et al., 1999; Wolf et al., 2006). The motivation for using this type of treatment is based on the evidence that stroke and other neurological damages cause a partial destruction of cortical tissue, with the involvement of sensorimotor areas, that can determine incorrect motor programmes. However, CIMT requires a significant level of motor function and is not suitable to patients with severe weakness or spasticity due to neurological damage.
Other treatments based on high-intensity and task-oriented active upper limbs movements led to significant improvements in cortical reorganization and motor function in people with disabilities, more than a year after the stroke (Fasoli et al., 2003; Duncan et al., 2005). Unfortunately, these traditional treatments for post-stroke rehabilitation shows some drawbacks: they require a manual interaction by the therapists that must be provided on a daily basis for several weeks: the administration of an intensive treatment for each patient is proved to be difficult and costly.
Several robotic devices for rehabilitation have been recently developed to overcome these disadvantages: they are able to provide a safe and intensive therapy to patients with different degrees of motor impairment (Riener et al., 2005a).
Furthermore, the training with the support of the robot can be extremely precise, intense and prolonged. The robotic systems can also measure the progress of the patient in an objective way, increase the effectiveness of treatment and reduce the costs associated with the healthcare system.
Several reviews have shown the robot-assisted sensorimotor treatments and task-orients repetitive movements can improve muscle strength and motor coordination in patients with

neurological damage (Kwakkel et al., 2008; Mehrholz et al., 2008), although a limited number of clinical studies have examined the effect of upper limb robot-aided post-stroke rehabilitation using robust study designs (i.e., Randomized Controlled Trial)[7].
Among these, only three studies involved more than 30 subjects, and only two trials focused on pre- and post-treatment measures in experimental and control group (Lum et al., 2002; Krebs et al., 2000), of which only one was an RCT about the effects of the robot therapy provided to patients in the sub-acute phase (Krebs et al., 2000).
Among the most advanced robotic systems providing a growing amount of experimental data, the "MIT-Manus" (InMotion[2], Interactive Motion Technologies. Inc., Cambridge, MA, USA), designed at the Massachusetts Institute of Technology (MIT), holds a special role (Hogan et al., 1995) (Figure 1a).
The clinical trial started in 1994 at the Burke Rehabilitation Hospital, White Plains, NY, USA and recently many hospitals around the world use the "MIT-Manus" robotic system as tool for the upper extremities rehabilitation therapy in patients with neurological damage.
In Europe, the first example of this robotic system was purchased by the Scuola Superiore Sant'Anna in Pisa, Italy in 1995 and is currently at the Neurological Rehabilitation and Traumatic Brain Injury Unit at Auxilium Vitae Rehabilitation Centre in Volterra, Italy, where clinical trials on hemiparetic subjects are in progress.
The "MIT-Manus" is an operational-type robotic system characterized by a human-machine interface which is limited to the robot's end-effector[8] (Figure 1b). Unlike in industrial robots, where any contact with the human operator is usually excluded for safety reasons, the "MIT-Manus" was specifically designed to interact with the patient in a safe, stable and, when necessary, compliant way.
These characteristics are obtained through the use of a control scheme, called "impedance control", which modulates the movement of the robot in order to adapt itself to the upper limb's dynamics of the patient. The "MIT-Manus" can move, drive or disturb the upper movement of a subject and enables the recording of significant variables such as position, speed and the forces applied at the end-effector (Krebs et al., 1998). The distinctive characteristics of the "MIT-Manus" robotic system are: high reversibility mechanism (back-drivability), low mechanical impedance, which can readily adapt to the actions of the patient and the control system, which through the above mentioned "impedance control" modulates the reactions of the system to mechanical disturbances and ensure a compliant behaviour[9].
In fact, the machine was designed so as to have an intrinsic low impedance, low isotropic inertia (0:33 ± 1 kg, maximum anisotropy 2:1), low isotropic friction (0:28 ± 0.84 N,

---

[7] A Randomized Controlled Trial (RCT) represents the most rigorous way of determining the cause-effect relation between treatment and outcome. It is a study in which recruited participants are randomly allocated to receive one of several clinical interventions. One of these interventions is the control (e.g. a standard practice), the other is the experimental treatment (e.g. robot-aided therapy).
[8] A different class of robotic systems for rehabilitation is represented by the exoskeleton-type systems, where the contact between the patient and the machine is extended to the whole impaired limb (or part of it).
[9] The robot assists the patient's movements in order to make their execution easier: only if the patient is not able to perform a movement, the robot's control system assists and guides the paretic upper limb in order to complete it.

maximum anisotropy 2:1) and is capable of producing a definite range of forces (0-45 N) and impedance (0-2 N/mm).
The rehabilitation treatment using the "MIT-Manus" consists of a series of complex motor tasks to be performed by the patient, who is asked to move the robot's end-effector (Figure 1b) to perform movements aimed at hitting targets in a bidimensional space (reaching tasks).

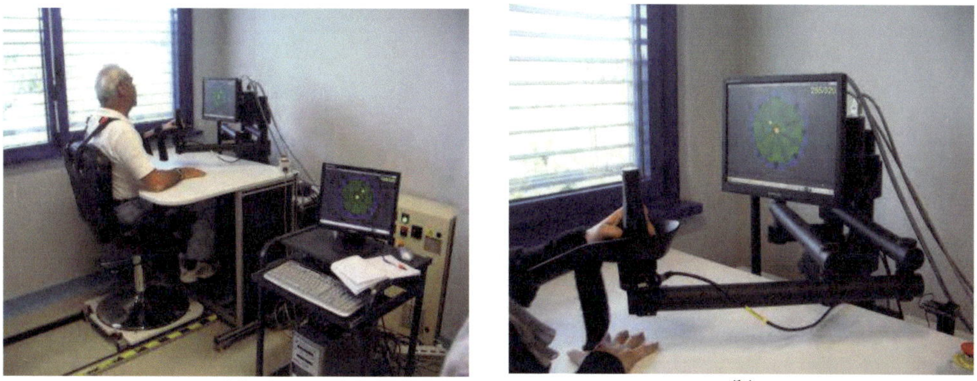

(a) (b)

Fig. 1. The "MIT-Manus" robotic system: a patient during the robotic therapy (a) end-effector of the robotic system

Specific biomechanical parameters, which are computed starting from the above mentioned physical variables recorded by the "MIT-Manus" system, can be used for the characterization of the patient's motor performance, together with other parameters calculated from the electroencephalographic signals (EEG) for the assessment of motor areas activation, before and after treatment (Figure 2).
The different clinical experimental trials performed using "MIT-Manus" were described in several articles. The results of a study performed on 96 patients, including 40 subjects in the control group and 56 subjects in the experimental group, showed an improvement of the latter over the former, according to the results of functional assessment performed by using different clinical scales, including the Motor Status Score for the shoulder and elbow (MSS-SE), the Motor Power Scale (MP) and the modified Ashworth scale (MAS) (Volpe et al., 2000).
The results of our clinical trials, in publication, confirm the significant improvement of motor damage by 1) an increase in MSS-SE values for both the shoulder and the elbow joint, 2) a decrease in the MAS values for the shoulder joint, 3) no increase in the MAS values for the elbow joint and 4) an increase of the ROM for the shoulder joint: the data are statistically significant and were recorded before and after the treatment in chronic hemiparetic subjects, which was administered for 6 weeks, 3 sessions per week, 45 minutes for each therapy session.
The three months follow-up has revealed a further reduction of the motor damage due to an increase in the ROM of the shoulder joint.
In conclusion, our results showed that the improvement in motor skills after a neurological damage may continue even a year after the acute event.

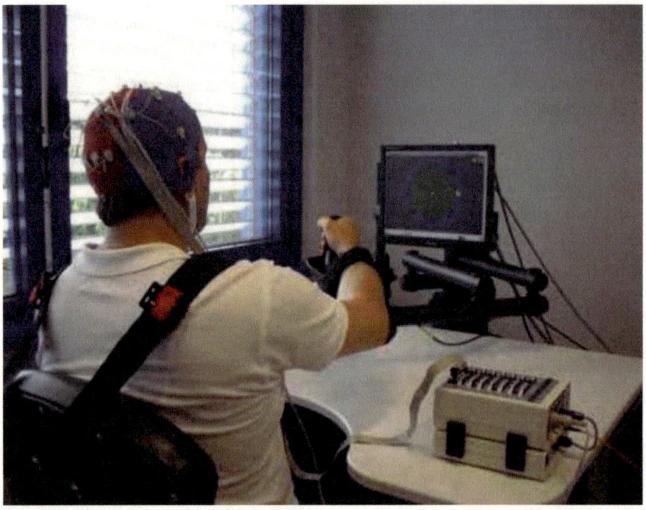

Fig. 2. A subject during the the robotic therapy using the MIT-Manus and recording of EEG signals for the functional assessment

The results of the questionnaires designed to measure the acceptability of the robotic therapy and distributed to the patients at the end of the experimental trial, showed a substantial acceptance of the robotic therapy, which can not considered as substitutive of the traditional rehabilitation treatments and the role of the therapist.

A clinical trial with chronic hemiparetic subjects in sub-acute phase, currently in progress, is aimed at assessing the effects of robotic therapy after a few weeks after the acute event and identifying prognostic parameters of motor recovery to be provided to physiatrists in order to early identify the most suitable rehabilitative treatments for each patient.

Another robotic system for the upper limb rehabilitation of post-stroke patients, the "MEMOS" (Mechatronic system for motor recovery after Stroke), was designed and developed at the Scuola Superiore Sant'Anna in Pisa, Italy (Figure 3). It is currently used at the Fondazione Salvatore Maugeri in Veruno, Italy, where the clinical trials has already shown excellent results: a significant reduction of the upper limb impairment in chronic hemiparetic patients was observed (Micera et al., 2005; Colombo et al., 2005; Colombo et al. 2008).

The "MEMOS" enables patients to perform reaching movements using a simple system of mechanical guides: the dimensions of its workspace are 70 cm x 70 cm. Before starting the therapy session, the worktable can be tilted by the therapist according to the needs of the patient through a manual adjustment. To a greater simplicity in terms of mechanical structure compared to the "MIT-Manus", the "MEMOS" adds the flexibility of a system that can be used for rehabilitation at home, supported in real-time by the remote control of physiatrists and therapists.

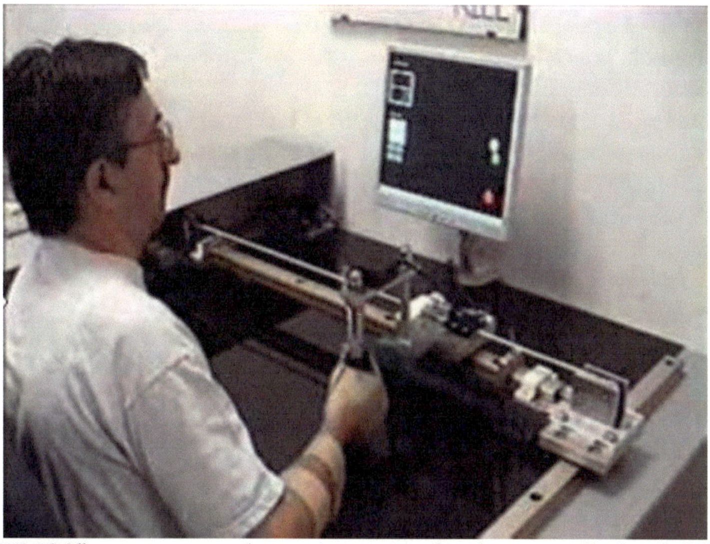

Fig. 3. The "MEMOS" system: a mechatronic system for the upper limb rehabilitation

4. Robotic systems for lower limbs rehabilitation

The objective of rehabilitation in paraplegic patients is to achieve maximum independence in the Activities of Daily Living (ADL). Over the past decades, the rehabilitation of locomotion in individuals with spinal cord injury (SCI) was largely developed and tested (Scivoletto & Di Donna, 2009; Barbeau & Rossignol, 1994; Barbeau et al., 1999; Barbeau et al., 2006).

Currently, the locomotion in these subjects is performed using standard wheelchairs, often without any training for the functional motor recovery of the lower limbs (i.e., the gait). An increasing evidence based on neurophysiological studies has demonstrated the ability to activate locomotion patterns in animals affected by a spinal cord injury through the use of systems for the body weight support (BWS) and a treadmill (Barbeau & Rossignol, 1987). Similar studies were performed on human subjects using a harness and variable BWS percentages on the treadmill: the locomotion activities in SCI subjects can be improved using such systems (Dietz & Colombo, 2004; Grasso et al., 2004; Edgerton et al., 2006; Scivoletto et al. 2007).

To activate the locomotor function using this approach, it is often necessary to help the progression of the lower limbs on a treadmill through a manual effort of two or more therapists. Therefore, the duration of the training is usually limited by the fatigue of the therapists and this causes a shorter duration of the sessions compared to the time needed to obtain a good rehabilitation outcome. To facilitate this type of rehabilitation and to ensure repeatability of the movements for locomotion, the "Lokomat" system (Hocoma AG, Volketswil, Switzerland), a bilateral robotic orthoses used in conjunction with a BWS system and a treadmill was developed (Figure 4) (Colombo 2000; Colombo 2001).

The joints in the "Lokomat" system, located in correspondence of the hip and knee joint, are moved by linear actuators which are integrated into the exoskeleton structure. A passive system of springs aimed at lifting the foot's sole induces an ankle's dorsiflexion during the swing phase. The patient's lower limbs, which are fixed to the exoskeleton structure through adjustable stripes and settings, are moved in the sagittal plane according to a control strategy with pre-defined hip and knee trajectories which enables a safe patient-machine interaction (Riener 2005b; Riener 2006).

Up to now, different studies about the use of the "Lokomat" system in subjects affected by neurological diseases were published (Colombo et al., 2000; Jezernik et al., 2003; Lünenburger et al., 2005; Hidler et al., 2005; Lünenburger et al., 2006; Hidler et al., 2009; Hornby et al., 2008; Israel et al., 2006; Wirz et al., 2005).

Recent experimental studies analyzed the activation of different EMG patterns in healthy and spinal cord injured subjects during the "Lokomat" training using the treadmill, in different experimental conditions (Hidler & Wall, 2005). The comparison of the EMG activity between healthy and SCI subjects show that in a complete spinal cord injury, the adaptation to different speeds is still present (Lünerberger 2006).

An experimental study currently in progress, carried out through a joint collaboration among Scuola Superiore Sant'Anna, the Neurorehabilitation Unit and the Spinal Cord Injuries Departmental Unit at Cisanello Hospital in Pisa, Italy is aimed at assessing the changes in the muscle activation after a period of rehabilitation using the "Lokomat" system in subjects with incomplete spinal cord injury.

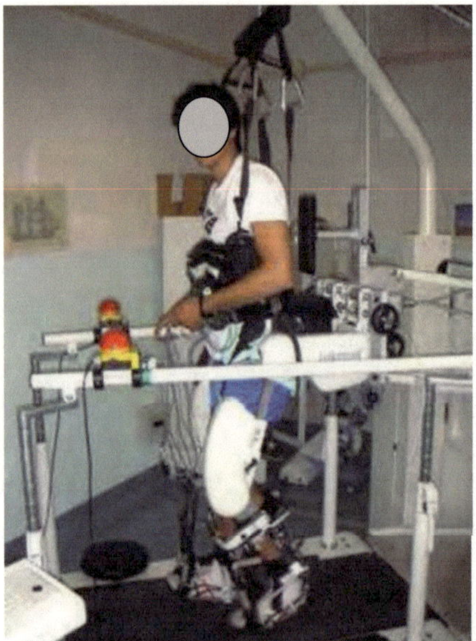

Fig. 4. The "Lokomat" robotic system for the lower limbs rehabilitation

To the current knowledge, the use of this device is still empirical: the choice of the variable parameters (BWS percentage, speed, driving force and duration) is not yet included in specific guidelines and the patterns of muscular activation of the robot-aided gait are not, for the time being, well defined.

Surface EMG signals from four muscles on each limb (rectus femoris, biceps femoris, anterior tibial and medial gastrocnemius) in healthy subjects and patients with incomplete spinal cord injury during the training with the "Lokomat" robotic system are recorded. In detail, three different treadmill speeds (1.0, 1.6 and 2.4 km/h) using two BWS percentages

(30% and 60%) and two different patient-cooperation modalities are used. The modalities are: (1) "passive", in which the subject does not contribute to the movement of the lower limbs that are mobilized by the robotic orthoses, and 2) "active", in which the subject accompanies the movement of the lower limbs according to his/her residual motor abilities. Subjects with a spinal cord injury follow, after a first recording (pre-training), a training program with the "Lokomat" system of variable duration (i.e., 4-8 weeks), after which a further EMG recording is performed (post-training). Follow-up recordings at 3 and 6 months will be performed as well.
So far, the results obtained on a limited sample of patients affected by an incomplete spinal cord injury, which carried out a physical therapy using the "Lokomat" robotic system, showed an improvement of the locomotor function, due to enhanced recruitment of the lower limb muscles in the "active" condition: the assessment was based on the analysis of EMG signals recorded before and after the treatment, using specific clinical scales (i.e., 10-Meter Walk Test, 6-Minute Walk Test, Timed Up & Go Test) (Mazzoleni et al., 2008).

5. Mechatronic systems for the functional assessment and the movement analysis

Mechatronics is defined as the synergistic integration of mechanical engineering, with electronics and intelligent computer control in the design and manufacturing of industrial products and processes (Harashima et al., 1996).
A mechatronic approach has several benefits. In fact, mechatronic systems have greater flexibility, a better performance, higher quality, wide areas of applications and are less expensive.
The present paragraph focuses on the applications of mechatronics to the rehabilitation's area, according two main research fields: functional assessment and movement analysis.
A mechatronic platform for the functional assessment of the rehabilitation treatment using isometric force/torque during the execution of ADL tasks, in post-stroke subjects was designed and developed within the "Alladin" project, funded by the European Commission under the 6th Framework Program (IST-2002-507424) (www.alladin-ehealth.org).
The platform (Figure 5) was validated in three clinical centers in Europe:

- Mary Middelares Algemeen Ziekenhuis Sint-Jozef (Gent, Belgium),
- Szent János Hospital (Budapest, Hungary),
- Adelaide & Meath Hospital (Tallaght, Ireland).

In these centres, an experimental clinical trial with 270 subjects (150 hemiparetic subjects and 120 healthy control subjects) was carried out: the results contributed to identify six parameters related to the motor recovery of post-stroke patients, which can be used to perform a functional assessment after a short time after the acute event, in order to promptly address the choices by physiatrists and therapists about the most appropriate rehabilitative treatment for each patient (Mazzoleni et al., 2005; Mazzoleni 2006; Mazzoleni 2007a; Mazzoleni 2007b).

Fig. 5. The ALLADIN mechatronic platform for functional assessment of post-stroke subjects

In recent years, different wearable mechatronic systems for the movement analysis were developed. The "MEKA" (MEchatronic device for the Knee Analysis) system, designed and developed through a collaboration between the ARTS Lab at Scuola Superiore Sant'Anna in Pisa, Italy and the company Humanware Inc., is a mechatronic device that performs a real-time monitoring of the knee's flexion-extension and varus-valgus angles (knee's dynamic analysis) during the gait and other ADL. The recorded data can be displayed in a numeric and graphic format.

The system is composed of three main units:

1. a modular mechatronic device for measuring values from the knee's flexion-extension and varus-valgus angles, through the use of Hall-effect sensors (Figure 6a);
2. a wireless system for the kinematic data acquisition and transfer to PC (Figure 6b);
3. a software tool for processing and displaying data.

One of the main advantages related to the use of the "MEKA" system is represented by the best cost/effectiveness ratio when compared to other similar devices aimed at performing a similar knee's dynamic analysis.

The results of experiments carried out by using such mechatronic device highlighted different locomotor strategies, depending on the age and the difficulty of the required task (Micera et al., 2003; Micera et al., 2004).

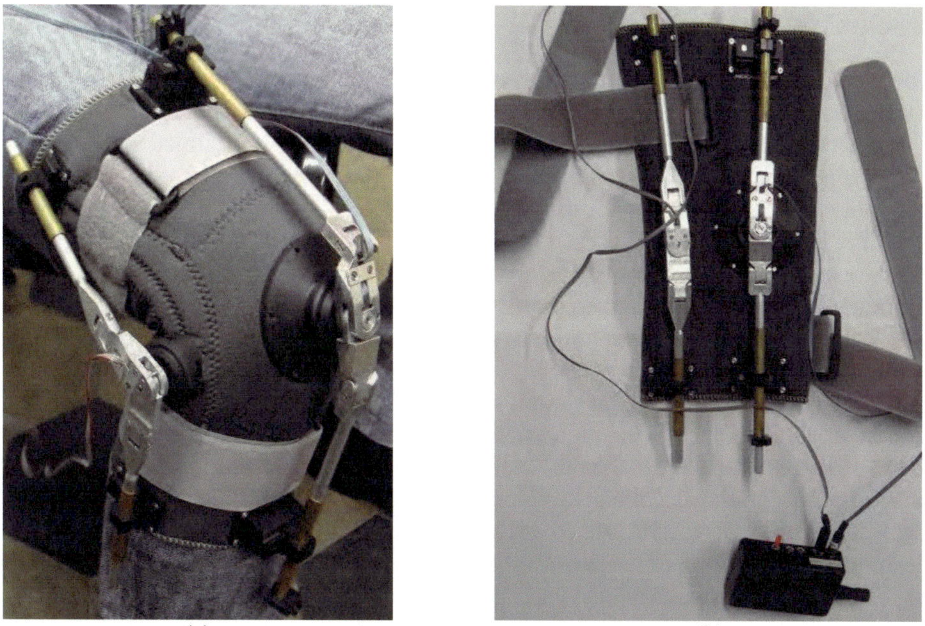

(a) (b)

Fig. 6. The "MEKA" mechatronic system for the knee movement analysis

6. Open issues

In the field of robotics and mechatronic systems for rehabilitation some problems are still unsolved and require further developments to be addressed. As regards the clinical trials, the main issues are:

- the need for a greater number of patients to be recruited in clinical trials;
- the use of homogeneous groups of patients and control groups (i.e., RCT);
- the development of innovative clinical protocols;
- the definition of quantitative parameters related to the rehabilitation treatment (milestones) and prognostic parameters (markers);
- the identification of neurophysiological models of the recovery process;
- the detailed analysis of scenarios and definition of the roles of involved subjects (patient, physiatrist and therapist);
- the incremental innovation of devices already available and the design of new technological systems for rehabilitation.

A key aspect concerns the active role that the patient has to perform during the administration of robotic therapy: it should be an integral part of the requirements to be taken into account when designing new technological systems for rehabilitation. Furthermore, the integration between robotic systems and the techniques of brain imaging for the functional evaluation represents a fundamental area of research on which focusing efforts in order to define therapeutic approaches, tailored to the specific needs of each patient and the degree of recovery.

7. Conclusion

Robotic and mechatronic systems presented in this chapter are increasingly used in hospitals and rehabilitation centres as technological tools for the clinical practice. These systems are used to administer intensive and prolonged treatments aimed at achieving the functional recovery of people affected by neurological impairments, in sub-acute and chronic stage, with a potential improvement of the cost/effectiveness ratio. They can evaluate the effects of rehabilitation treatments in a quantitative way and contribute to increase the knowledge of motor control and learning mechanisms in humans.

Such systems have to be designed according to the ever-increasing knowledge in the neurophysiologic field, in order to improve the motor function through stimulating, entertaining exercises, able to promote patients' motivation.

The integrated use of robotic system for rehabilitation, mechatronic systems for the movement analysis and functional assessment and advanced brain imaging techniques enables to analyse the mechanisms of motor recovery for each patient and define customized rehabilitation treatments, towards an optimization of hospitalization's period and healthcare resources (Figure 7).

Furthermore, the development of low-cost robotic systems together with safe, reliable and robust telecommunications networks can also contribute to the validation of innovative tele-rehabilitation protocols, leading to potential high advantages:

(i) to continue the rehabilitation process at home or resident care (for the patient);

(ii) to reduce the costs associated with hospitalization and rehabilitation at the hospital (for the healthcare system).

Finally, the collaboration between physiatrists, therapists, patients and engineers, which has already shown important clinical and scientific results, has to be strengthened through research projects and specific programmes, at local and national level, as it is essential to guide the development of safe and effective technological systems for a patient-oriented robotics aimed at improving the quality of life of disabled persons.

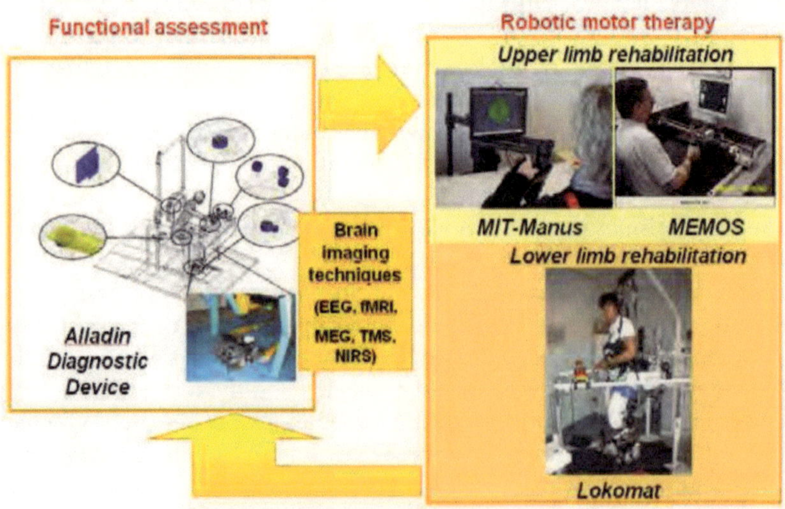

Fig. 7. Robotic and mechatronic systems for rehabilitation: an integrated approach

8. References

Barbeau, H. & Rossignol, S. (1987). Recovery of locomotion after chronic spinalization in the adult spinal cat. *Brain Res*, Vol. 412, pp. 84-95.

Barbeau, H. & Rossignol, S. (1994). Enhancement of locomotor recovery following spinal cord injury. *Current Opinion Neurol*, Vol. 7, pp. 517-524.

Barbeau, H.; Ladouceur, M.; Norman, K. E. & Pepin, A. (1999). Walking after spinal cord injury: Evaluation, treatment, and functional recovery. *Arch Phys Med Rehabil*, Vol. 80, pp. 225-235.

Barbeau, H.; Nadeau, S. & Garmeau, C. (2006). Physical determinants, emerging concepts, and training approaches in gait of individuals with spinal cord injury. *J Neurotrauma*, Vol. 23, pp.571-85.

Colombo, R.; Pisano, F.; Micera, S.; Mazzone, A.; Delconte, C.; Carrozza, M. C.; Dario, P. & Minuco, G (2005). Robotic techniques for upper limb evaluation and rehabilitation of stroke patients. *IEEE Trans Neural Sys Rehab Eng*, Vol. 13, No. 3, pp. 311-324.

Colombo, R.; Pisano, F.; Micera, S.; Mazzone, A.; Delconte, C.; Carrozza, M. C.; Dario, P. & Minuco, G. (2008). Assessing mechanisms of recovery during robot-aided neurorehabilitation of the upper limb. *Neurorehabil Neural Repair*, Vol. 22, No. 1, pp. 50-63.

Colombo, G.; Joerg, M.; Schreier, R. & Dietz, V. (2000). Treadmill training of paraplegic patients using a robotic orthosiS. *J Rehabil Res Dev*, Vol. 37, pp. 693-700.

Colombo, G.; Wirz, M. V. Dietz (2001). Driven gait orthosis for improvement of locomotor training in paraplegic patients", *Spinal Cord*, Vol. 39, pp. 252-255.

Dario, P.; Guglielmelli, E.; Carrozza, M. C.; Micera, S.; Dipietro L. & Pisano F. (2003). Sistemi meccatronici e robotici per la neuro riabilitazione. In *Bioingegneria della Postura e del Movimento*, Bressanone (Italy), ISBN 9788855527170.

Dietz, V. & Colombo, G. (2004). Recovery from spinal cord injury: underlying mechanisms and efficacy of rehabilitation. *Acta Neurochir Suppl*, Vol. 89, pp. 95-100.

Donoghue, J. P.; Hess, G. & Sanes, J. N. (1996). Motor cortical substrates and mechanisms for learning. In *Acquisition of Motor Behavior in Vertebrates*, JR Bloedel (Ed.), TJ Ebner, SP Wise, pp. 363–86. Cambridge, MA: MIT Press.

Duncan, P. W.; Goldstein, L. B.; Matchar, D.; Divine, G. W. & Feussner, J. (1992). Measurement of motor recovery after stroke. *Stroke*, Vol. 23, pp. 1084-89.

Edgerton, V. R.; Kim, S. L.; Ichiyama, R. M.; Gerasimenko, Y. P. & Roy, R. R. (2006). Rehabilitative therapies after spinal cord injury. *J Neurotrauma*, Vol. 23, No. 3-4, pp. 560-570.

Feigin, V. L.; Lawes, C. M.; Bennett, D. A. & Anderson, C. S. (2003). Stroke epidemiology: a review of population-based studies of incidence, prevalence, and case-fatality in the late 20th century. *Lancet Neurol*, Vol. 2, pp. 43-53.

Grasso, R.; Ivanenko, Y. P.; Zago, M.; Molinari, M.; Scivoletto, G.; Castellano, V.; Macellari, V. & Lacquaniti, F. (2004). Distributed plasticity of locomotor pattern generators in spinal cord injured patients. *Brain*, Vol. 127, pp. 1019-1034.

Harashima, F.; Tomizuka, M. & Fukuda, T. (1996). Mechatronics—what is it, why, and how? An editorial. *IEEE/ASME Transactions on Mechatronics*, Vol. 1, pp. 1–4.

Heart Disease and Stroke Statistics 2009 Update: A Report From the American Heart Association Statistics Committee and Stroke Statistics Subcommittee. Available at: http://circ.ahajournals.org/cgi/reprint/119/2/e21.pdf

Hendricks, H. T.; Van Limbeek, J.; Geurts, A. C. & Zwarts, M. J. (2002). Motor recovery after stroke: a systematic review of the literature. *Arch Phys Med Rehabil*, Vol. 83, pp. 1629-37.

Hidler, J. M. & Wall, A. E. (2005). Alterations in muscle activation patterns during robotic-assisted walking, *Clinical Biomechanics*, Vol. 20, pp. 184-193.

Hidler, J.; Nichols, D.; Pelliccio, M.; Brady, K.; Campbell, D. D.; Kahn, J. H. & Hornby, T. G. (2009). Multicenter randomized clinical trial evaluating the effectiveness of the Lokomat in subacute stroke. *Neurorehabil Neural Repair*, Vol. 23, pp. 5-13.

Hornby, T. G.; Campbell, D. D.; Kahn, J. H.; Demott, T.; Moore, J. L. & Roth, H. R. (2008). Enhanced gait-related improvements after therapist- versus robotic-assisted locomotor training in subjects with chronic stroke: a randomized controlled study. *Stroke*, Vol.39, pp. 1786-92.

Israel, J. F.; Campbell, D. D.; Kahn, J. H. & Hornby, T. G. (2006). Metabolic costs and muscle activity patterns during robotic- and therapist-assisted treadmill walking in individuals with incomplete spinal cord injury. *Physical Therapy*, Vol.86, pp. 1466-78.

Jezernik, S.; Schärer, R.; Colombo, G. & Morari, M. (2003). Adaptive robotic rehabilitation of locomotion: a clinical study in spinally injured individuals. *Spinal Cord*, Vol. 41, No. 12, pp. 657–666.

Krebs, H. I.; Hogan, N.; Aisen, M. L. & Volpe, B. T. (1998). Robot-Aided Neurorehabilitation. *IEEE Trans. Rehabilitation Engineeering*, Vol. 6, pp. 75-87.

Krebs, H. I.; Volpe, B. T.; Aisen, M. L. & Hogan, N. (2000). Increasing productivity and quality of care: Robot-aided neuro-rehabilitation. J *Rehabil Res Dev*, Vol. 37, No. 6, pp. 639–52

Kwakkel, G.; Kollen, B. J. & Krebs, H. I. (2008). Effects of robot-assisted therapy on upper limb recovery after stroke: a systematic review. *Neurorehabilitation and Neural Repair*, Vol. 22, No. 2, pp. 111-121.

Lum, P. S.; Burgar, C. G.; Shor, P. C.; Majmundar, M. & Van der Loos M. (2002). Robot-assisted movement training compared with conventional therapy techniques for the rehabilitation of upper-limb motor function after stroke. *Arch Phys Med Rehabil*, Vol. 83, No. 7, pp. 952–59.

Lünenburger, L.; Colombo, G.; Riener, R. & Dietz, V. (2005). Clinical assessments performed during robotic rehabilitation by the gait training robot Lokomat. In *Proceedings of the 9th International Conference on Rehabilitation Robotics*, pp. 4888-4490, Chicago, USA.

Lünenburger, L.; Bolliger, M.; Czell, D.; Müller, R. & Dietz, V. (2006). Modulation of locomotor activity in complete spinal cord injury. *Exp Brain Res*, Vol. 174, pp. 638-646.

Marini, C.; Baldassarre, M.; Russo, T.; De Santis, F.; Sacco, S.; Ciancarelli I. & Carolei, A., (2004). Burden of first-ever ischemic stroke in the oldest old: evidence from a population-based study. *Neurology*, Vol. 62, pp. 77-81.

Mazzoleni, S; Stampacchia, G.; Cattin, E.; Lefevbre, O.; Riggio, C.; Troncone, M.; Bradaschia, E.; Tolaini, M.; Rossi, B. & Carrozza, M. C. (2008). Effects of a robot-mediated locomotor training in healthy and spinal cord injured subjects. In *Proceedings of the 1st National Conference on Bioengineering*, pp.245-246, Pisa, Italy.

Mazzoleni, S.; Van Vaerenbergh, J.; Toth, A.; Munih, M.; Guglielmelli, E. & Dario, P. (2005). ALLADIN: a novel mechatronic platform for assessing post-stroke functional recovery. *Proceedings of the 9th IEEE International Conference on Rehabilitation Robotics*, pp. 156-159, Chicago, IL, USA.

Mazzoleni, S.; Micera, S.; Romagnolo, F.; Dario, P. & Guglielmelli, E. (2006). An ergonomic dynamometric foot platform for functional assessment in rehabilitation. In *Proceedings of the 1st IEEE/RAS-EMBS International Conference on Biomedical Robotics and Biomechatronics*, pp. 619-624, Pisa, Italy.

Mazzoleni, S.; Cavallo, G.; Munih, M.; Cinkelj, J.; Jurak, M.; Van Vaerenbergh, J.; Campolo, D.; Dario, P. & Guglielmelli, E. (2007a). Towards application of a mechatronic platform for whole-body isometric force-torque measurement to functional assessment in neuro-rehabilitation. In *Proceedings of the IEEE International Conference on Robotics and Automation*, pp. 1535-1540, Rome, Italy.

Mazzoleni, S.; Van Vaerenbergh, J.; Toth, A.; Munih, M.; Guglielmelli, E. & Dario, P. (2007b). The ALLADIN diagnostic device: an innovative platform for assessing post-stroke functional recovery, In: *Rehabilitation Robotics*, ARS Scientific book, pp. 535-554, I-Tech Education and Publishing, Vienna, Austria.

Mehrholz, J.; Platz, T.; Kugler, J. & Pohl, M. (2008). Electromechanical and robot-assisted arm training for improving arm function and activities of daily living after stroke. *Cochrane Database of Systematic Reviews*, Vol. 4, CD006876.

Micera, S.; Mazzoleni, S.; Guglielmelli, E. & Dario, P. (2003). Assessment of gait in elderly people using mechatronic devices: preliminary results. *Gait & Posture*, Vol. 18, Supplement 1, pp. S22.

Micera, S.; Carpaneto, J.; Scoglio, A.; Zaccone, F.; Freschi, C.; Guglielmelli, E. & Dario, P. (2004). On the analysis of knee biomechanics using a wearable biomechatronic device. *Proceedings of the International Conference on Intelligent Robots and Systems*, vol. 2, pp. 1674 - 1679, Sendai, Japan.

Micera, S.; Carrozza, M. C.; Guglielmelli, E.; Cappiello, G.; Zaccone, F.; Freschi, C.; Colombo, R.; Mazzone, A.; Delconte, C.; Pisano, F.; Minuco, G. & Dario, P. (2005). A simple robotic system for neurorehabilitation. *Autonomous Robots*. Vol. 19, pp. 1-11.

Murray, C. J. L. & Lopez, A. D. (1997). Global mortality, disability and the contribution of risk factors. Global burden of the disease study. *Lancet*, Vol. 349, pp. 1436-1442.

Nichols-Larsen, D. S.; Clark, P. C.; Zeringue, A.; Greenspan, A. & Blanton, S. (2005). Factors influencing stroke survivors' quality of life during sub-acute recovery. *Stroke*, Vol. 36, pp. 1480-84.

Riener, R.; Nef, T. & Colombo, G. (2005a). Robot-aided neurorehabilitation of the upper extremities. *Med. Biol. Eng. Comput.*, Vol. 43, pp. 2-10.

Riener, R.; Lünenburger, L.; Jezernik, S.; Anderschitz, M. & Colombo, G. (2005b). Patient-cooperative strategies for robot-aided treadmill training: first experimental results. *IEEE Trans Neural Syst Rehabil Eng*, Vol. 13, pp. 380-394.

Riener, R.; Lünenburger, L. & Colombo, G. (2006). Human-centered robotics applied to gait training and assessment, *J Rehab Res Dev*, Vol. 43, pp. 679-694.

Schmidt, H.; Werner, C.; Bernhardt, R.; Hesse, S. & Krüger, J. (2007). Gait rehabilitation machines based on programmable footplates. *J Neuroeng Rehabil*, Vol. 4:2.

Scivoletto, G.; Ivanenko, Y.; Morganti, B; Grasso, R.; Zago, M.; Lacquaniti, F.; Ditunno, J. & Molinari, M. (2007). Plasticity of spinal centers in spinal cord injury patients: new concepts for gait evaluation and training. *Neurorehabil Neural Repair*, Vol. 21, No. 4, pp. 358-365.

Scivoletto, G. & Di Donna, V. (2009). Prediction of walking recovery after spinal cord injury. *Brain Res Bull*, Vol. 78, No. 1, pp. 43-51.

SPREAD, Stroke Prevention and Educational Awareness Diffusion (2007), In: *Ictus cerebrale: linee guida italiane di prevenzione e trattamento*, 5th Edition, Catel (Ed.), Milano.

Volpe, B. T.; Krebs, H. I.; Hogan, N.; Edelstein, L.; Diels, C. M. & Aisen, M. (2000). A Novel Approach to Stroke Rehabilitation: Robot Aided Sensorymotor Stimulation. *Neurology*, Vol. 54, pp. 1938-1944.

Wirz, M.; Zemon, D. H.; Rupp, R.; Scheel, A.; Colombo, G.; Dietz, V. & Hornby, T. G. (2005). Effectiveness of automated locomotor training in patients with chronic incomplete spinal cord injury: a multicenter trial. *Arch Phys Med Rehabil*, Vol. 86, pp. 672-80.

WHO (World Health Organization), The atlas of Heart Disease and Stroke. Available at: http://www.who.int/cardiovascular_diseases/resources/atlas/en/

8

Wearable Sensor System for Human Dynamics Analysis

Tao Liu, Yoshio Inoue, Kyoko Shibata and Rencheng Zheng
Kochi University of Technology
Japan

1. Introduction

In clinical applications the quantitative characterization of human kinematics and kinetics can be helpful for clinical doctors in monitoring patients' recovery status; additionally, the quantitative results may help to strengthen confidence during their rehabilitation. The combination of 3D motion data obtained using an optical motion analysis system and ground reaction forces measured using a force plate has been successfully applied to perform human dynamics analysis (Stacoff et al., 2007; Yavuzer et al., 2008). However, the optical motion analysis method needs considerable workspace and high-speed graphic signal processing devices. Moreover, and if we use this analysis method in human kinetics analysis, the devices are expensive, while pre-calibration experiments and offline analysis of recorded pictures are especially complex and time-consuming. Therefore, these devices is limited to the laboratory research, and difficult to be used in daily life applications. Moreover, long-term, multi-step, and less restricted measurements in the study of gait evaluation are almost impossible when using the traditional methods, because a force plate can measure ground reaction force (GRF) during no more than a single stride, and the use of optical motion analysis is limited due to factors such as the limited mobility and line-of-sight of optical tracking equipment. Recently, many lower-cost and wearable sensor systems based to multi-sensor combinations including force sensitive resistors, inclinometers, goniometers, gyroscopes, and accelerometers have been proposed for triaxial joint angle measurement, joint moment and reaction force estimation, and muscle tension force calculation.

As for researches of wearable GRF sensors, pressure sensors have been widely used to measure the distributed vertical component of GRF and analyze the loading pattern on soft tissue under the foot during gait (Faivre et al., 2004; Zhang et al., 2005), but in these systems the transverse component of GRF (friction forces) which is one of the main factors leading to fall were neglected. Some flexible force sensors designed using new materials such as silicon or polyimide and polydimethyl-siloxane have been proposed to measure the normal and transverse forces (Valdastri et al., 2005; Hwang et al., 2008), but force levels of these sensors using these expensive materials were limited to the measurements of small forces (about 50N). By mounting two common 6-axial force sensors beneath the front and rear boards of a special shoe, researchers have developed a instrumented shoe for ambulatory

measurements of CoP and triaxial GRF in successive walking trials (Veltink et al., 2005; Liedtke et al., 2007), and an application of the instrumented shoe to estimate moments and powers of the ankle was introduced by Schepers et al. (2007).
About researches on body-mounted motion sensors, there are two major directions: one is about state recognition on daily physical activities including walking feature assessment (Sabatini et al., 2005; Aminian et al., 2002), walking condition classification (Coley et al., 2005; Najafi et al., 2002) and gait phase detection (Lau et al., 2008; Jasiewicz et al., 2006), in which the kinematic data obtained from inertial sensors (accelerometer or gyroscope) were directly used as inputs of the inference techniques; and another direction is for accurate measurement of human motion such as joint angles, body segment's 3-D position and orientation, in which re-calibration and data processing by fusing different inertial sensors are important to decrease errors of the quantitative human motion analysis. In our research, a wearable sensor system which can measure human motion and ground reaction force will be developed and applied to estimate joint moment and muscle tension force, so we are focusing on the second direction for quantitative human motion analysis. There are growing interests in adopting commercial products of 3D motion sensor system, for example a smart sensor module MTx (Xsens, Netherlands) composed of a triaxial angular rate sensor, a triaxial accelerometer and a triaxial magnetic sensor, which can reconstruct triaxial angular displacements by means of a dedicated algorithm. However, it is sensitive to the effect of magnetic filed environment, and the dynamic accuracy of this sensor is about two degrees, which depends on type of experimental environments.
In this chapter, a developed wearable sensor system including body-mounted motion sensors and a wearable force sensor is introduced for measuring lower limb orientations, 3D ground reaction forces, and estimating joint moments in human dynamics analysis. Moreover, a corresponding method of joint moment estimation using the wearable sensor system is proposed. This system will provide a lower-cost, more maneuverable, and more flexible sensing modality than those currently in use.

2. Wearable GRF Sensor

2.1 Mechanical Design and Dimension Optimization

We developed a wearable multi-axial force sensor with a parallel mechanism to measure the ground reaction forces and moments in human dynamics analysis. First, the parallel mechanism for sensing triaxial forces and moments was introduced. As shown in Fig.1, the sensor is composed of a bottom plane, x-, y- and z-axial load cells, and four balls. When forces and moments are imposed to the bottom plane, they are transferred onto the four support balls. The support balls are connected with three load cells by point contacts. Therefore, only translational forces can be transferred to the corresponding load cells and measured using the strain gauges attached on the load cells. The x-axial load cell can measure F_{X1} and F_{X2}. Similarly, the y-axial load cell measures F_{Y1} and F_{Y2}, while the z-axial load cell measures F_{Z1}, F_{Z2}, F_{Z3} and F_{Z4}. Based on these measured values, the three-axis forces and moments can be calculated by the use of the following equations:

$$F_x = F_{x1} + F_{x2} \quad (1)$$

$$F_y = F_{y1} + F_{y2} \quad (2)$$

$$F_z = F_{z1} + F_{z4} + F_{z2} + F_{z3} \quad (3)$$

$$M_x = (F_{z2} + F_{z3} - F_{z1} - F_{z4})L/2 \quad (4)$$

$$M_y = (F_{z3} + F_{z4} - F_{z1} - F_{z2})L/2 \quad (5)$$

$$M_z = (F_{x2} + F_{y2} - F_{x1} - F_{y1})L/2 \quad (6)$$

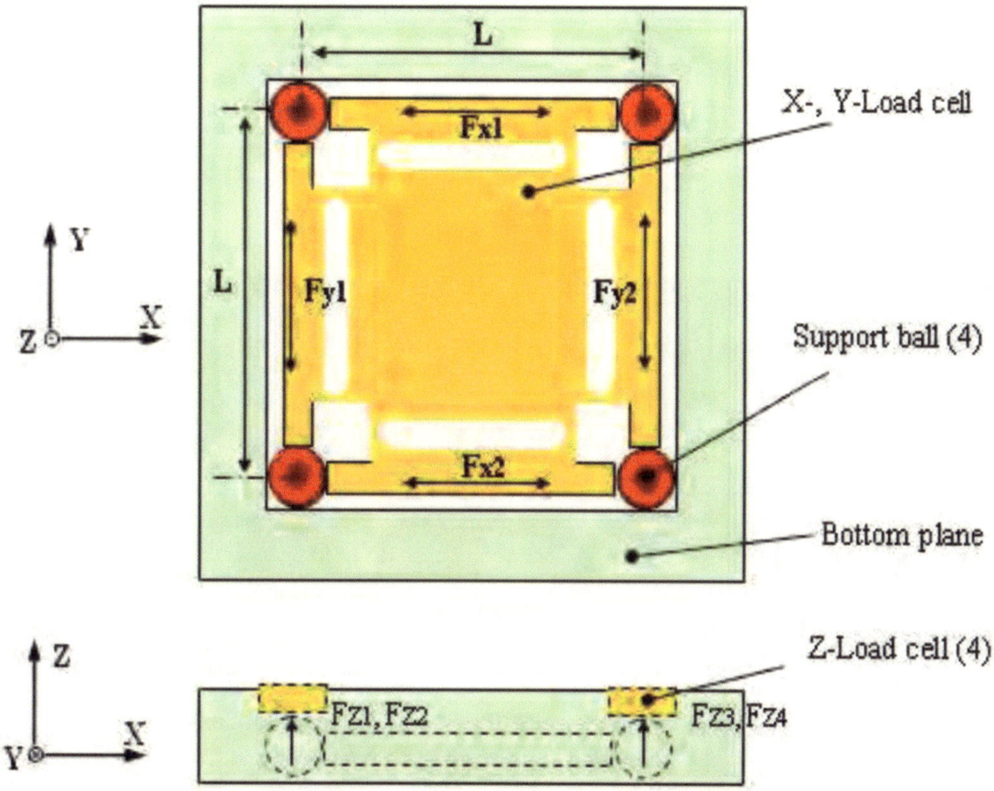

Fig. 1. Schematic picture for the new Sensor with a parallel Support mechanism. The transverse load cells are composed of two x-axial load cells for measuring F_{x1} and F_{x2} and two y-axial load cells for measuring F_{y1} and F_{y2} respectively. The z-load cells under the four support ball at the four corners (L=100mm) can measure four z-directional forces including F_{z1}, F_{z2}, F_{z3} and F_{z4}.

Figure 2 shows the detail of the load cells. Two strain gauges are attached on the load cell to sense a uniaxial translational force. In order to obtain a high sensitivity, the strain gauges should be distributed on the points where the maximum strains occur. ANSYS, FEA software, was used to perform the static analysis of the load cell. Based on the sensitivity limitation of the strain gauge, the optimal dimensions of the load cell were determined by ANSYS simulations. Figure 3 shows representative results of the static analysis for the load cell.

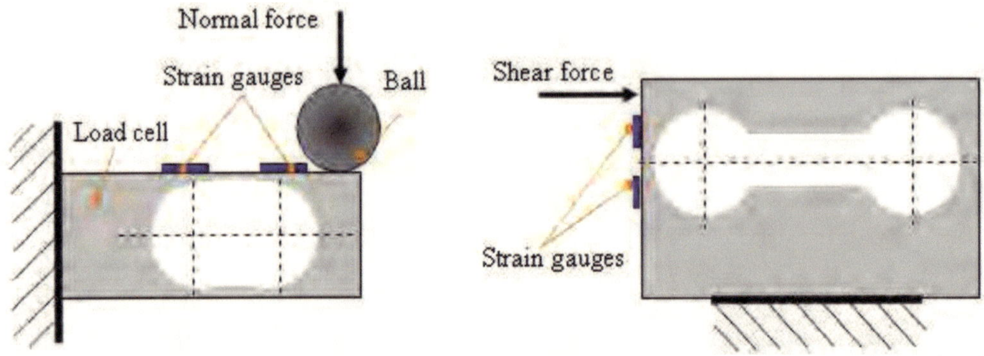

Fig. 2. Schematic of the design of load cell. We put two strain gauges on each load cell's flexible mechanical-body, and a set of two strain gauges is only sensitive to single directional translational force.

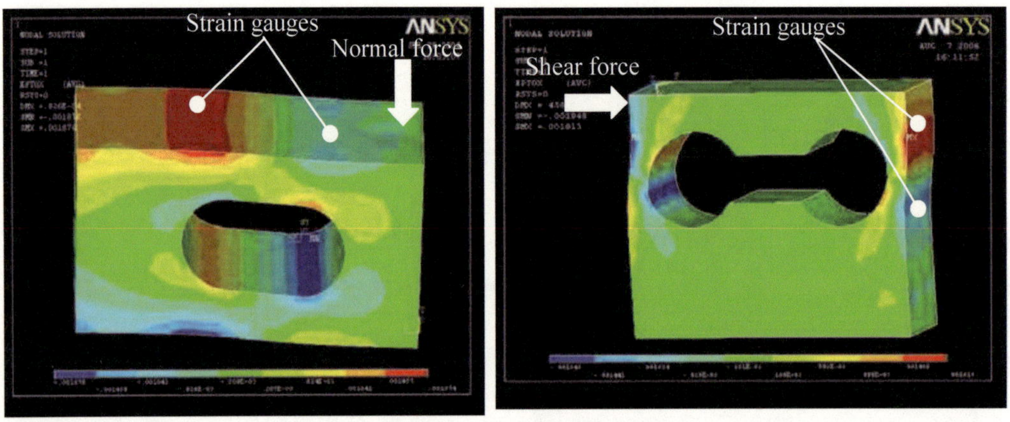

Fig. 3. Result graph of FEA. Finite element method was adopted to optimize the mechanism dimension of the load cells' flexible mechanical-body, and to improve the sensitivity of the force sensor.

As shown in Fig. 4, based on the single load cell obtained by the optimal design, we designed a prototype of the sensor, and the 3D model was constructed using an engineering modeling software of Pro/E. Figure 5 shows the prototype of the load cells in the wearable force sensor, and the flexible beams were made of ultra hard duralumin. Four groups of the strain gauges were used to construct the x- and y-axial load cells, and another four groups were used to make the z-axial load cell. In order to implement a more compact structure, hybrid measurement load cells were adopted for x- and y-directional translational force measurements. This new design can decrease the number of strain gauges and simplify amplifier modules.

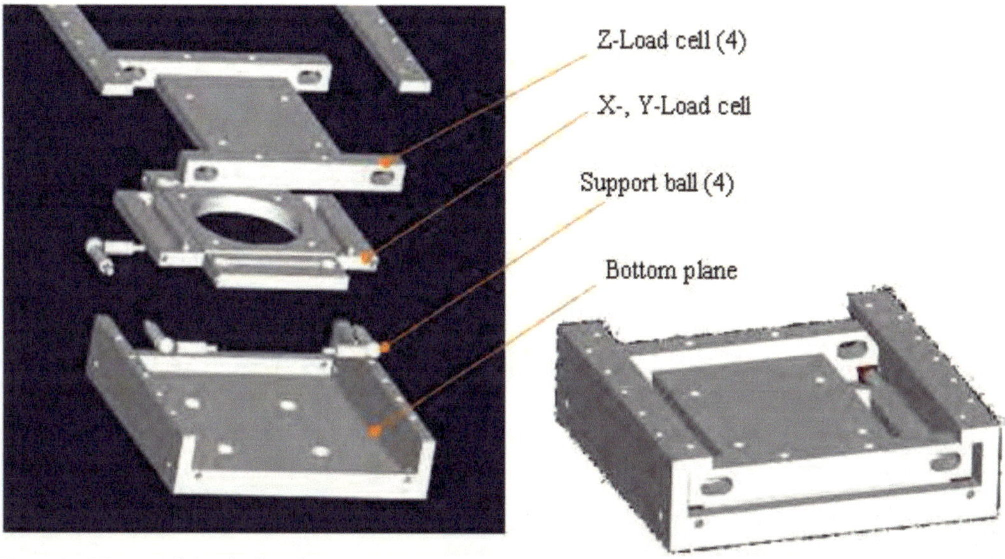

Fig. 4. 3D model of the force sensor using the stimulation model of the force sensor. According to the 3D model, we designed the mechanical structure of the parts in the sensor.

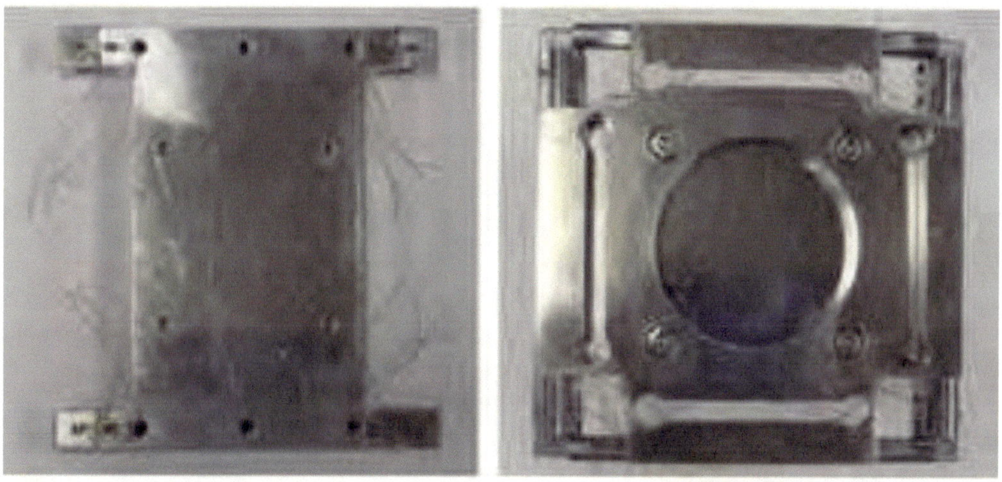

(a) (b)

Fig. 5. Mechanical structure of the load cells. (a) The mechanical structure of z-load cell with four sub-load cells which can measure z-direction vertical forces at the four support points. (b) The picture of the x-, y-load cell for the measurements of the horizontal forces.

2.2 Electrical System Design and Integrated Sensor System

As shown in Fig. 6, an integrated electrical system was developed and incorporated into the force sensor. The strains due to forces applied on the flexible body are converted to the resistance changes. Then the resistance changes are converted to the voltage signals by the conditioning modules, and are amplified by the amplifier modules. The amplified voltage signals X_i (i=1, 2, 3…8) are input into a personal computer through serial port (RS232) after A/D conversion using a micro-computer system. Since eight channels of the strain gauges were used (four groups for x- and y-directional forces and another four groups z-directional forces), there are eight channels of the voltage signals. A program specially developed in MATLAB was used to sample the eight channels of the voltage signals and calculate the forces and the moments.

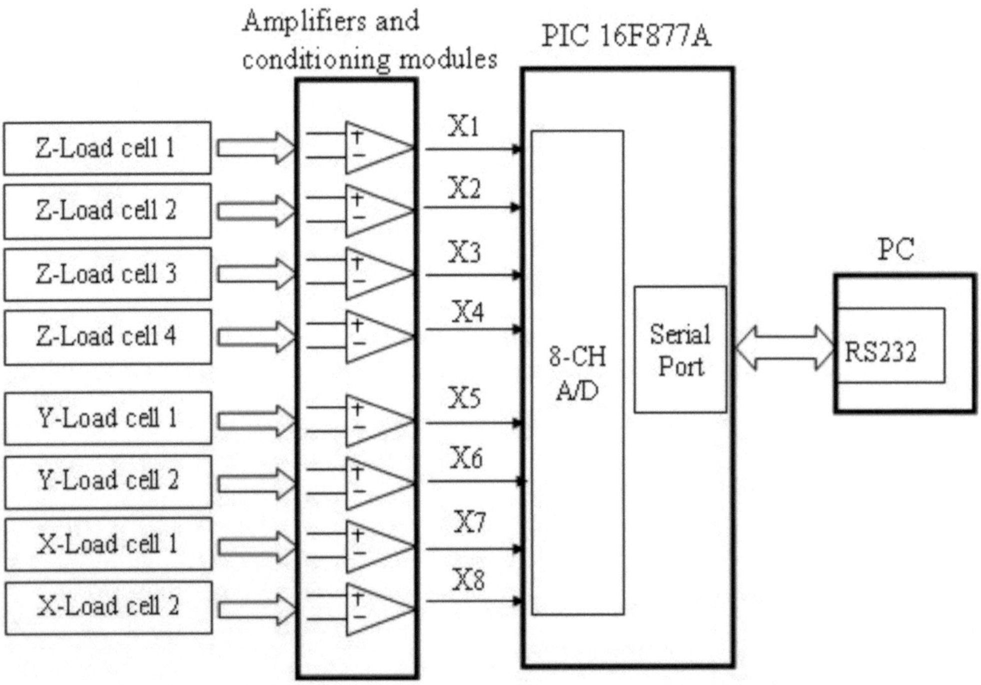

Fig. 6. Electrical hardware system of the sensor. The amplifier modules, conditioning circuits and microcomputer system were integrated on a based board, which was fixed in the mechanical structures of the sensor. The outputs of the amplifiers and conditioning modules (X_i) were used to calculate triaxial forces and moments applied on the sensor.

2.3 Prototype of Force Sensor

In order to achieve a high signal to noise ratio, amplifier modules, conditioning circuits and interface program were integrated into the force sensor. The large resistance strain gages (5000 ohm) of Vishay Micro-measurements were used, so the sensor system is low power consumed and can be powered using a small battery. Figure 7 shows the integrated sensor system and an interface software developed specially for monitoring the data obtained from the force sensor.

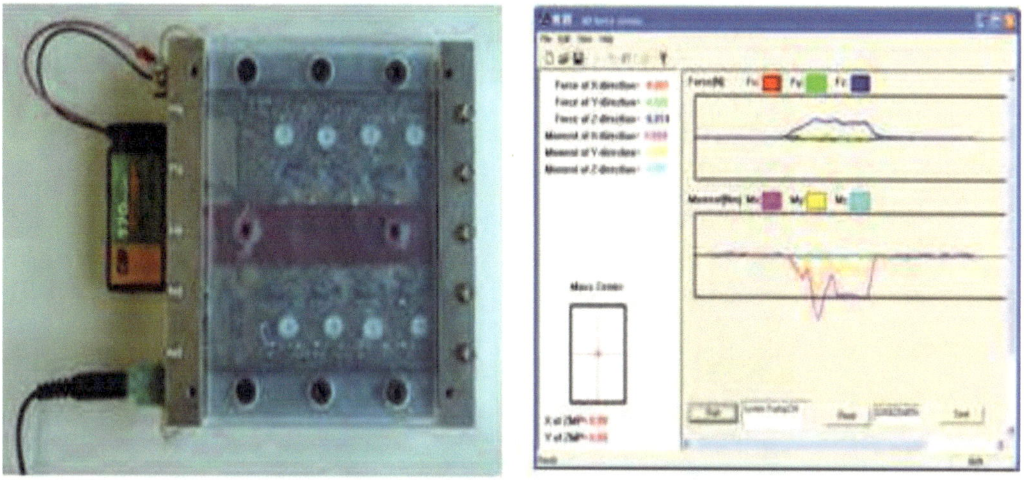

(a) (b)

Fig. 7. Sensor system including a mechanical system, an electrical system and an interface software system. (a) The sensor hardware system can be power using a battery and communicate with a personal computer through a serial port of a micro-computer system; (b) An interface software for operation of the senor and sampling data from the sensor.

3. Wearable Motion Sensor

3.1 Motion Sensor System

As shown in Fig. 8, we developed a wearable motion sensor system which includes an eight-channel data recorder, a gyroscope and accelerometer combination unit, and two gyroscope units. The two gyroscope units are attached on the foot and thigh respectively, and the gyroscope and accelerometer combination unit is fixed on the shank, which is near to the ankle. The data-logger can be pocketed by subjects. The principle operation of the gyroscope is measurement of the Coriolis acceleration which is generated when a rotational angular velocity is applied to the oscillating piezoelectric bimorph. The inertial sensors can work under lower energy consumption (4.6 mA at 5V), so it is appropriate for ambulatory measurements. The signals from the gyroscopes and accelerometer are amplified and low-pass filtered (cutoff frequency: 25Hz) to remove the electronic noise. The frequencies outside the pass-band are filtered out because they are invalid for the study of human kinetics.

As shown in Fig. 9, three local coordinate systems were defined for the three sensor units, in which the sensing axis of the gyroscopes is along y-axis, and the z-axis is along the leg-bone. Three gyroscopes are used to measure angular velocities of leg segments of the foot, shank and thigh (ω_1, ω_2 and ω_3). The sensing axis (y-axis) of the gyroscopes is vertical to the medial-lateral plane so that the angular velocity in the sagittal plane can be detected. A bio-axial accelerometer is attached on the side of shank to measure two-directional accelerations along the tangent direction of x-axis (a_t) and the sagittal direction of z-axis (a_r). In this system the data obtained from accelerometer are fused with data collected from gyroscopes for a cycle re-calibration, through supplying initial angular displacements of the attached leg segment.

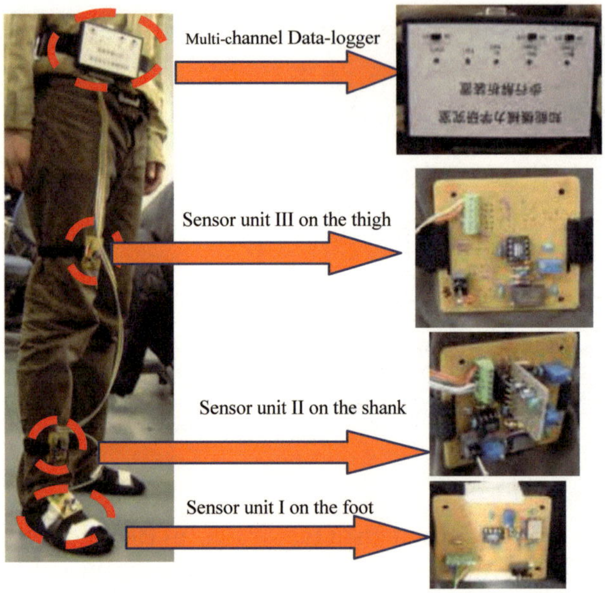

Fig. 8. Wearable motion sensor system. A strap system is designed for binding the sensor units to lower limb segments. Each sensor unit is attached to a strap. During walking, the strap is tied around the limb to secure the position of the sensor unit.

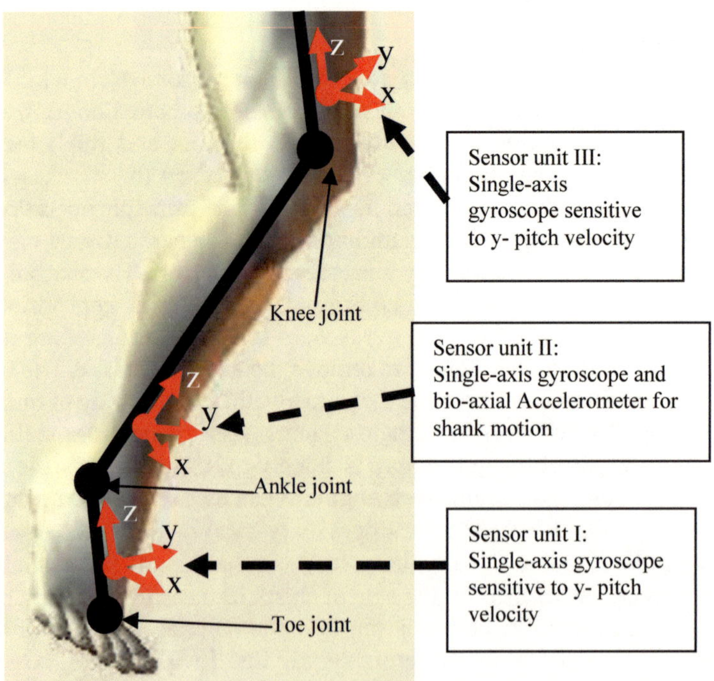

Fig. 9. Position and coordinates of the sensor units. In the local orientation coordinate of the sensor unit (x-, y- and z- axis), the y-axis denotes each joint's rocker axis, which is parallel to

the sensitive axis of the gyroscope, while x-axis and z-axis denote the unity vectors in the radial and tangential direction respectively.

A multi-channel data-logger was also specially designed for the wearable motion sensor system. A micro-computer (PIC 16F877A) was used to develop the pocketed data-logger, and the sampled data from the inertial sensors could be saved in a SRAM which can keep recording for five minutes. An off-line motion analysis can be performed by feeding data saved in the SRAM to a personal computer through a RS232 communication module. Since gyroscope (ENC-03J), accelerometer-chip (ADXL202) and the PIC system are all devices with a lower energy consumption, the wearable sensor system could be powered using a battery of 300mAh (NiMH 30R7H).

3.2 Calibration of Sensor Units

Complete architecture of a calibration system is showed in Fig. 10(a), and hardware devices of the system mainly include a data-logger (A/D card: Keyence NR-110), a potentiometer, a reference angle finder and a clamp (Fig. 10(b)).

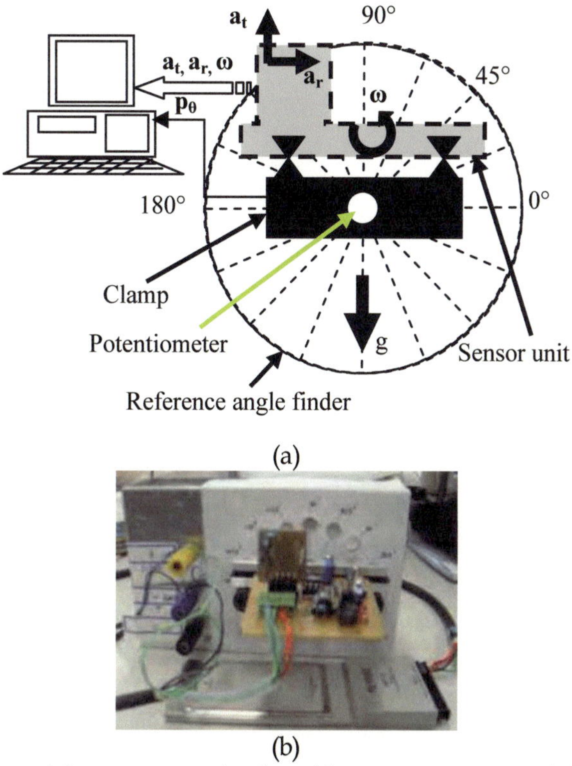

Fig. 10. (a) Architecture of the equipment for the calibration experiment of the sensor unit. A bio-axial accelerometer is used to measure two-directional accelerations of a_t and a_r, and p_θ is the output signal of the potentiometer which measures imposed rotational quantities and provides reference angular velocity quantities through a differential computing. Signal of the gyroscope in the sensor unit is defined as ω, and its positive direction is anticlockwise. The four signals of p_θ, ω, a_t and a_r are sampled into computer through the 12-bit A/D card. (b) Hardware devices with a mechanical case and an interface for the calibration of the sensor unit.

The sensor units are calibrated in static state and dynamic state respectively. First, the calibration of the accelerometer sensor is carried out during the static state. The accelerometer in sensor unit is subjected to different gravity vectors by rotating a based axis that is connected with a potentiometer. Second, the dynamic calibration is completed to calibrate the gyroscopes and test the accelerometer in a dynamic condition. In both cases the calibration matrixes are computed using the least squares method.
$[C_\theta]$ is calibration matrix for the angle position in (7). The matrix of the imposed quantities $[\theta]$ in a specific case can be recorded when the sensor unit is rotated to different positions on the angle finder plane; $[p_\theta]$ is the matrix of the quantities acquired from the potentiometer, in the specific positions of a sensor unit. Angular displacement $[\theta_r]$ can be calculated using (8), and angular velocity of the sensor unit is obtained through the differential computing of the angular positions in a serial time. Gravity acceleration g projected to the two sensitive axes (a_t and a_r) of the bio-axial accelerometer is estimated using (9) and (10).

$$[C_\theta] = [\theta][p_\theta]^T([p_\theta][p_\theta]^T)-1 \quad (7)$$

$$[\theta_r] = [C_\theta]\cdot[p_\theta] \quad (8)$$

$$A_t = -g\cdot\cos(\theta_r) \quad (9)$$

$$A_r = -g\cdot\sin(\theta_r) \quad (10)$$

$[C_{at}]$ and $[C_{ar}]$ are the calibration matrixes for the bio-axis accelerometer in (11) and (12), where $[A_t]$ and $[A_r]$ are the matrixes of the imposed quantities which were obtained when the sensor unit is subjected to different directional gravity vectors by rotating the sensor unit on the angle finder plane; $[a_t]$ and $[a_r]$ are the matrixes of the quantities assessed by the accelerometer in the sensor unit. (13) and (14) give summations of the subjected gravity g and the segment motion acceleration on the two sensitive axes using the output signals of the bio-axial accelerometer. Table 1 shows the data of a sensor unit from a representative calibration experiment.

$$[C_{at}] = [A_t][a_t]^T([a_t][a_t]^T)-1 \quad (11)$$

$$[C_{ar}] = [A_r][a_r]^T([a_r][a_r]^T)-1 \quad (12)$$

$$\lfloor A^{re}{}_r \rfloor = [C_{ar}]\cdot[a_r] \quad (13)$$

$$\lfloor A^{re}{}_t \rfloor = [C_{at}]\cdot[a_t] \quad (14)$$

$[C_g]$ in (15) is the calibration matrix for gyroscope sensor in the sensor units, where $[V_g]$ and $[p_\theta]$ are the matrixes of the quantities respectively acquired from the gyroscope and potentiometer in a serial time (t); $[C_\theta]$ is the calibration matrix for the angle position in (7). Angular position $[\theta_r]$ can be calculated using (8), and angular velocity of the sensor unit is obtained through difference computing of the angular positions in the serial time. Gravity acceleration g projected to the two sensitive axes (a_t and a_r) of the bio-axial accelerometer is estimated using (9) and (10).

$$[C_g] = [C_\theta][p_\theta]\left[\int V_g dt\right]\left(\left[\int V_g dt\right]\left[\int V_g dt\right]^T\right)-1 \tag{15}$$

Angle finder θ (Degrees)	Potentiometer p_θ (V)	Accelerometer	
		a_t (V)	a_r (V)
0	1.438	2.541	3.004
-22.5	1.355	2.798	2.943
-45	1.284	2.985	2.813
-67.5	1.209	3.108	2.602
-90	1.134	3.138	2.369
22.5	1.512	2.294	2.969
45	1.595	2.096	2.849
67.5	1.671	1.937	2.663
90	1.748	1.882	2.457

Table 1. Calibration results of a sensor unit.

3.3 Estimation of Segment Orientations

The loop frequency of the gait record is 100 Hz which is equal to the sensors sampling frequency, and the number of sampling time point is counted by an integer value i (i = 1, 2, 3....). The orientation of leg segment ($\theta(i)$) can be calculated by integral operator of the angular velocity ($\omega(i)$) of leg segment ((16) and (17)), which is directly measured using the wearable sensor units. The inclination of shank and thigh is set to zero in the initial period, while the inclination of foot is set to 90° at initial setting. Because the gyroscope in the sensor unit is a kind of inertial sensor that is affected by the drift errors when it is worn on the human body, the integral calculation in (16) must produce cumulated errors in the motion analysis of a multi-step walking trial.

$$\theta(i) = \theta(i-1) + (\omega(i-1) + \omega(i))\Delta t / 2 \tag{16}$$

where

$$\theta(0) = \theta_0; \quad \mathrm{i} = 1, 2, 3.... \tag{17}$$

We define the gait cycle (walking gait cycle number k = 1, 2....) as the period from one stance phase of a foot to the next stance phase of the same foot. In every walking cycle, the time points of transition from loading response phase to mid stance phase, and transition from pre-swing to initial swing phase are defined (Parry, 1992). Based on the analysis of transition of the gait phases, the human motion analysis is implemented by calculating body segments' angular displacements using inertial sensors of gyroscopes and accelerometers. As shown in Fig. 11, we can primarily detect the mid stance phase just using gyroscope signals and raw integration results of gyroscope signals from the three sensor units ($\omega_t < 0$, $\omega_f = \varepsilon_\omega$, $\omega_s < 0$ and $\theta_s < 0$). Moreover, we find that the rotational angular velocities of the shank and thigh are very small in the later interval of this phase, because the ankle is in a state of dorsiflexion, and the shank's rotational velocity is limited. Therefore, the accelerometer can be used for inclination measurement with respect to the gravity acceleration, when shank's sagittal direction A_r (18) and A_t (19) are mainly affected by the gravity acceleration's projection. Hence, we can make cycle calibration by measuring the initial angular orientation of the attached segment (the shank) using (20), and the foot orientation ($\theta_f{}^m = 0$) and the thigh orientation ($\theta_t{}^m = \theta_s{}^m$) are the initial calibration quantities for calculation of the foot and thigh orientations. Integral calculations are performed in every gait cycle, which can decrease the cumulated errors in longtime walking trials.

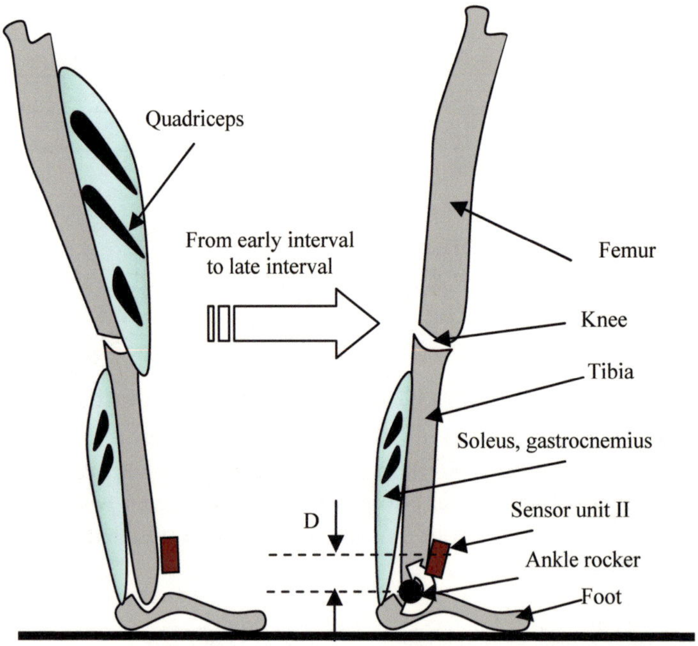

Fig. 11. Mid stance phase including the early and late intervals. The orientation calibration is implemented in the late interval of each mid stand. Early interval has body over mid foot with climb vertical, ankle neutral and foot flat, in which quadriceps and soleus muscles are in activity. Later interval has body over forefoot with continued heel contact, while ankle is in the state of dorsiflexion which limits the shank rotational velocity. A small distance between sensor unit II and ankle rocker is denoted using D (it is about 50mm). Moreover in the later interval, soleus and gastrocnemius are the only extensor muscles around tibia, which produces the least vibration effect on the accelerometer.

$$A_r = -g \cdot \sin(\theta_s{}^m) + D \cdot (\omega_s{}^m)^2 \tag{18}$$

$$A_t = -g \cdot \cos(\theta_s{}^m) + D \cdot \dot{\omega}_s{}^m \tag{19}$$

$$\theta_s^m = \arctan(A_r / A_t) \tag{20}$$

However, the heel of some subjects (e.g. paralytic patient) may never contact the ground, and in this case the proposed direct inference algorithm can not be used for the cycle recalibration to decrease the integration errors. A linear regression method was developed to calculate drift errors coming from the digital integration, in which we supposed the error increase in a linear function. We let θ_i be the direct integration results by (16), and θ_m be reference static inclination angle measured using accelerometer by (20). The summary error ($\theta_e = \theta_i - \theta_m$) is obtained from a single walking trial during which subjects are required to walk along a straight leading line. Error estimation function ($\theta_e(i)$) is given as following equation based to a linear regression.

$$\theta_e(i) = \Delta t \cdot i \cdot (\theta_i - \theta_m) / t, \quad \mathrm{i = 1, 2, 3....t} \tag{21}$$

$$\theta_c(i) = \theta(i) - \theta_e(i) \tag{22}$$

where Δt is the sampling time and $\theta_c(i)$ was defined as the estimated angle after an error correction.

4. Calculation of Joint Moment

Based on the measurements of GRF and segment orientations using the two sensor systems, we can calculate lower limb's joint moments which are useful for evaluating in vivo forces of human body during gait. For calculation purposes, such as estimating joint moments of the ankle during loading response and terminal stance phases (Parry, 1992), all vectors including joint displacement vector, GRF vector and gravity vector have to be expressed in the same coordinate system, being the global coordinate system. The *y*-axis of the global coordinate system was chosen to represent the anterior-posterior direction of human movement, and the *z*-axis was made vertical, while the *x*-axis was chosen such that the resulting global coordinate system would be right-handed. The origin of the global coordinate system was fixed to a point around the anatomical center of the ankle when the wearable force sensor was worn under the foot. The GRF and the moments measured using the wearable force sensor are expressed by the vectors in (23), and the coordinates of center of pressure (CoP) ${}^g x_{CoP}$ in the global frame is calculated using the following equation:

$${}^g F_{GRF} = \begin{pmatrix} {}^g F_x \\ {}^g F_y \\ {}^g F_z \end{pmatrix} \quad {}^g M_{GRF} = \begin{pmatrix} {}^g M_x \\ {}^g M_y \\ {}^g M_z \end{pmatrix} \quad {}^g x_{CoP} = \begin{pmatrix} \frac{{}^g M_y}{{}^g F_z} \\ \frac{{}^g M_y}{{}^g F_z} \\ 0 \end{pmatrix} \tag{23}$$

The ankle, knee and hip joint moments in the global coordinates system can be calculated using the inverse dynamic method (Hof, 1992; Zheng et al., 2008)

$$
{}^{g}M_{k,k+1} = -({}^{g}r_{CoP} - {}^{g}r_{k,k+1}) \times F_{GRF} - \sum_{i=1}^{k} \left[({}^{g}r_i - {}^{g}r_{k,k+1}) \times m_i \cdot g \right] + \sum_{i=1}^{k} \left[({}^{g}r_i - {}^{g}r_{k,k+1}) \times m_i \cdot a_i \right] + \sum_{i=1}^{k} \frac{d}{dt}\left(I_i \cdot \dot{\theta}_i \right) \quad , k=1, 2, 3, \tag{24}
$$

where ${}^{g}M_{1,2}$, ${}^{g}M_{2,3}$ and ${}^{g}M_{3,4}$ represent the ankle, knee and hip joint moments respectively. ${}^{g}r_{CoP}$ is referred as the position vector pointing from the origin of the global coordinate system to the CoP. ${}^{g}r_1$, ${}^{g}r_2$ and ${}^{g}r_3$ are the position vectors from the origin of the global coordinates system to the centers of the ankle, knee and hip joints. ${}^{g}r_{1,2}$, ${}^{g}r_{2,3}$ and ${}^{g}r_{3,4}$ are the position vectors from the origin of the global coordinates system to the center of mass (CoM) of the foot, shank and thigh respectively. The mass of the foot, shank and thigh was defined as m_1, m_2 and m_3 respectively. a_1, a_2 and a_3 are the accelerations of the CoM of the foot, shank and thigh respectively. I_1, I_2 and I_3 are the inertia moments of the foot, shank and thigh respectively. The angular velocities of the foot, shank and thigh were obtained by the differential operators ($\dot{\theta}_1$, $\dot{\theta}_2$ and $\dot{\theta}_3$).

5. Experiment Study

5.1 Experiment Method

To validate the sensor system performance we have compared the quantitative results of the sensor system with the measurements obtained with a commercial optical motion analysis system Hi-DCam (NAC image technology. Japan). The commercial motion analysis system can track and measure three-dimensional (3-D) trajectories of retro-reflective markers placed on the subject's body, as shown in Fig. 12. The cameras with a sampling frequency of 100 Hz were used to track the marker positions with an accuracy of about 1 mm. A stationary force plate of EFP-S-2KNSA12 (Kyowa co. Japan) was also used as a reference sensor to validate the measurement of the developed force sensor. In our experiment, these sensor systems simultaneously worked in the measurements of human force and motion. The data from the reference sensor systems and our developed sensor systems were sampled at the same time, and were compared.

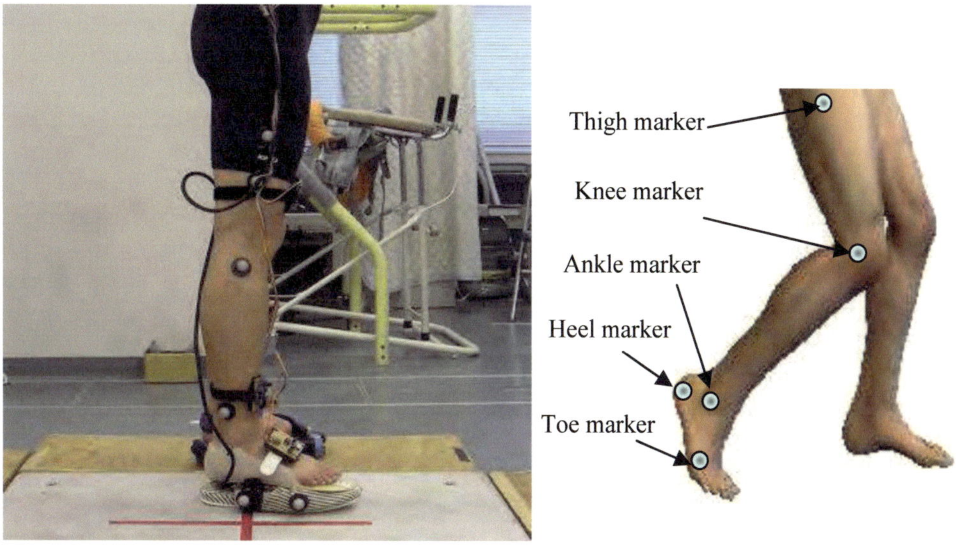

Fig. 12. Verification experiment system and positions of the retro-reflective markers

The root of the mean of the square differences (RMS) was used to compare the closeness in amplitude of the two sensor measurement results. The percent error (PE) was calculated as the ratio between the RMS errors to the average peak-to-peak amplitude of the reference measurements.

$$RMS = \sqrt{\left(\frac{1}{n}\sum(F - F_r)^2\right)} \tag{25}$$

Ten healthy subjects (eight men and two women: age=28.1±2.0years, height=169.2±4.2cm, weight=66.3±9.2kg) were required to wear the developed sensor system in experiment tests. Segmental mass and inertial moment parameters of the lower extremities of the subjects could be estimated using the empirical regression method (Zatsiorsky and Seluyanov, 1983). The length of the thigh, shank and foot of the subjects is 0.4165±0.0224m, 0.3715±0.03m, and 0.2453±0.008m respectively. The mass of the foot, shank and thigh is 0.7355±0.0635kg, 3.1043±0.3902kg and 7.6924±0.8436kg respectively. The ratio of the center of gravity was calculated as the mean value of the percentage of the segment length measured from the proximal end. The mean of the ratio of the foot, shank and thigh is 59.5%, 40.6% and 0.475% respectively for the male subjects, while 59.4%, 41% and 45.8% respectively for the female subjects. The inertial moment of the foot, shank and thigh is 0.00037732±0.00000365kgm^2, 0.0302±0.0101kgm^2 and 0.0973±0.0139kgm^2 respectively.

5.2 Experiment Results: GRF and Segmental Orientations

As shown in Fig. 13 (a)-(c), the comparisons of the three components of GRF measured using the wearable sensor systems and the stationary force plate system were demonstrated in a representative trial. The results of GRF were normalized with respect to the body weight.

The results show a good correspondence between two methods, which is confirmed by comparison analysis of the GRF (see Fig. 13(d)) and errors analysis of the GRF (see Fig. 13 (e)), and RMS differences by the ten subjects' trials of 0.045±0.003 N/N (mean ± standard deviation), which corresponds to 4.26±0.34% of the maximal GRF magnitude. A separate analysis of each component of the GRF results for the vertical component in an RMS difference of 0.046±0.002N/N, being 4.8±0.2% of the maximal GRF magnitude, or 4.2±0.2% of the maximal vertical components. For the x-directional component of the horizontal GRF, RMS difference is 0.011±0.008N/N, being 1.07±0.91% to the maximal GRF magnitude, or 10.3±2.2% to the maximal x-directional component. For the y-directional component of the horizontal GRF, RMS difference is 0.014±0.002N/N, being 1.17±0.12% to the maximal GRF magnitude, or 10.1±3.6% to the maximal y-directional component of GRF.

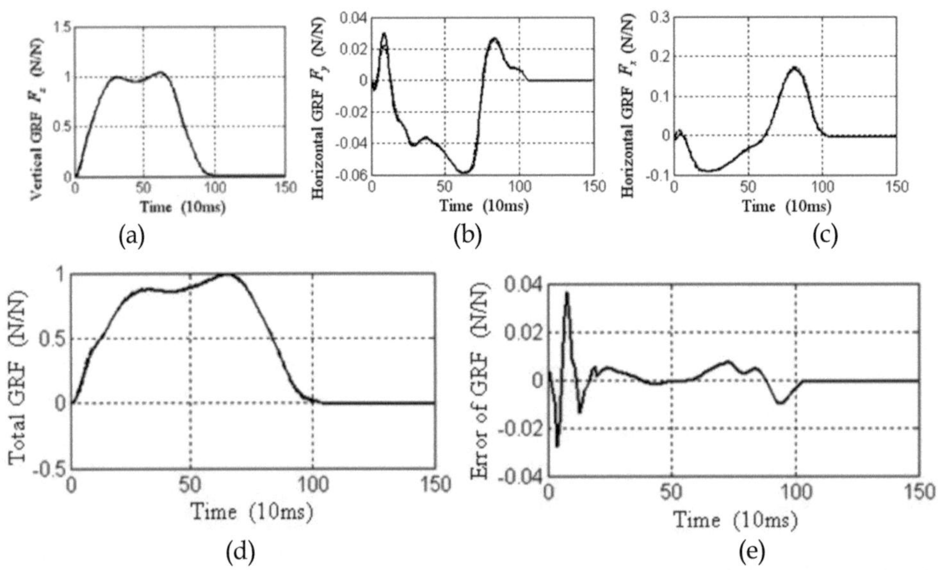

Fig. 13. Triaxial GRF measured by the wearable sensor systems (solid line) and the force plate (dashed line). (a) The vertical components: F_z. (b) The horizontal component: F_y. (c) Horizontal component: F_x. (d) The total GRF. (e) Error of GRF between the two measurement systems.

The estimation of the position of the CoP is shown in Fig. 14. The trajectories agree well, and RMS difference between the two methods was 10.4±1.2mm, corresponding to 3.7±0.5% of the length of the instrumented shoe.

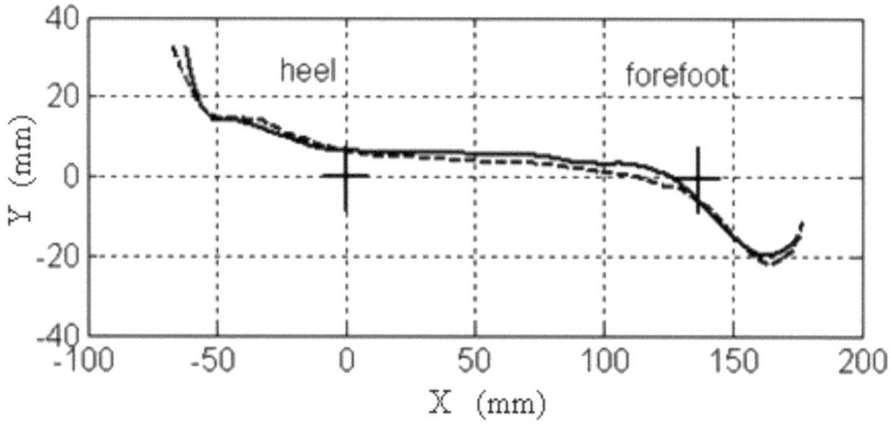

Fig. 14. Center of pressure (CoP) measured by the wearable sensor system (solid line) and the stationary force plate (dashed line) in the global coordinate system.

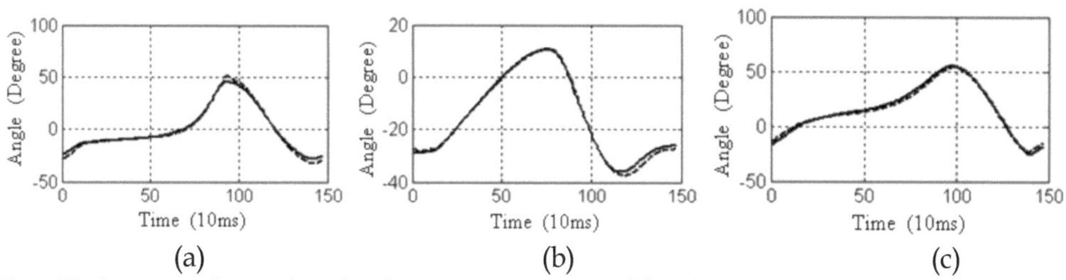

Fig. 15. Segmental angular displacements measured by the wearable motion sensor system (solid line) and the Hi-DCam camera system (dashed line). (a) Foot angular displacements. (b) Shank angular displacements. (c) Thigh angular displacements.

As shown in Fig. 15, the segmental angular displacements estimated using the wearable sensor system and Hi-DCam camera system were also compared. For the foot angular displacement, the RMS difference between the results estimated using the two systems was calculated as 2.91±0.12degrees, being 5.6±0.35% of the maximal magnitude of the foot angular displacements. For the shank angular displacement, the RMS difference between the two methods is 2.61±0.93degrees, being 4.71±1.4% of the maximal magnitude of the shank angular displacement. For the thigh angular displacement, the RMS difference between the two methods is 1.3±0.39degrees, being 4.71±2.05% of the maximal magnitude of the thigh angular displacement.

5.3 Experiment Results: Joint Moments

The comparison results of the joint moments of the ankle, knee and hip in the sagittal plane between the two measurement systems are shown in Fig. 16. The RMS difference of the ankle joint moments calculated using the two systems is 2±0.34Nm, being 5.4±0.7% of the maximal magnitude of the ankle joint moments during a whole gait cycle. The RMS difference of the knee joint moments calculated using the two systems is 7.2±1.34Nm, being 6±0.32% of the maximal magnitude of the knee joint moments. The RMS difference of the hip joint moments calculated by using the two systems is 11.2±1.3Nm, being 6.1±0.25% of the maximal magnitude of the hip joint moments.

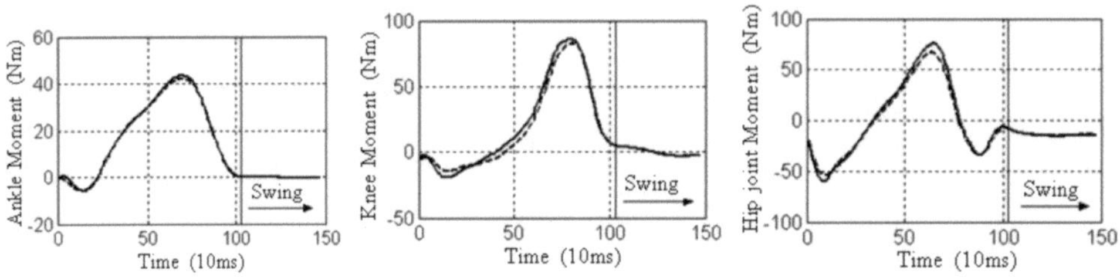

(a) (b) (c)

Fig. 16. Joint moments estimated by the wearable sensor systems (solid line), and the force plate & Hi-DCam camera systems (dashed line). Ankle plantarflexor, knee extensor, and hip flexor are positive, and ankle dorsiflexion, knee flexor, and hip extensor are negative. (a) Ankle joint moments. (b) Knee joint moments. (c) Hip joint moments.

6. Conclusion

A new wearable sensor system was developed for measuring tri-directional ground reaction force (GRF) and segment orientations. A stationary force plate can not measure more than one stride; moreover, in studies of stair ascent and descent measurements, a complex system consisting of many stationary force plates and a data fusion method must be constructed (Stacoff et al., 2005; Della and Bonato, 2007). The wearable sensor system proposed in this chapter can be applied to successive walking trials in a number of non-laboratory environments, because of its lower-cost, small size and lower effects on human movements. The joint moment estimations of the ankle, knee and hip joints using the measurements of the wearable sensor system is an application study. The joint moment data should be helpful for understanding the cause of certain gait abnormalities and the motion mechanism of human walking, when we combine it with the other measurements, such as 3D joint angles and electromyography. In the verification experiments, we measured the triaxial GRF, coordinates of CoP, and lower limbs' angular displacements, and caluculated joint moments using the measurments of the wearable sensor system and the force plate & optical camera system. The analysis of the RMS differences between the two systems was implemented, and the results of the RMS differences demonstrated that the force and motion measurement results and the joint moment estimation results have high consistence between the two systems. In the next step, based on the estimated joint moments, we will use the wearable sensor systems to estimate muscle tension forces by a static optimization method and a musculoskeletal model of the lower limb, because muscle tension forces provide further details about human physiological data during gait.

7. References

Aminian, K.; Najafi, B.; Bula, C.; Leyvraz, P.F. & Robert, P. (2002). Spatio-temporal parameters of gait measured by an ambulatory system using miniature gyroscopes. *Journal of Biomechanics*, vol. 35, pp. 689-699, ISSN: 0021-9290.

Coley, B.; Najafi, B.; Paraschiv-Ionescu, A. & Aminian, K. (2005). Stair climbing detection during daily physical activity using a miniature gyroscope. *Gait and Posture,* vol. 22 (4), pp. 287-294, ISSN: 0966-6362.

Della, U. & Bonato, P. (2007). A novel design for an instrumented stairway. *Journal of Biomechanics,* vol. 40, pp. 702-704, ISSN: 0021-9290.

Faivre, A.; Dahan, M.; Parratte, B. & Monnier, G. (2004). Instrumented shoes for pathological gait assessment. *Mechanics Research Communications,* vol. 31, pp. 627-632, ISSN: 0093-6413.

Hof, A. L. (1992). An explicit expression for the moment in multibody systems, *Journal of Biomechanics,* Vol. 25, No. 10, pp. 1209-1211, ISSN: 0021-9290.

Hwang, E.S.; Yoon, Y.R.; Yoon, H.R.; Shin, T.M. & Kim, Y.J. (2008). Flexible contact force sensing device using metal/polymer multilayer structure for robotic applications. *Sensors and Materials,* vol. 20, pp. 55-69, ISSN: 0914-4935.

Jasiewicz, J.M.; Allum, J.H.J.; Middleton, J.W.; Barriskill, A.; Condie, P.; Purcell, B. & Li, R.C.T. (2006). Gait event detection using linear accelerometers or angular velocity transducers in able-bodied and spinal-cord injured individuals. *Gait and Posture,* vol. 24, pp. 502-509, ISSN: 0966-6362.

Lau, H. & Tong, K. (2008). The reliability of using accelerometer and gyroscope for gait event identification on persons with dropped foot. *Gait and Posture,* vol. 27, pp. 248-257, ISSN: 0966-6362.

Liedtke, C.; Fokkenrood, S.A.W.; Menger, J.T.; Kooij, H. & Veltink, P.H. (2007). Evaluation of instrumented shoes for ambulatory assessment of ground reaction forces. *Gait and Posture,* vol. 26, pp. 39-47, ISSN: 0966-6362.

Najafi, B.; Aminian, K.; Loew, F.; Blanc, Y. & Robert, P.A. (2002). Measurement of stand-sit and sit-stand transitions using a miniature gyroscope and its application in fall risk evaluation in the elderly. *IEEE Transactions on Biomed. Eng.,* vol. 49, pp. 843-851, ISSN: 0018-9294.

Parry, J. (1992). *Gait analysis normal and pathological function,* Slack Incorporated, pp. 149-158, ISBN: 9781556421921.

Sabatini, A.M.; Martelloni, C.; Scapellato, S. & Cavallo, F. (2005). Assessment of walking features from foot inertial sensing. *IEEE Transactions on Biomed. Eng.,* vol. 52 (3), pp. 486-494, ISSN: 0018-9294.

Schepers, H.M.; Koopman, H. F. J. M. & Veltink, P.H. (2007). Ambulatory assessment of ankle and foot dynamics. *IEEE Trans. Biomed. Eng.,* vol. 54, pp. 895-900, ISSN: 0018-9294.

Stacoff, A.; Diezi, C.; Luder, G.; Stussi, E. & Quervain, I.A. (2005). Ground reaction forces on stairs: Effects of stair inclination and age. *Gait and Posture,* vol. 21, pp. 24-38, ISSN: 0966-6362.

Stacoff, A.; Quervain, I.A.K.d.; Luder, G.; List, R. & Stussi, E. (2007). Ground reaction forces on stairs. Part II: Knee implant patients versus normal. *Gait and Posture,* vol. 26, pp. 48-58, ISSN: 0966-6362.

Valdastri, P.; Roccella, S.; Beccai, L.; Cattin, E.; Menciassi, A.; Carrozza, M.C. & Dario, P. (2005). Characterization of a novel hybrid silicon three-axial force sensor. *Sensors & Actuators: A. Physical,* vol. 123-124, pp.249-257, ISSN: 0924-4247.

Veltink, P.H.; Liedtke, C.; Droog, E. & Kooij, H. (2005). Ambulatory measurement of ground reaction forces. *IEEE Trans Neural Syst Rehabil Eng.*, vol. 13, pp. 423–527, ISSN: 1534-4320.

Yavuzer, G., Oken, O., Elhan, A., Stam, H.J., 2008, "Repeatability of lower limb three-dimensional kinematics in patients with stroke," *Gait and Posture*, 27(1), pp. 31-35, ISSN: 0966-6362.

Zatsiorsky, V. M. & Seluyanov, V. N. (1983). The mass and inertia characteristics of the main segments of the human body, In: *Biomechanics VIII-B,* Edited by Matsui, H. & Kobayashi, K., pp. 1152-1159. Human Kinetics Publishers, Chamapaign, IL.

Zhang, K.; Sun, M.; Lester, D.K.; Pi-Sunyer, F.X.; Boozer, C.N. & Longman, R.W. (2005). Assessment of human locomotion by using an insole measurement system and artificial neural networks. *Journal of Biomechanics*, vol. 38, pp. 2276-2287, ISSN: 0021-9290.

Zheng, R.; Liu, T.; Inoue, Y.; Shibata, K.; Liu, K.; (2008). Kinetics analysis of ankle, knee and hip joints using a wearable sensor system, *Journal of Biomechanical Science and Engineering*. Vol. 3(3), pp.343-355, ISSN: 1880-9863.

9

Postural Mechatronic Assistant for Laparoscopic Solo Surgery (PMASS)

Arturo Minor Martínez and Daniel Lorias Espinoza
CINVESTAV I.P.N. Center for Research and Advanced Studies / ITNL Technological Institute of Nuevo León
Monterrey, México

1. Abstract

Laparoscopes used in laparoscopic surgery are manipulated by human means, passive systems or robotic systems. All three methods accumulate downtime when the laparoscope is cleaned and the optical perspective is adjusted. This work proposes a new navigation system that autonomously handles the laparoscope, with a view to reducing latency, and that allows real-time adjustment of the visual perspective.

Methods. The system designed is an intuitive mechatronic system with three degrees of freedom and a single active articulation. The system uses the point of insertion as the invariant point for navigation and has a work space that closely resembles an inverted cone.

Results. The mechatronic system has been tested in a physical trainer, cutting and suturing chicken parts, as well as in laparoscopic ovariohysterectomies in dogs and pediatric surgeries. In all the procedures, surgeons were able to auto-navigate and there was no visual tremor while using the system. Surgeons performed visual approaches in real time and had both hands free to carry out the procedure.

Conclusion. This new mechatronic system allows surgeons to perform solo surgery. Cleaning and positioning downtime are reduced, since it is the surgeon him/herself who handles the optics and selects the best visual perspective for the surgery.

2. Introduction

Laparoscopic surgery, which is at the vanguard of technology, has encompassed various technological fields. Given its characteristics, this type of surgery demands that the specialty surgeon acquire new abilities and quickly adapt to new technology. Technological advances in assistance with holding and handling the laparoscope during surgery focus on solo surgery, in which the surgeon is provided with the technological means to perform the surgery alone. At present, in addition to "passive" electromechanical systems, such as Tiska[1], Endofreeze[2] and Passist[3], there are various "active" robotic positioning systems, such as Aesop (Sackier et al., 1994), Endoassist (Dagan et al., 1982), FIPS (Buess et al., 2005) and Tonatiuh (Minor et al., 2002), with which to perform solo laparoscopic surgery. The work space of all these systems is an inverted cone. However, depending on the

application or subspecialty, the real work space is limited to half or less than half of an inverted cone, which is normally located in front of the surgeon.
A tool such as a robotic assistant is very advantageous if it can be used in all laparoscopic specialties. However, considering transportation, installation and maintenance costs, the tool becomes very expensive if its use is limited to a few specialties. In addition, downtime due to manual or voice-activated repositioning of the laparoscope is cumulative and generally not taken into account during the evaluation process. Nevertheless, downtime can be reduced if the laparoscope is handled intuitively and, preferably, in real time. We therefore propose an optimized tool, based on the Postural Mechatronic Assistant for Laparoscopic Training (PMAT) (Minor et al., 2005), which can be applied selectively to solo surgeries, easily installed and transported, and used to provide functional assistance during surgery.

3. Material and Methods

The PMAT comprises a harness, an electrically activated linear guide and a pair of passive articulations to which the laparoscope is coupled. Before the surgeon scrubs his/her hands, he/she places the harness on his/her body. The passive articulations are sterilized in a solution before the procedure and the linear guide is covered in sterile plastic.
The surgeon, having scrubbed his/her hands and assisted by the nurse, places the surgical dress over the harness. Port positions are located and a manual visual exploration is carried out. Once the entry ports are in place, the linear guide of the mechatronic system is coupled to the harness, with a quick push over the surgical dress, and secured. Lastly, the passive module of the mechatronic assistant is secured to the linear guide and the 30° laparoscope is attached. Once the optical system is coupled to the mechatronic system, it is introduced into the cavity to carry out the exploration and begin the procedure Figure 1.
Development of the mechatronic system evolved from a modified PMAT which was conceptualized for selftraining in the specialty. This assistant requires the postural movements of the surgeon to position the laparoscope within the training cavity. Although the PMAT offers significant advantages for self-training, it is limited to use in the operating room. The PMAT uses a linear guide to switch between the up and down visual perspectives, but the guide has two main limitations. Firstly, if the work space is to be enlarged, the guide needs to be lengthened, which implies increased weight and a poor fit on the chest of the surgeon. Secondly, the guide, which must be sterilized for routine procedures and fitted after the surgeon has donned his/her surgical coveralls, may rip surgical wear, thus compromising sterility. In order to remove these limitations, the new design still uses the harness and postural movements to position the laparoscope, but the linear guide is replaced by a rotary articulation.

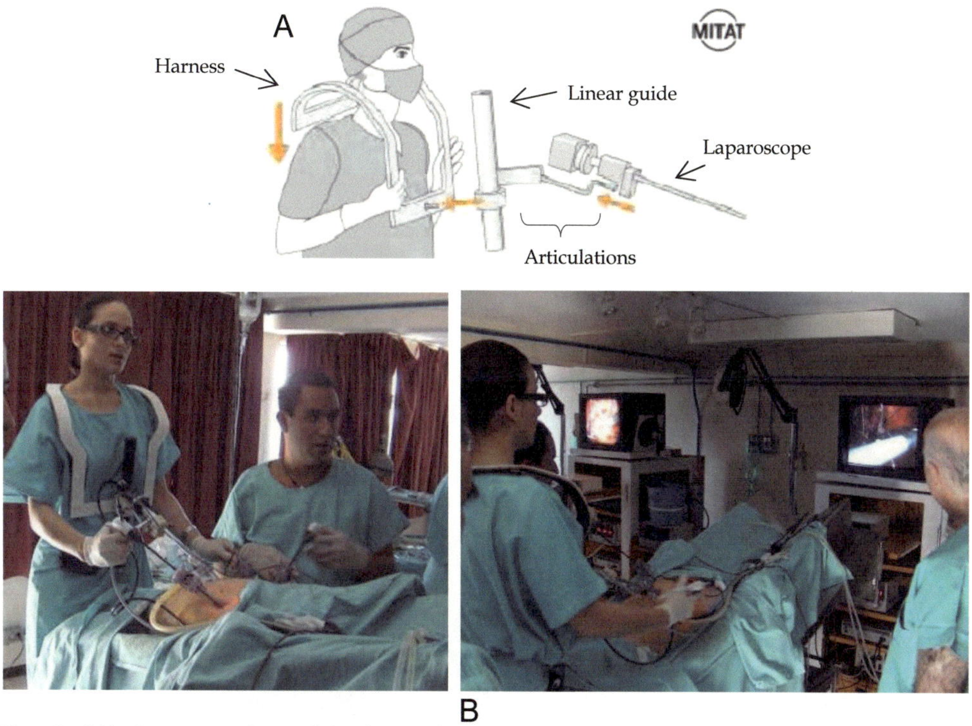

Fig. 1. (A) Conceptual model of non human laparoscopic assistant (B) Surgery without human laparoscopic assistant in pig.

The new system consists of three articulations, as shown in Fig. 2a. Two articulations are passive and one is active. The first articulation is active and rotary (1). The second articulation (2) is rotary, but passive; it operates in the same plane as the first articulation. The third articulation (3) is passive and rotary; it operates in a plane perpendicular to that of the first two articulations.

The system uses the entry point as the fixed and invariant point for exploration and navigation. Laparoscopic navigation with 0° optics requires six basic movements: right, left, up, down, in, and out. To pan horizontally right or left, the surgeon turns his/her torso right or left. The third articulation (3) and the point of insertion complete this movement (Fig. 2b). To establish the entry angle, which corresponds to the change in optical orientation above and below the laparoscope, the system uses the first articulation, which is active and rotary, together with the second articulation, which is passive. This mechanism has an almost linear displacement, as shown in Fig. 2c. The surgeon activates this articulation by means of two proximal switches.

To move the laparoscope in or out, the surgeon moves his/her body towards or away from the patient. The trajectory is almost linear, so there is no visual loss when moving in or out from tissues and organs during the procedure (Fig. 2d).

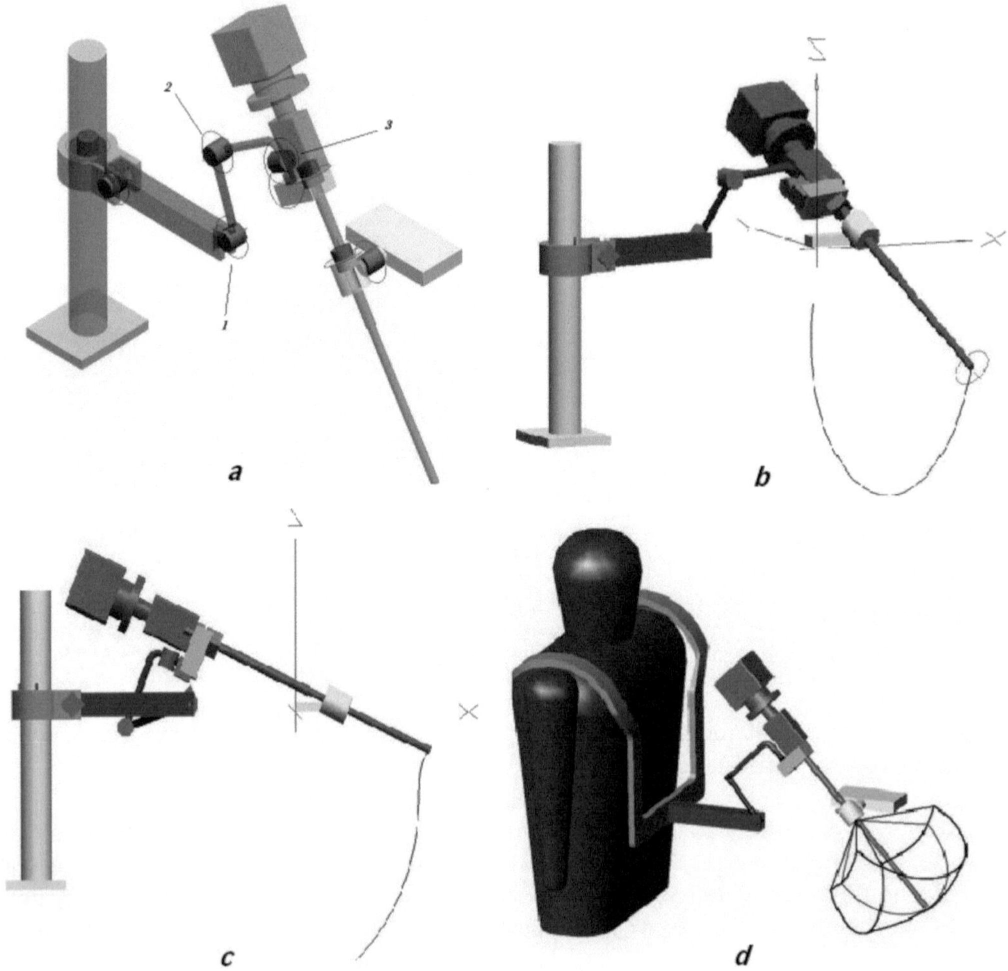

Fig. 2 (A) Concept design of the mechatronic assistant. (B) Model of the system for right-left perspective change. (C) Model of the system for above-below perspective change. (D) Work space of the mechatronic system using Visual Nastram 4D and Mechanical Desktop. The point of entry is invariant throughout the entire navigation

The simulation also determines the position in all moment of the laparoscope tip and the displacement magnitude inside the abdominal cavity for the movements right-left and in-out figure 3.

Once the mechatronic system had been modeled, a multidisciplinary group met to establish the following criteria:

- The system should be mountable in modules, so setup in the operating room is quick and trans-operatory sterilization is maintained.
- The system should allow quick disconnection of the laparoscope, so the surgeon can perform unforeseen explorations of anatomic spaces and maneuvers during any standard surgery.

- The system should be manufactured from a material that can be sterilized.
- The system weight should be kept to a minimum.
-

The system was manufactured from medical-grade steel, together with Teflon-steel parts, to allow for sterilization. The system, which weighs half a kilogram, is set up in a three-step procedure.

A

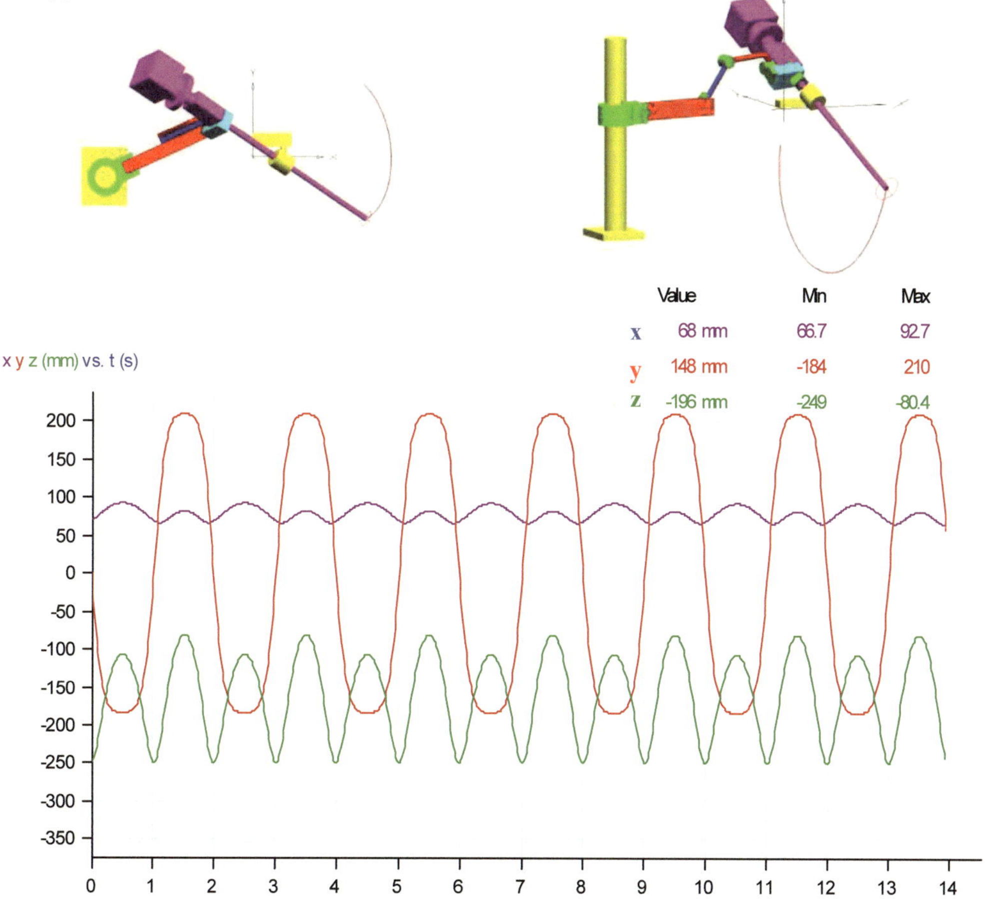

B

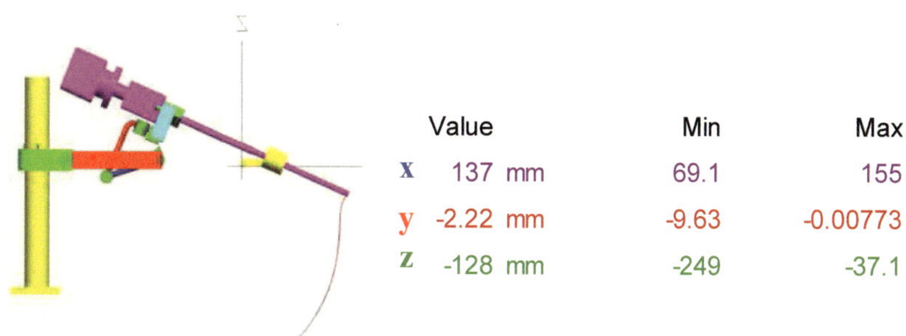

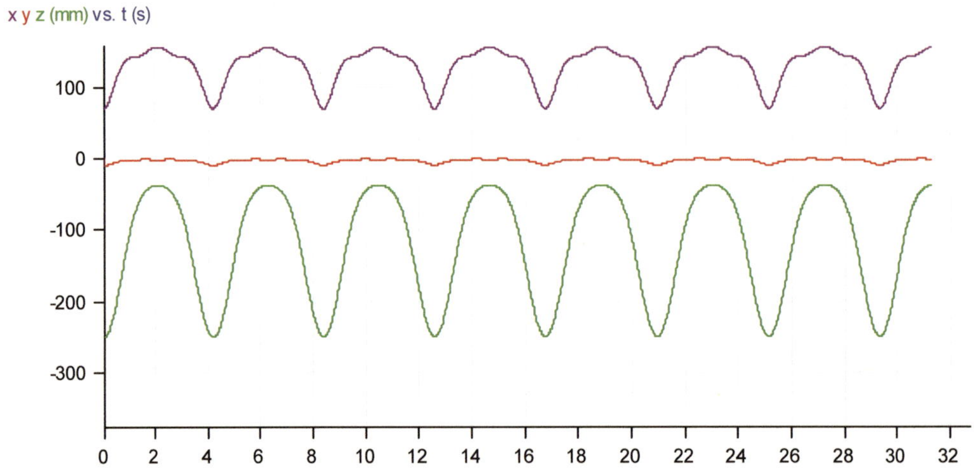

Fig. 3. Space location of the laparoscopy tip during the sailing (A) right–left perspective change. (B) above–below perspective change

Step one is to fit the harness onto the surgeon, under his/her surgical coveralls. Step two is to place on the harness, but over the coveralls, the motor which drives the active articulation and is wrapped in sterile plastic. Step three is to connect to the motor the passive articulations, which are sterilized by normal procedures. The mechatronic system is fitted onto the surgeon in two minutes. The laparoscope connects to and disconnects from the system with ease

4. Test

The PMASS was evaluated by experienced doctors, who first dissected, cut, and sutured chicken parts using 0° optics and performed maneuvers to change optical perspective and explore surrounding areas (Fig. 4 a, b). Surgeries were chosen so that the surgeon approach was frontal to the patient and ergonomic. The next evaluation consisted of three ovariohysterectomies in dogs, using 0° optics (Fig. 4 c, d). The final evaluation consisted of two pediatric surgeries for which the selected procedure was a Nissen fundoplication to correct gastroesophageal reflux disease, not corrected by pharmacological treatment and with persistent esophagitis and alimentary tract bleeding. The surgeries were performed on two 1-year-old patients, whose cases were fully studied prior to the surgery and for whom, in accordance with the Helsinki Treaty, there was informed parental consent. These surgeries were performed solo and using 0° and 30° optics. Also five appendicectomies, four ovarian cystectomies and four laparoscopic sterilizations - were performed by three experienced surgeons (Fig. 5).

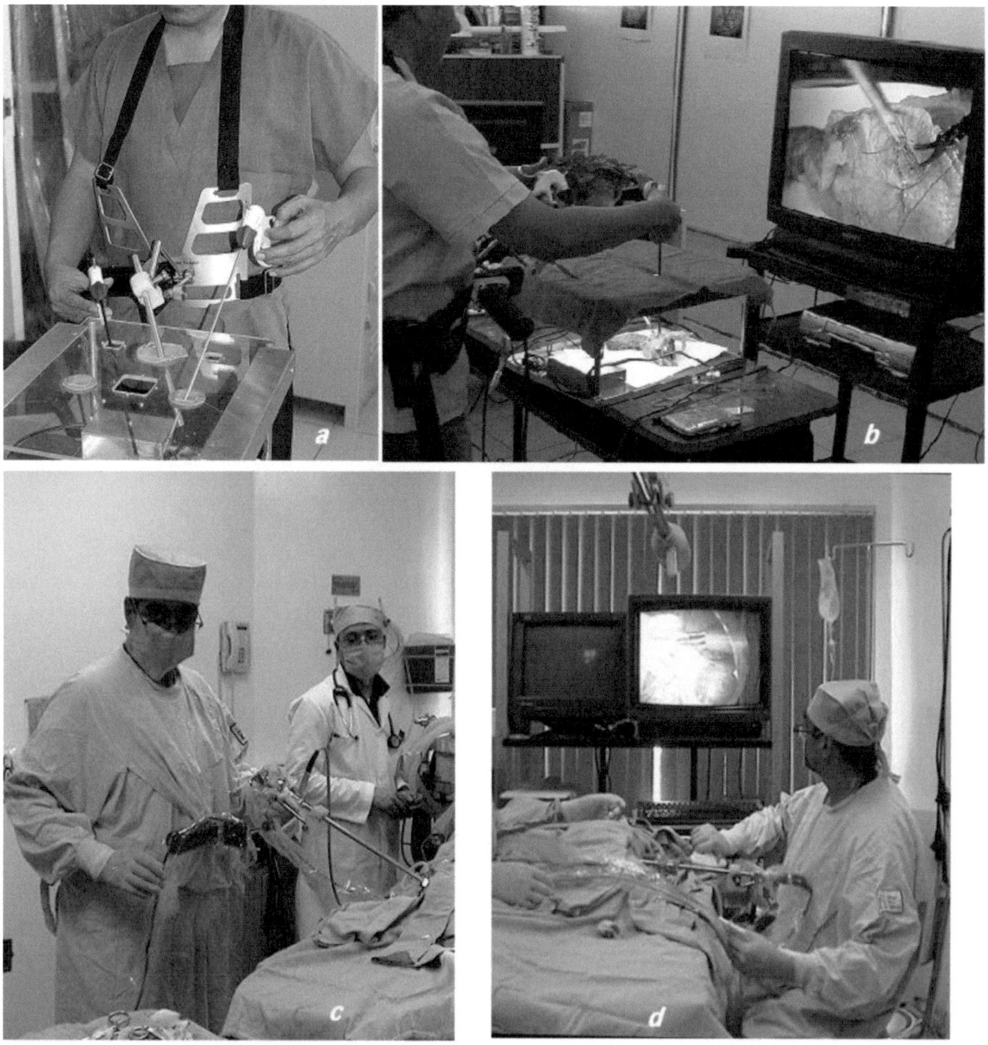

Fig. 4. (a) From concept to design. (b) Adaptation to suturing of chicken parts. (c) Coupling of the PMASS prior to the laparoscopic ovariohysterectomy. (d) Solo laparoscopic ovariohysterectomy surgery in dog.

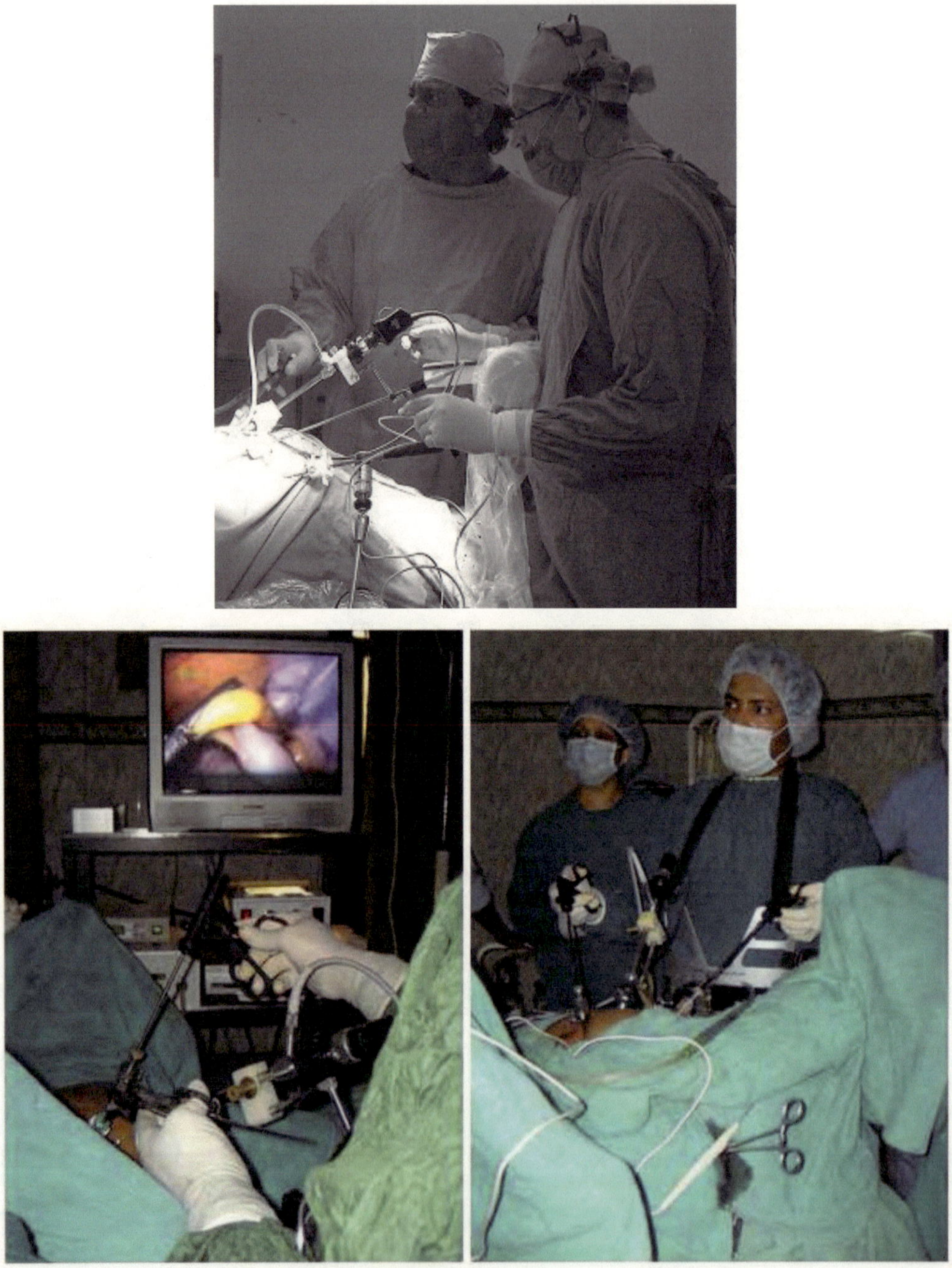

Fig. 5. Configuration and arrangement of the PMASS in the surgical scenario.

5. Result

The mechatronic system was fitted onto the surgeons in an average time of 1 min. The surgeons required an average of 5 min for adaptation to establish hand-eye feedback with the system. Surgeons used postural movements and visual feedback to achieve the required positions while keeping both hands free to perform their procedures. Optics showed no sign of tremor, and exploration, while moving in and out, was intuitive and took place in real time. It was also observed that the system worked equally well whether the surgeon stood or sat, and that postural changes did not limit visual perception. Simulations using the prototype showed that its in and out movements are not exactly linear. However, when operated with visual feedback, the mechatronic system does not have this limitation. The mechatronic system does not afford a view of the roof of the pneumoperitoneum when using 0° optics, so some surgeries, such as the Nissen surgery, are better performed using 30° optics, although the visual correction is manual. In order to clean the laparoscope, the surgeon steps back from the patient, cleans the device, and then returns to optics that have naturally taken up their last visual position. The average time required for cleaning and repositioning was 1 min, which is less than that required by the human assistant (Arezzo et al., 2005), (Buesset al., 2005), (Arezzo al., 2000) and much less than that required by the Aesop robot, Endoassist, Tiska, etc. (Arezzo et al., 2005), (Buesset al., 2005), (Arezzo al., 2000) or Tonatiuh (Minor et al., 2002). With respect to comfort, the subjective evaluation of the surgeons is that the harness is not uncomfortable to wear (average surgery time was 1.5 h), but that initially they feel anchored to the patient.

6. Discussion

The use of new technology in surgery is enabling surgeons to operate solo in some procedures. The advantages are adequate work space on the operating table and maneuverability. However, although active and passive systems offer these advantages, the time required to relocate or reaccommodate optics and clean the laparoscope between operations has not been reduced. In addition, there is a perceptual difference between what the surgeon wishes to see and the proximity he/she achieves with the active or passive system via the robot. There will always be the feeling that there exists a greater visual perspective to perform the procedure. The new mechatronic system shows that it is possible to have solo surgery in which the surgeon auto-manipulates the laparoscope to obtain the best optical perspective and has both hands free to perform the procedure.

One of the maneuvers that causes most delay during surgery is cleaning the laparoscope, whether the laparoscope is soiled or steamed up. Such delays, which are cumulative, are greater for robotic systems than for human assistants, given the set of verbal orders that must be given to the robot and the time required to mechanically couple and uncouple the laparoscope. The new design reduces cleaning time without affecting surgical quality. This assistant facilitates work on the operating table, as do active and passive assistants. Still pending are tests in the various laparoscopic subspecialties and a randomized clinical study to measure downtime under standard conditions. The clinical study will also seek to determine the advantages and limitations of the system, as well as the possibility of selectively applying the system to certain types of surgery in which it offers advantages over other known assistants. Presently under development is a prospective study for application to cholecystectomy, given that these surgeries are among those with the highest demand at first-level hospitals in Mexico, and are well suited to the PMASS approach.

7. References

Arezzo A, Schurr MO, Braun A, Buess GF (2005) Experimental assessment of a new mechanical endoscopic solo surgery system: endofreeze. *Surg Endosc* 19:581–588

Arezzo A, Ulmer F, Weiss O, Schurr MO, Hamad M, Buess GF (2000) Experimental trial on solo surgery for minimally invasive therapy. *Surg Endosc Ultrasound Intervent Tech* 14:955–959

Sackier JM, Wang Y (1994) Robotically assisted laparoscopic surgery from concept to development. *Surg Endosc* 8:63–66

Buess GF, Arezzo A, Schurr MO, Ulmer F, de Pescador H, Gumb L, Testa T, Nobman C (2000) A new remote-controlled endoscope positioning system for endoscopic solo surgery. The FIPS endoarm. *Surg Endosc* 14(4):395–399

Dagan J, Bat L (1982) Endoassist, a mechanical device to support an endoscope. *Gastrointest Endosc* 28(2):97–98

Minor A, Mosso JL, Domınguez A, Martınez RC, Muñoz R, Lara V (2002) Robot para cirugıa laparoscopica. *Revista Mexicana Ingenierıa Biomedica* 23(1):27–32

Minor Martinez A, Muñoz Guerrero R, Nieto J, Ordorica Flores R (2005) Postural mechatronic assistant for laparoscopic training. *Minimally Invasive and Their Allied Technol* 14(6):357–359

10

Model-Based Fault Detection and Isolation for a Powered Wheelchair

Masafumi Hashimoto, Fumihiro Itaba and Kazuhiko Takahashi
Doshisha University
Japan

1. Introduction

In mobile robotics and vehicle automation, the demand for fault detection and isolation (FDI) of sensors, actuators, and system components is growing significantly to assure system reliability and safety (Carlson, 2004; Petterson, 2005). FDI techniques are broadly classified into hardware redundancy and analytical redundancy approaches (Simani et al., 2003). The typical hardware redundancy approach for FDI is based on majority voting logic; outputs from redundant sensors are compared with each other, and the sensor whose output does not match the others is determined to be the faulty one. The major problem with hardware redundancy is the extra equipment required.

Unlike hardware redundancy, analytical redundancy requires no additional hardware components, and its algorithm can be implemented using software on a computer. Analytical redundancy makes use of mathematical models of the system under investigation, and it is therefore often referred to as a model-based method.

The typical model-based FDI is based on Bayesian filtering. The traditional FDI method uses the multiple-model adaptive estimation (MMAE) algorithm (*i.e.*, a bank of Kalman filters) (Magill, 1965); its application to FDI of mechanical failures (flat tires) and internal sensor faults has been presented (Roumeliotis et al., 1998a & 1998b). The MMAE algorithm is a noninteracting multiple-model algorithm. Therefore, it is not suitable for situations where system faults occur abruptly. To make the MMAE algorithm a better fit for such situations, various ad hoc techniques have been investigated (Maybeck & Hanlon, 1998; Goel et al., 2000) . To cope with the weakness of the MMAE-based approach, the interacting multiple-model (IMM) estimator (Blom & Bar-shalom, 1988; Zhang & Li, 1998) was applied to sensor FDI in mobile robots (Hashimoto et al., 2001 & 2003). Another typical model-based approach for FDI is the use of particle filters (Verma et al., 2004; Duan et al., 2006) ; this approach is suitable for FDI in nonlinear systems or systems with multimodal distribution or non-Gaussian noise. However, the computational cost increases rapidly with the number of state dimensions, even though there are means for increasing efficiency.

The major problems encountered with the model-based method are due to the use of imperfect models. In practical systems, model plant mismatches always exist; unexpected faults and disturbances also occur. These conditions may cause false alarms or missed alarms in FDI. Thus, model-based FDI needs to be made more robust (Simani et al., 2003).

In our previous paper, we presented model-based FDI for mobile robots and powered wheelchairs (Hashimoto et al., 2001; 2003 & 2007); external sensors and actuators were assumed to be always fault free, and only the faults of internal sensors were handled. The IMM estimator and Kalman filter were applied to FDI of the internal sensors.
In this chapter, we present model-based FDI for a powered wheelchair handling faults of both the internal and external sensors, as well as actuators. This chapter is organized as follows: Section 2 overviews our experimental wheelchair. Sections 3 and 4 present the FDI method for internal sensors and actuators, and Section 5 presents the FDI method for external sensors. In Section 6, we describe experiments conducted in a typical indoor environment to evaluate the FDI method, followed by our conclusions.

2. Experimental Wheelchair

Figure 1 shows the powered wheelchair used in the experiments; it has wheel units on the left and right sides to achieve differential steering. The wheel units each consist of a wheel motor and a wheel resolver. The wheel resolver measures the wheel velocity. A yaw-rate gyro is attached to the chassis of the wheelchair to sense the turn velocity. A laser range sensor (LRS) is mounted at the front side of the wheelchair to sense environments. It takes a laser scan image, which is represented by a sequence of distance samples in the horizontal half plain. The angular resolution of the LRS is 0.5 [deg], and one scan image contains 361 distance samples.
We handle the faults of three internal sensors (two wheel resolvers and one gyro), one external sensor (LRS) and two wheel motors.

3. Detection and Isolation of Internal Sensor and Wheel Motor Hard Faults

3.1 Internal Sensor FDI

First, we consider the hard faults of the wheel resolver and the gyro—the sensor fails completely, and the sensor output is stuck at a constant value.
A dynamic model related to the left and right wheel units is derived by

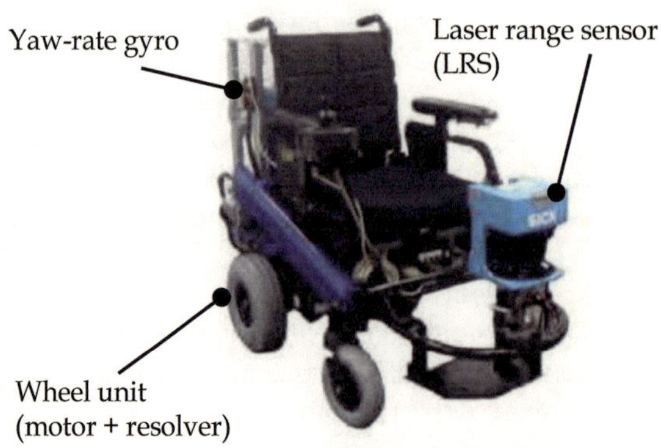

Fig. 1. Experimental wheelchair.

$$v_i(t+1) = A_i v_i(t) + B_i u_i(t) + \Delta v_i(t) \tag{1}$$

where i = left (wheel) and right (wheel). v_i is the wheel velocity and Δv_i is the disturbance. u_i is the input of the wheel unit. A_i and B_i are constants.

We assume that the wheelchair moves stably at an almost constant velocity through fault-tolerant operation, even though sensor faults have occurred. The dynamic model related to the turning motion of the wheelchair is then assumed to be

$$\dot{\phi}(t+1) = \dot{\phi}(t) + \Delta\dot{\phi}(t) \tag{2}$$

where $\dot{\phi}$ is the turning velocity and $\Delta\dot{\phi}$ is the disturbance.

When the sensor output of the hard fault is assumed to be stuck at zero, the measurement model related to the wheel unit is

$$\text{Fault-free model } (m_0)\text{: } z_i(t) = v_i(t) + \Delta z_{i0}(t) \tag{3}$$

$$\text{Hard fault model } (m_1)\text{: } z_i(t) = 0 + \Delta z_{i1}(t) \tag{4}$$

where z_i is the output of the wheel resolver. Δz_{i0} and Δz_{i1} are the sensor noises.

Because the wheel resolver is assumed to be stuck at zero on the hard fault, the measurement is modeled by zeroing out the kinematic function, as in Eq. (4). If it is stuck at a nonzero value on the hard fault, the measurement is modeled by $z_i(t) = C + \Delta z_{i1}(t)$, where C is the stuck value.

The gyro output is also modeled by

$$\text{Fault-free model } (m_0)\text{: } z_{gyro}(t) = \dot{\phi}(t) + \Delta z_{gyro0}(t) \tag{5}$$

$$\text{Hard fault model } (m_1)\text{: } z_{gyro}(t) = 0 + \Delta z_{gyro1}(t) \tag{6}$$

We assume that the disturbances and the sensor noises are both zero means and white Gaussian sequences, and we estimate the mode probability based on Eqs. (1)-(6) via the IMM estimator (Zhang & Li, 1998; Hashimoto et al., 2001). The mode probability provides an indication of FDI of the internal sensor; for example, we consider FDI of the gyro and denote the mode probability estimates of fault free and hard fault by μ_0 and μ_1, respectively. The fault decision is made based on simple logic: If $\mu_0 > \mu_1$, then the gyro is fault free; else, a hard fault occurs.

The stop condition of the wheelchair results in wheel resolver FDI making an incorrect fault decision, because it makes the sensor output zero. In addition, the forward motion and stop condition of the wheelchair make it difficult to discriminate between fault-free and hard fault conditions of the gyro, because in both these conditions, the gyro output is almost zero. To cope with these problems, we switch the following model sets, according to the motion of the wheelchair:

$$\left.\begin{aligned} MG_1 &= \{m_0\} \\ MG_2 &= \{m_0, m_1\} \end{aligned}\right\} \tag{7}$$

In FDI for the wheel resolver, when the wheelchair is considered to stop (*i.e.*, $u_{left} \approx 0$ and $u_{right} \approx 0$), model set MG_1 is selected; else, MG_2 is selected. In FDI for the gyro, when the wheelchair stops (*i.e.*, $u_{left} \approx 0$ and $u_{right} \approx 0$) or moves straight (*i.e.*, $u_{left} \approx u_{right}$), model set MG_1 is selected; else, MG_2 is selected. The model-set-switching-based FDI algorithm is formulated based on a variable-structure-interacting-multimodel (VSIMM) estimator (Li, 2000; Hashimoto et al., 2003).

3.2 Fault Isolation of the Wheel Resolver and Wheel Motor

We consider a hard fault of the wheel motor as well as that of the wheel resolver. For a hard fault of the wheel motor, wheel rotation might stop completely. The hard fault of the wheel motor and that of the wheel resolver make the wheel-resolver output zero; thus, the IMM-based FDI mentioned in 3.1 cannot identify the faulty component (the wheel resolver or the wheel motor). In this subsection, when the IMM-based FDI detects the fault of the wheel unit, we identify the faulty component.

A hard fault of the wheel motor and that of the wheel resolver cause a difference in the turning velocity of the wheelchair. When the hard fault of the wheel motor occurs, the turning velocity sensed by the gyro is almost the same as that calculated by the wheel resolver. On the other hand, when a hard fault in the wheel resolver occurs, the turning velocity sensed by the gyro is much different from that calculated by the wheel resolver. We therefore perform fault identification based on the following logic:

$$\text{If } \left| z_{gyro} - \left(\frac{z_{right} - z_{left}}{b} \right) \right| < h \tag{8}$$

then the wheel motor is faulty;
else, the wheel resolver is faulty.

where b is the tread length of the wheelchair. h is the threshold; in our experiment, we set h as 0.01 [rad/s].

In principle, when a gyro fault occurs, it is impossible to discriminate between faults of the wheel resolver and the motor. For simultaneous faults of the wheel resolver and the motor, our FDI decides that the wheel resolver is faulty.

4. Detection and Isolation of Soft Faults of the Internal Sensor

4.1 Scan Matching Based Velocity Estimate

In this section, we handle soft faults of the internal sensors that appear as changes in the sensor gain. A change in the sensor gain is caused by not only sensor faults but also various system component failures and environmental interactions. For example, a flat tire, wheel slippage, and soft wheel motor faults change the gain of the wheel resolver; therefore, their conditions are considered to be soft faults of the wheel resolver.

In principle, it is necessary to determine the accurate linear and turning velocity of the wheelchair to detect soft faults of the internal sensors. We make use of laser scan matching with the fault-free LRS to estimate the velocity of the wheelchair. Our laser scan matching is based on typical point-to-point scan matching using an iterative closest point (ICP) algorithm (Besl & Mckay, 1992; Lu & Milios, 1997). In ICP-algorithm-based scan matching, the distance samples in the new laser scan are matched with those in the previous scan, so that the sum of squared distance can be minimized, and the velocity of the wheelchair is estimated.

Before scan matching, we preprocess the scan image to remove erroneous distance samples (Diosi & Kleeman, 2007). This preprocessing consists of noise reduction, segmentation, and interpolation. The noise reduction is based on a median filter, and it replaces outliers of distance samples with suitable samples. The segmentation prevents incorrect interpolation between two disjoint distances. The interpolation allows estimation of distance samples at every angular resolution of 0.5 [deg] of the LRS.

We determine the linear and turning velocities, $\dot{\boldsymbol{x}} = (\dot{x}, \dot{y}, \dot{\phi})^T$, of the wheelchair by the weighted least-squares method; the cost function is

$$J = \sum_{i=1}^{361} w_i \{\boldsymbol{q}_j - (\boldsymbol{R}\boldsymbol{p}_i + \boldsymbol{T})\}^2 \tag{9}$$

Here, $\boldsymbol{p}_i = (p_{ix}, p_{iy})^T$, where i = 1, 2,..., 361, is the distance sample in the current scan, while $\boldsymbol{q}_j = (q_{jx}, q_{jy})^T$, where j = 1, 2,..., 361, is that in the previous scan. Each sample $\boldsymbol{p}_i$ corresponds with the minimum distance sample $\boldsymbol{q}_j$ of all samples in the previous scan. w_i is the weight. $\boldsymbol{R}$ and $\boldsymbol{T}$ are the rotational matrix and the translational vector, respectively, given by

$\boldsymbol{R} = \begin{pmatrix} \cos\tau\dot{\phi} & -\sin\tau\dot{\phi} \\ \sin\tau\dot{\phi} & \cos\tau\dot{\phi} \end{pmatrix}$ and $\boldsymbol{T} = \begin{pmatrix} \tau\dot{x} \\ \tau\dot{y} \end{pmatrix}$. τ is the scanning period of the LRS; in our experimental system, τ = 0.1 [s].

The correspondence errors of distance samples of two successive scans affect the velocity estimate. To reduce the effect, we determine the weight w_i according to the errors between the correspondence points:

$$w_i = \frac{1}{1 + \frac{d_{ij}}{C}} \tag{10}$$

where d_{ij} is the distance error between the new and previous scans. C is a constant; in our experiments, C = 0.1.

From Eq. (9), the iterative least-squares method updates the wheelchair velocity $\dot{\boldsymbol{x}}^{(m-1)}$ as follows:

$$\dot{\boldsymbol{x}}^{(m)} = \dot{\boldsymbol{x}}^{(m-1)} + (\boldsymbol{H}^T \boldsymbol{W} \boldsymbol{H})^{-1} \boldsymbol{H}^T \boldsymbol{W} (\boldsymbol{q} - \boldsymbol{p}) \tag{11}$$

where $\boldsymbol{p} = (\boldsymbol{p}_1^T, \boldsymbol{p}_2^T, \ldots, \boldsymbol{p}_{361}^T)^T$, $\boldsymbol{q} = (\boldsymbol{q}_1^T, \boldsymbol{q}_2^T, \ldots, \boldsymbol{q}_{361}^T)^T$, $\boldsymbol{W} = \mathrm{diag}(w_1, w_2, \ldots, w_{361})$, and $\boldsymbol{H} = \partial\boldsymbol{p} / \partial\dot{\boldsymbol{x}}$. In our experiments, the maximum iterative number of the velocity update is 20. The velocity estimate calculated by the iterative least-squares method is noisy, and it is smoothed using Kalman filter. One disadvantage of the ICP algorithm is that it converges slowly. To accelerate the convergence, we use dead-reckoning information as the initial velocity $\dot{\boldsymbol{x}}^{(0)}$.

4.2 FDI Algorithm

A soft fault appears as a change in the sensor gain, which deviates it from the fault-free condition; thus, we detect a soft fault by estimating the sensor gain. Sensor gain α is defined by

$$\alpha = \frac{\text{Actual output of the sensor}}{\text{Sensor output in the fault - free condition}} \tag{12}$$

From this definition, when a sensor is fault free, the sensor gain equals unity.
We estimate the sensor gain based on the wheelchair velocity estimated by laser scan matching. We assume that the state model of the sensor gain is given by

$$\alpha_i(t+1) = \alpha_i(t) + \Delta\alpha_i(t) \tag{13}$$

where i = left (wheel), right (wheel), and gyro.
From the inverse kinematics of the wheelchair velocity, the measurement model is derived:

$$\begin{pmatrix} z_{left}(t) \\ z_{right}(t) \\ z_{gyro}(t) \end{pmatrix} = \begin{pmatrix} \alpha_{left}(t)\left(\sqrt{\dot{x}^{(M)2}(t)+\dot{y}^{(M)2}(t)} + \dfrac{b\dot{\phi}^{(M)}(t)}{2}\right) \\ \alpha_{right}(t)\left(\sqrt{\dot{x}^{(M)2}(t)+\dot{y}^{(M)2}(t)} - \dfrac{b\dot{\phi}^{(M)}(t)}{2}\right) \\ \alpha_{gyro}(t)\dot{\phi}^{(M)}(t) \end{pmatrix} + \begin{pmatrix} \Delta z_{left}(t) \\ \Delta z_{right}(t) \\ \Delta z_{gyro}(t) \end{pmatrix} \tag{14}$$

where $(\dot{x}^{(M)}, \dot{y}^{(M)})$ and $\dot{\phi}^{(M)}$ are the linear and turning velocities of the wheelchair, which are estimated by laser scan matching.
The sensor gain is estimated based on Eqs. (13) and (14) via Kalman filter. A small change in the gain estimate always occurs due to wheel slippage, sensor noise, and so on. Thus, if the estimate of the sensor gain is far away from one (*e.g.*, 0.7), the sensor is determined to be faulty. The stop condition of the wheelchair yields an incorrect estimate of the sensor gain, because it makes the right-handed side of Eq. (14) small. When the wheelchair moves straight, the turning velocity is very small, making it impossible to estimate the sensor gain. To better estimate the sensor gain, the following heuristic rules are incorporated into the fault decision making:
Rule 1: If the control inputs of the two drive wheels are very small (*i.e.*, $u_{left} \approx 0$ and $u_{right} \approx 0$), the gain estimates of all internal sensors in the previous cycle are held for the current cycle.
Rule 2: If the control inputs of the two drive wheels are almost the same (*i.e.*, $u_{left} \approx u_{right}$), the gain estimate of the gyro in the previous cycle is held for the current cycle.
Because a hard sensor fault leads to $\alpha = 0$, we can detect the hard fault by estimating the sensor gain. However, the time delay in the hard fault detection by the sensor gain estimate is longer than that in the IMM-based hard fault detection. For this reason, we exploit the IMM estimator for hard fault detection.

5. Fault Detection of the LRS

5.1 Fault Detection Algorithm

An LRS fault can be detected based on an error related to scan matching (Hashimoto et al., 2008a); if the current scan cannot match the previous scan, the LRS is determined to be faulty. For fault detection, we define the following cost function:

$$J_{LRS} = \sum_{i=1}^{361} \{ \boldsymbol{q}_i - (\boldsymbol{R}^{(M)} \boldsymbol{p}_i + \boldsymbol{T}^{(M)}) \}^2 \tag{15}$$

where $\boldsymbol{R}^{(M)}$ and $\boldsymbol{T}^{(M)}$ are the rotational matrix and the translational vector, respectively, which contain the velocity estimate $\dot{\boldsymbol{x}}^{(M)}$ from laser scan matching.

A fault is determined when $J_{LRS} > h_{LRS}$, where h_{LRS} is a threshold. In detecting an LRS fault, Eq. (9) removes the correspondence points of large errors in two successive scan points. For this reason, we use Eq. (15) for detecting LRS faults. Noisy distance samples of the LRS may cause false detection, even though the LRS is fault free. To cope with this problem, the LRS is determined to be faulty, when $J_{LRS} > h_{LRS}$ is always satisfied in several successive scans.

In featureless environments (*e.g.*, in long corridors), the corresponding pair between distance samples in two successive laser scans does not necessarily represent the same physical point in the environment, and the estimate of the wheel velocity becomes inaccurate, even though the LRS is fault free. To cope with this problem, we evaluate the distance samples by means of the Nieto approach (Nieto et al., 2007) and diagnose the LRS fault only in featured environments.

In dynamic environments, moving objects and backgrounds occluded by the objects cause J_{LRS} to be large, and false detections may occur. To cope with this problem, we detect the moving objects by the occupancy grid method (Hashimoto et al., 2006). As shown in Fig. 2, a grid map is represented in a reference coordinate frame Σ_w ($O_w;X_wY_w$). The LRS measurements are mapped in Σ_w and marked on the grid map. For each new cell of occupation in the current scan, the corresponding cell in the previous scan is checked. If that cell is marked, the cell in the current scan is considered a stationary object cell; else, it is a moving object cell. An object consisting of many moving object cells is determined as a moving object, and we remove the distance samples related to that moving object and backgrounds occluded by it when calculating the cost function J_{LRS}. In our experiments, the cell size is set as 0.3 [m] × 0.3 [m]. For mapping the laser image on the grid map, we use the wheelchair velocity estimated with laser scan matching.

5.2 Selection of Initial Velocity

In scan matching, we use dead-reckoning information as the initial velocity $\dot{\boldsymbol{x}}^{(0)}$ to accelerate the convergence. A fault of an internal sensor thus causes J_{LRS} to be large, even though the LRS is fault free. To cope with this problem, we define the following cost function:

$$J_{DR} = \sum_{i=1}^{361} \{ \boldsymbol{q}_i - (\boldsymbol{R}^{(0)} \boldsymbol{p}_i + \boldsymbol{T}^{(0)}) \}^2 \tag{16}$$

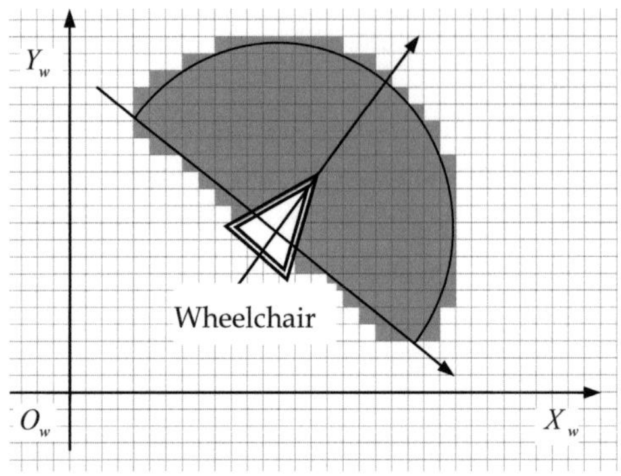

Fig. 2. Grid map; gray cells show the field of view of the LRS.

where $\boldsymbol{R}^{(0)}$ and $\boldsymbol{T}^{(0)}$ are the rotational matrix and the translational vector, respectively, which contain the initial velocity $\dot{x}^{(0)}$ resulting from dead reckoning.

If $J_{DR} < h_{DR}$, where h_{DR} is a threshold, the internal sensor is assumed to be fault free, and we use dead-reckoning information as the initial velocity for scan matching. However, if $J_{DR} \geq h_{DR}$, the internal sensor is assumed to be faulty, and the velocity estimate in the previous scan is used as the initial velocity for scan matching in the new scan.

When abrupt faults occur in the LRS, our algorithm detects the fault. However, incipient faults (a slow degradation of LRS performance), which may occur in the LRS in the real world, would allow scan matching, but estimate the wheelchair velocity inaccurately, causing incorrect fault detection. To cope with this problem, an alternative approach (Sundvall & Jensfelt, 2005) can be considered; the internal sensors estimate the wheelchair velocity by dead reckoning. The LRS also estimates it by scan matching. We thus can consider that the wheelchair is equipped with two different velocity providers (dead reckoning and laser scan matcher). Velocity estimates by fault-free providers yield similar values; whereas those of fault-free and faulty providers are inconsistent. This approach might be suitable for various fault patterns. However, this indicates only that something is going wrong in the two velocity providers; it cannot identify which provider is failing. If a wheelchair is equipped with three or more velocity providers, we can perform majority-voting-logic-based FDI (Hashimoto et al., 2008b). The velocity estimates from the providers are compared, and the provider whose velocity estimate does not match the others is determined to be the faulty one.

6. Experimental Results

6.1 Hard Fault of the Wheel Unit

The wheelchair was moved in a typical indoor environment. We conducted experiments in the following three scenarios:

Scenario 1: All the internal sensors and LRS are fault free.

Scenario 2: A hard fault of the left wheel resolver occurs at 8 [s].

Scenario 3: A hard fault of the left wheel motor occurs at 8 [s].

Figures 3–5 show the experimental results; subfigures (a) and (b) show the velocity of the left and right wheels, respectively. Subfigure (c) shows the turning velocity of the wheelchair; the bold and broken lines indicate the results calculated by the wheel resolvers and sensed by the gyro, respectively. Subfigures (d)–(f) show the mode probability of the left wheel unit, the right wheel unit, and the gyro, respectively; the bold and broken lines indicate the fault-free and hard fault modes, respectively.

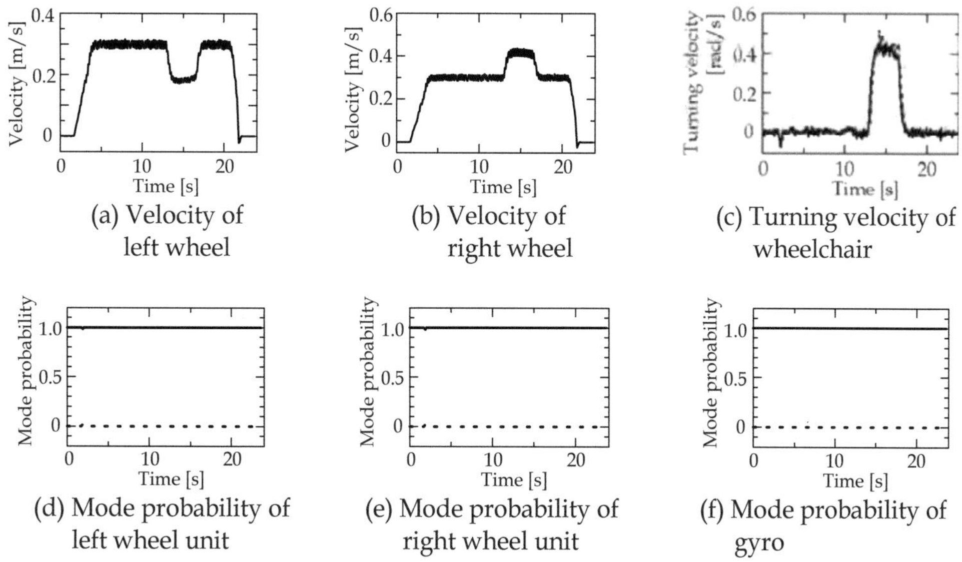

(a) Velocity of left wheel (b) Velocity of right wheel (c) Turning velocity of wheelchair

(d) Mode probability of left wheel unit (e) Mode probability of right wheel unit (f) Mode probability of gyro

Fig. 3. Experimental result (Scenario 1).

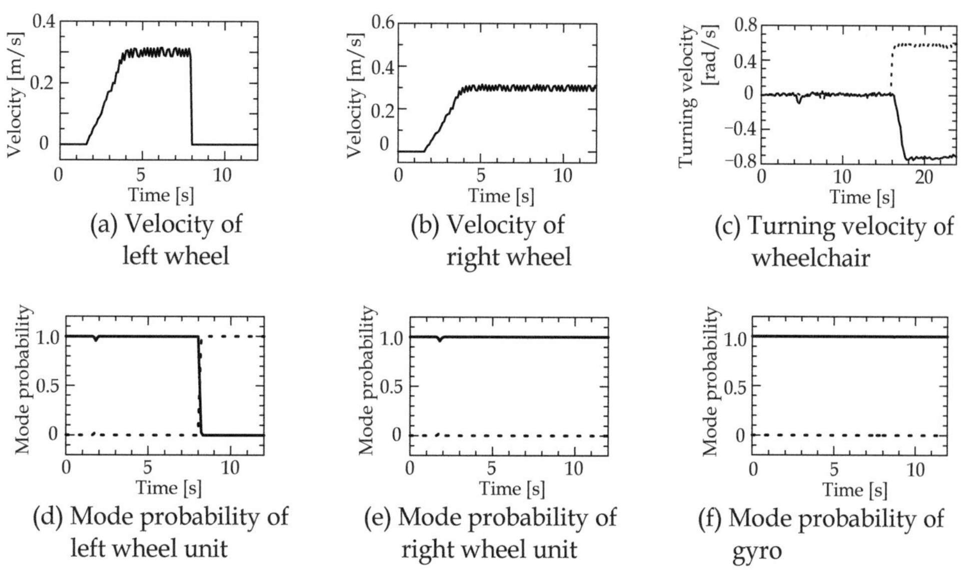

(a) Velocity of left wheel (b) Velocity of right wheel (c) Turning velocity of wheelchair

(d) Mode probability of left wheel unit (e) Mode probability of right wheel unit (f) Mode probability of gyro

Fig. 4. Experimental result (Scenario 2).

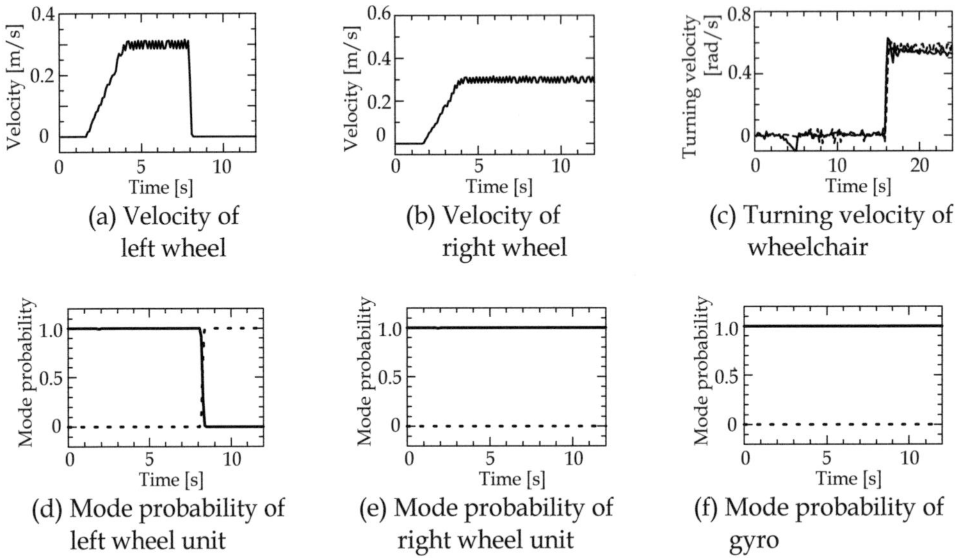

(a) Velocity of left wheel (b) Velocity of right wheel (c) Turning velocity of wheelchair

(d) Mode probability of left wheel unit (e) Mode probability of right wheel unit (f) Mode probability of gyro

Fig. 5. Experimental result (Scenario 3).

It is clear from subfigure (d) that in scenarios 2 and 3, we can detect a hard fault of the left wheel unit exactly, based on the mode probability. In scenario 2, the turning velocity sensed by the gyro is much different from that calculated by the wheel resolvers. On the other hand, the turning velocity sensed by the gyro is almost the same as that calculated by the wheel resolvers in scenario 3. This difference enables identification of the component (resolver or motor) that fails.

6.2 Soft Fault of Internal and External Sensors

The wheelchair was moved in a typical indoor environment. Figure 6 shows the velocity profile of the wheelchair. Two people walked around in the environment. We set the thresholds related to scan matching at $h_{LRS} = h_{DR} = 15$ [m^2]. We conducted experiments in the following four scenarios:

Scenario 4: All the internal sensors and LRS are fault free.

Scenarios 5 and 6: The sensor gain of the right wheel resolver changes abruptly to 0.5 (scenario 5) and 0.8 (scenario 6), respectively, from 1.0 at 21 [s].

Scenarios 7: The sensor gain of the LRS changes abruptly to 0.8 from 1.0 at 30 [s].

Figures 7–10 show the experimental results. Because in scenarios 4, 5, and 6, the cost function J_{LRS} is always less than the threshold h_{LRS}, the LRS is determined to be fault free. Then, we estimate the sensor gain of the internal sensor, and then, based on the estimate, we detect a soft fault of the right wheel resolver, as shown in Figs. 8(c) and 9(c). In scenario 7 (Fig. 10), the cost function J_{DR} is larger than the threshold h_{DR}; the internal sensor is assumed to be faulty, and the velocity estimate in the previous scan is used as the initial velocity for laser scan matching in the new scan. We also check the value of the cost function J_{LRS}. Because $J_{LRS} > h_{LRS}$, the LRS is determined to be faulty.

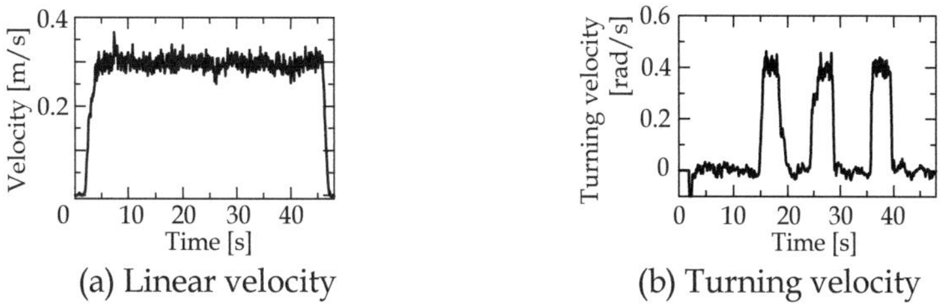

Fig. 6. Profile of wheelchair velocity.

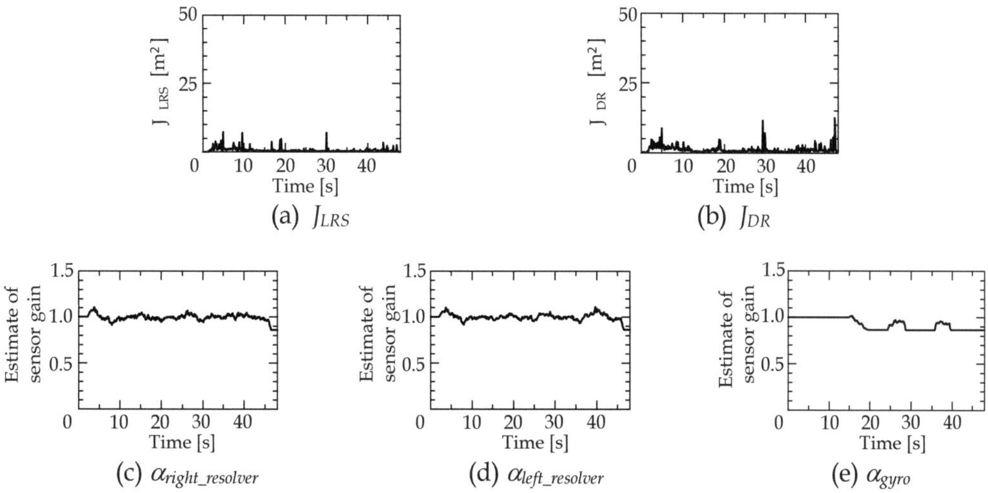

Fig. 7. Experimental result (Scenario 4).

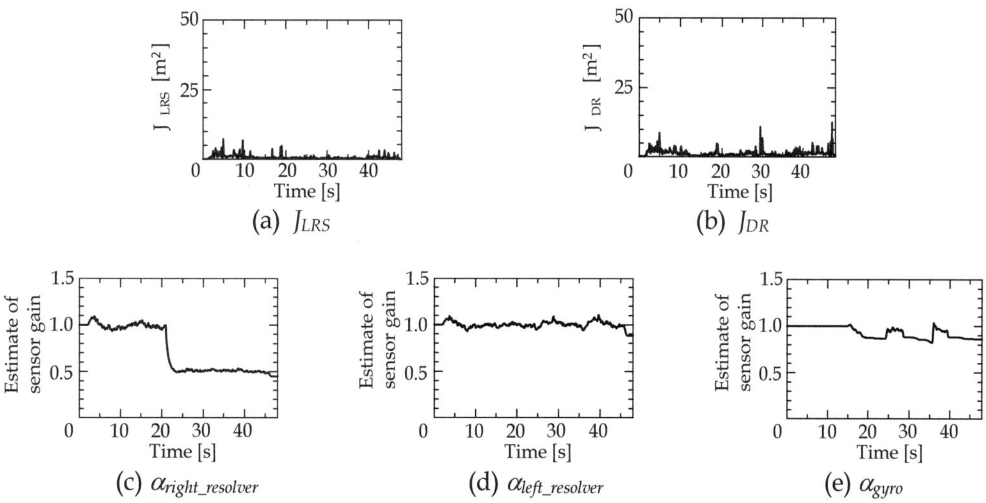

Fig. 8. Experimental result (Scenario 5).

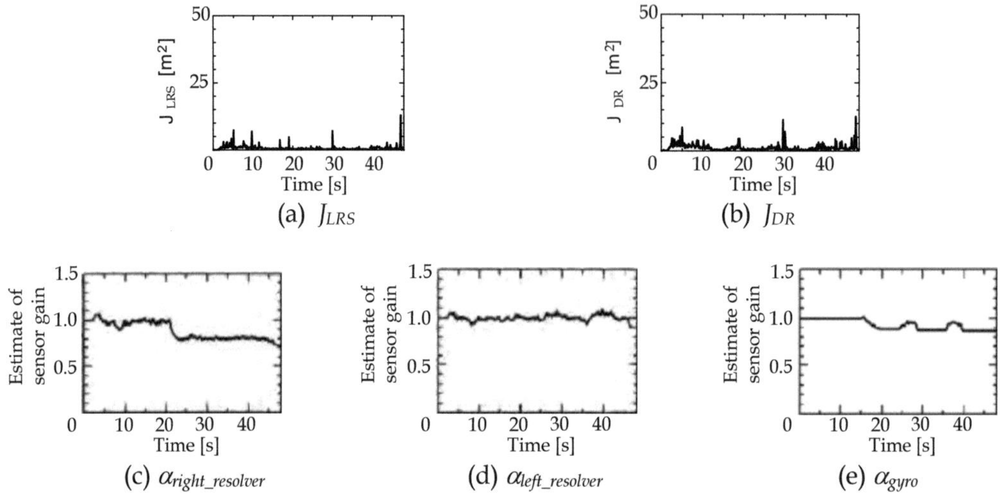

(a) J_{LRS} (b) J_{DR}

(c) $\alpha_{right\_resolver}$ (d) $\alpha_{left\_resolver}$ (e) α_{gyro}

Fig. 9. Experimental result (Scenario 6).

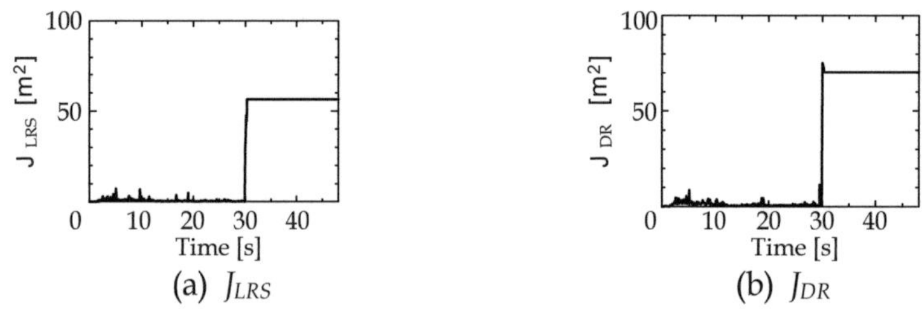

(a) J_{LRS} (b) J_{DR}

Fig. 10. Experimental result (Scenario 7).

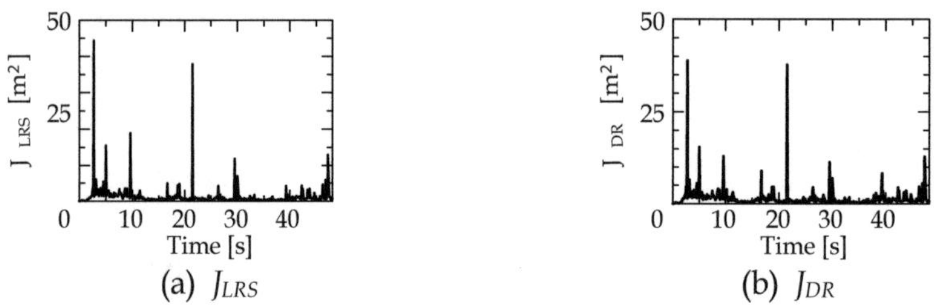

(a) J_{LRS} (b) J_{DR}

Fig. 11. Experimental result (without moving-object detector).

For comparison, we conducted experiments using the FDI system without the moving-object detector, based on the occupancy grid method. Figure 11 shows the results (the values of the cost functions) in a fault-free condition of both the internal sensors and LRS. The FDI system without the moving-object detector causes false detection of the LRS fault because the values of the cost functions sometimes become larger than the thresholds. On the other hand, as shown in Fig. 7(a) and (b), our FDI system with the moving-object detector, provides the correct detection of the sensor state.

7. Conclusions

This chapter presented a model-based FDI for wheelchair sensors and actuators. Hard faults of the internal sensors and the wheel motors were detected based on the IMM estimator. Fault isolation of the wheel resolvers and the motors was achieved based on the information of the fault-free gyro. A soft fault of the internal sensors was diagnosed based on the velocity estimate of the wheelchair from laser scan matching, using the fault-free LRS. The LRS fault was detected based on errors related to scan matching.

Abrupt faults of the LRS can be detected by our algorithm. However, incipient faults (a slow degradation of LRS performance), which can occur in the real world, allows scan matching, but estimates the wheelchair velocity inaccurately, and causes incorrect LRS fault detection. Our research effort is directed toward FDI for sensors and actuators in various fault patterns.

8. Acknowledgments

This work was partially supported by the Academic Frontier Research Project on "New Frontier of Biomedical Engineering Research" of the Ministry of Education, Culture, Sports, Science and Technology, Japan.

9. References

Besl, P. J. & Mckay, N. D. (1992). A Method of Registration of 3-D Shapes, *IEEE Transactions on Pattern Analysis and Machine Intelligence*, Vol. 14, No. 2, pp. 239-256.

Blom, H. A. P. & Bar-Shalom, Y. (1988). The Interacting Multiple Model Algorithm for Systems with Markovian Switching Coefficient, *IEEE Transactions on Aerospace and Electronic Systems*, Vol. 33, No. 8, pp. 780-783.

Carlson, J. (2004). *Technical Report for the Safety Security Rescue Research Center, ICRA 2004 Robot Fault Diagnosis Workshop*, http://www.csee.usf.edu/~jcarlsol/papers/Fault DiagnosisWorkshop Report.pdf.

Diosi, A. & Kleeman, L. (2007). Fast Laser Scan Matching Using Polar Coordinates, *The International Journal of Robotics Research*, Vol. 26, No.10, pp.1125-1153.

Duan, Z.; Cai, Z. & Yu, J. (2006). Adaptive Particle Filter for Unknown Fault Detection of Wheeled Mobile Robots, *Proceedings of IEEE/RSJ International Conference on Intelligent Robots and Systems*, pp.1312-1315.

Goel, P.; Dedeoglu, G.; Roumeliotis, S.I. & Sukhatme, G.S. (2000). Fault Detection and Identification in a Mobile Robot Using Multiple-Model Estimation and Neural Network, *Proceedings of 2000 IEEE International Conference on Robotics and Automation*, pp.2302-2307.

Li, X.R. (2000). Engineer's Guide to Variable-Structure Multiple-model Estimation for Tracking, In: *Multitarget-Multisensor Tracking: Applications and Advances volume III* , Bar-Shalom, Y. & Blair, W.D. (Ed.) Artech House, pp.499-567.

Lu, F. & Milios E. (1997). Robot Pose Estimation in Unknown Environments by Matching 2D Range Scans, *Journal of Intelligent and Robotic Systems*, Vol.20, pp.249-275.

Hashimoto, M.; Kawashima, H.; Nakagami, T. & Oba, F. (2001). Sensor Fault Detection and Identification in Dead-Reckoning System of Mobile Robot: Interacting Multiple Model Approach, *Proceedings of IEEE/RSJ International Conference on Intelligent Robots and Systems.*

Hashimoto, M.; Kawashima, H. & Oba, F. (2003). A Multi-Model Based Fault Detection and Diagnosis of Internal Sensor for Mobile Robot, *Proceedings of IEEE/RSJ International Conference on Intelligent Robots and Systems.*

Hashimoto, M.; Ogata, S.; Oba, F. & Murayama, T. (2006). A Laser Based Multi-Target Tracking for Mobile Robot, *Intelligent Autonomous Systems 9*, pp.135-144.

Hashimoto, M.; Watanabe, S. & Takahashi, K. (2007). Sensor Fault Detection and Isolation for a Powered Wheelchair, *Proceedings of IEEE/ASME International Conference on Advanced Intelligent Mechatronics.*

Hashimoto, M.; Itaba, F. & Takahashi, K. (2008a). Fault Diagnosis of Internal and External Sensors with Scan Matching for Mobile Robot, *Proceedings of IEEE International Conference on Systems, Man and Cybernetics*, pp.3096-3101.

Hashimoto, M.; Kitade, R.; Itaba, F. & Takahashi, K. (2008b). Voting Based Fault Isolation of In-Vehicle Multi-Sensors, *Proceedings of International Conference on Instrumentation, Control and Information Technology.*

Magill, D.T. (1965). Optimal Adaptive Estimation of Sampled Stochastic Processes, *IEEE Transactions on Automatic Control*, Vol. AC-10, pp. 434-439.

Maybeck, P. S. & Hanlon, P. D. (1998). Performance Enhancement of a Multiple Model Adaptive Estimator, *IEEE Transactions on Aerospace and Electronic Systems*, Vol. 31, pp. 1240-1254.

Nieto, J.; Bailey, T. & Nebot, E. (2007). Recursive Scan-Matching SLAM, *Robotics and Autonomous Systems*, Vol. 55, pp. 39-49.

Petterson, O. (2005). Execusion Monitoring in Robotics: A Survey, *Robotics and Autonomous Systems*, Vol.53, pp.73-88.

Roumeliotis, S. I. ; Sukhatime, G. S. & Bekey, G.A. (1998a). Fault Detection and Identification in a Mobile Robot using Multiple-Model Estimation, *Proceedings of* IEEE *International Conference on Robotics and Automation*, pp. 2223-2228.

Roumeliotis, S. I.; Sukhatime, G. S. & Bekey, G.A. (1998b). Sensor Fault Detection and Identification in a Mobile Robot," *Proceedings of* IEEE/RSJ *International Conference on Intelligent Robots and Systems*, pp. 1383-1388.

Simani, S.; Fantuzzi, C. & Patton, R.J. (2003). *Model-based Fault Diagnosis in Dynamic Systems Using Identification Techniques*, Springer.

Sundvall, P. & Jensfelt, P. (2005). Fault Detection for Mobile Robots Using Redundant Positioning Systems, *Proceedings of IEEE International Conference on Robotics and Automation.*

Verma, V.; Gordon, G.; Simmons, R. & Thrun, S. (2004). Real-Time Fault Diagnosis, *IEEE Robotics and Automation Magazine*, pp. 56-66.

Zhang, Y. & Li, X.R. (1998). Detection and Diagnosis of Sensor and Actuator Failures Using IMM Estimator, *IEEE Transactions on Aerospace and Electronic Systems*, Vol.34. No.4, pp. 1293-1313.

11

Electric Wheelchair Navigation Simulators: why, when, how?

Patrick Abellard, Iadaloharivola Randria, Alexandre Abellard, Mohamed Moncef Ben Khelifa and Pascal Ramanantsizehena
HANDIBIO EA4322, IUT, Université du Sud Toulon Var / Ecole Supérieure Polytechnique d'Anatananarivo
France / Madagascar

1. Introduction

It is now widely admitted that every assistive technical system has to be evaluated so as to fit the user's needs. Tests in realistic situation provide pertinent indications concerning a user's adaptation to a new technical assistance.
However, they are often a costly solution as their usual drawback is to be very specific to each handicap or each person. Furthermore, the acceptability and appropriaton of any new technical assistance has to be insured. In fact, it is not only a question of proposing technological solutions to a disabled person so that it can gain some autonomy, but also to integrate a human dimension in their choice (Pan et al., 2006). In this context, one challenge that is faced by the international community to bring it answers is the evaluation of motor and cognitive capacitied in the choice of a technical help. Some of these works propose evaluation methods based on questionnaires or usability clinical tests (Kirby et al., 2002).
The main drawback of these solutions is that they are often subjective, dedicated to a specific pathology and are not sufficient to make the final choice adapted to the motor and psychological user specificities. The decision then relies on doctors expertise, particularly ergotherapists. In the case of an electric wheelchair, this choice is all the more delicate that it makes up a complex motorized technical help, inevitably driving to safety, ergonomy and penibility problems. The necessity to propose an evaluation tool that does not subsitute to the ergotherapist expertise, but rather complete it, becomes therefore pertinent, all the more that tests in real situation can by costly since often very specific (Roussel, 2002). In this context, three point must be studied with a particular attention :

- disabled person fatigability
- material financial cost
- risks for the disabled person and the environment (accident, collision...)

In fact, concentration asked to a disabled person during navigation experiments quickly implies a fatigue, and these are precisely the most disabled persons that generally need the more assistive tools which are the most costly. Moreover, as the more disabled person have the highest risks when driving a wheelchair, the prescription of a wheelchair required many cares from the team made of a prescribing doctor, an ergotherapist (and/or

psychomotrician) and the user. In this aim, the team has to globally evaluate the user capacities so as to determine the most appropriated wheelchair to his handicap and his navigation environment. This evaluation includes various exercises similar to those encountered in a daily wheelchair use.
Sometimes, the prescription of an electric wheelchair can be refused to a disabled person for safety or apprehension reasons. It becomes therefore impossible to evaluate the help that could be brought by technical assistance. Thus, simulation exercises in a virtual environment are an interesting solution, since they enable :

- to reduce previous constraints
- to bring a solution to the safety problem
- to diversify experiments
- to evaluate driving capacities
- to quantify needs in terms of functionnalities

Several research teams have worked on wheelchair driving simulators prototypes developments.

2. Electric wheelchair simulators

Scientific and research communities have often tried to put technologies for human self being for decades. In the early 80s, Pronk and his mates have been the first to develop a simulation tool for electric wheelchair driving (Pronk *et al.*, 1980). In this aim, they used computer assisted simulation. A global "bird's-eye" view of the wheelchair movement was shown on a terminal screen. Authors concluded that the simulator could constitute in the future a pertinent help for evaluation and/or adaptation of electric wheelchairs. Since these first works, several other projects have exploited this idea to use virtual reality for electric wheelchair driving learning.

2.1 PWMS (Powered Wheelchair Mobility Simulator): State University of New York (Buffalo), USA

This electric wheelchair simulator project is based on original works from « Turtle Trainer » (Fig. 1), (Bresler, 1990), developed by a team from the Center for Assistive Technology at the State University of New York (Buffalo), so as to build an evaluation and formation tool for people with important physical disabilities and cognitive and/or sensorial deficiencies (Schmeler et al., 1993). This realization consists in a motorized platform on which a manual wheelchair is brought in order to evaluate driving capacities of a real electric wheelchair.
A performance validation study has been done on this prototype so as to determine its viability as a movement and command simulator. Performance has been compared to the one obtained with a real wheelchair and experienced users on a trajectory composed of several tasks and measurements. Results proved statistical and clinical similarity between the two methods, justifying that the simulator would enable a identical behaviour.
This work led to the realization of PWMS platform for clinicians and patients. It is devoted to become an evaluation and learning tool for determining electric wheelchair driving capacities of people with physical, cognitive and/or sensorial disabilities, having a manual wheelchair but unable to use it (Schmeler, 1993).

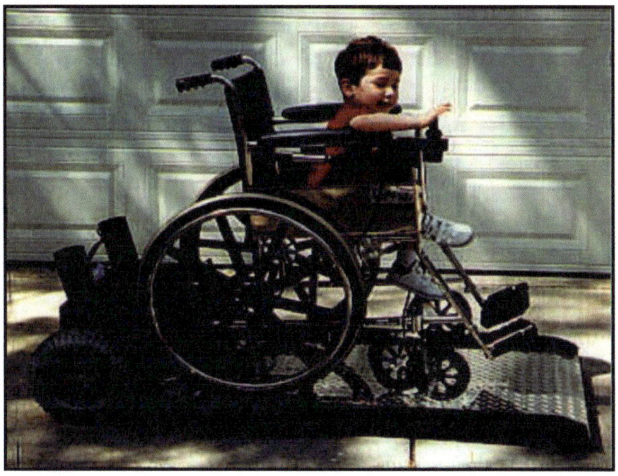

Fig. 1. Turtle trainer

2.2 Virtual Electric Power Wheelchair Driving: Pittsburgh University & VA Healthcare System, Pennsylvania, USA

This is a study in which real-time speed control of an electric wheelchair has been analyzed facing perturbations, internal parameters variations and external uncertaintities (Ding et al., 2003). Researchers have done this study since simple Proportional-Integral controller used in motorized wheelchairs has been considered insufficient and does not behave correctly with perturbations, sensors uncertainties and load variations. This team thus proposed a robust adaptive controller with which system stability and speed errors convergence have been examined. Controller efficiency has been validated by simulation studies. Authors think this study could lead to driving improvements, more fluidity sensations and few collisions.

They are actually investing many efforts to develop environments and virtual systems in order to improve evaluation and mobility learning, as well as environment perception (Fig. 2).

The resulting system brings wheelchair users a formation tool in various environments without any risk (no physical constrains such as walls, furniture, stairs, no surfance unevenness, no road traffic...).

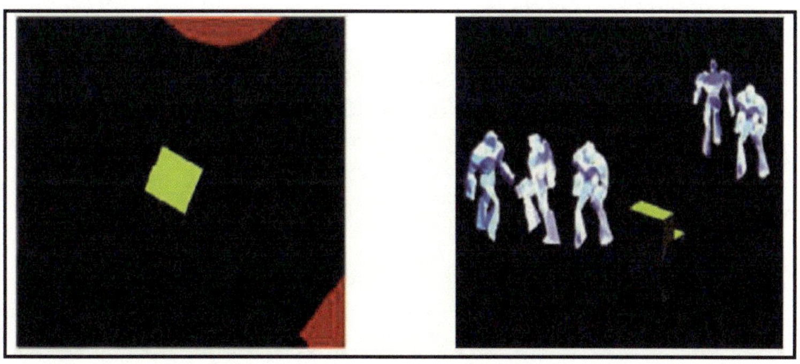

Fig. 2. 3D environment static (left) and dynamic (right) views

A following work on virtual reality potentialities for rehabilitation, formation and motorized wheelchair driving training has been developped (Cooper et al., 2005). The aim is to enable users to drive with any risk and efficiently in a virtual environment using a lever.
The context of virtual learning includes simple scenarii simulating basic motor takes, and complex dynamic scenarii simulating movements on a pavement with people walking around. Clinicians can change virtual scenarii in order to define specific tasks and can record formation progression.
Virtual formation platform is made of 3 modules : static environment, dynamic environment, evaluation. The first module enables the virtual wheelchair movement in a nempty 3D environment or with simple static scenarii simulating basic motor tasks. The second one simulates the wheelchair evolving on a sidewalk surrounded by walkers and examinates attention and complex context training capacities. The third module records virtual wheelchair speed and position, as well as virtual walkers. Number of collisions, formation scenario execution time and trajectory performed are also recorded (Marchuk et al., 2007). Data can be used to analyze formation effects or as evaluation criteria.

2.3 Wheelchair User Proficiency Through Virtual Simulation: OSC (Ohio Supercomputer Center) and Ohio State University, USA

The basic idea was to conceive a system enabling a user to drive a motorized wheelchair in a virtual architectural environment (Swan et al., 1994). In fact, lack of evaluation methods for disabled person for defining the best suited appropriated command were often reported. However :

- technology evolutions then enabled to make control mechanisms more performant,
- the « Americans with Disabilities Act » (ADA, 1990) legally made an obligation to almost all public structures to become accessible to disabled persons.

In order to be free from difficulties linked to the limitation of users to real wheelchair training, the developed system uses virtual reality to evaluate the user competency evolving into virtual structured environments. It is composed of an instrumented electric wheelchair linked to a workstation to simulate real speed and orientation from an architectural database. Display is done inside and outside via stereoscopic vision using polarized lens (Fig. 3). A hierarchical data structure enables to detect collisions between the virtual wheelchair and the environment. This system can be useful for wheelchair users going to public structures as well as architects and conceivers, since it provides structures visualization and a mean to study its ADA conformity (Carlson et al., 1994).
The Rehabilitation Engineering Center for the Quantification of Physical Performance from Ohio State University and the OSC (Ohio Supercomputer Center) in Colombus collaborated in order to take part in the development of the virtual world prototyping system enabling the electric wheelchair navigation (Fig. 4) under the direction of Dr. Wayne Carlson, "Advanced Computing Center for the Arts and Design" director, et de Don Stredney from « Ohio Supercomputer Center ».

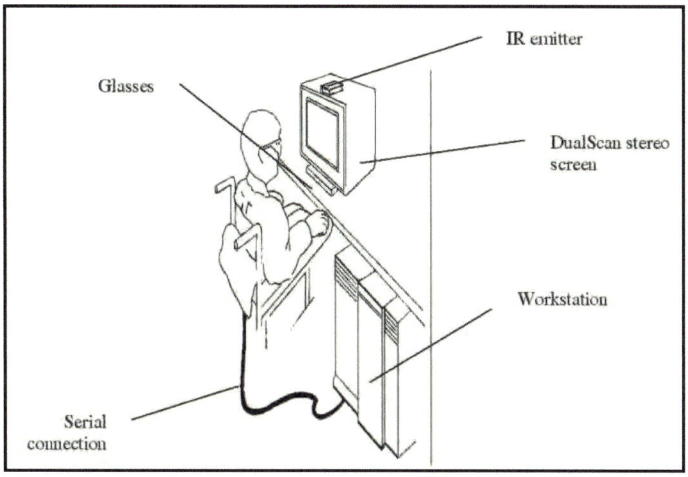

Fig. 3. System configuration

A sub-project named " The determination of wheelchair user proficiency and environmental accessibility through virtual simulation" (Swan et al., 1994) is designed to examine human activity in free environments navigation, by using virtual simulations generated with computers, so as to provide pertinent information in order to define the most suited technological assistances to a specific disabled person. Furthermore, this research aims at validating direct applications of developped optimized method for the design of secured environments.

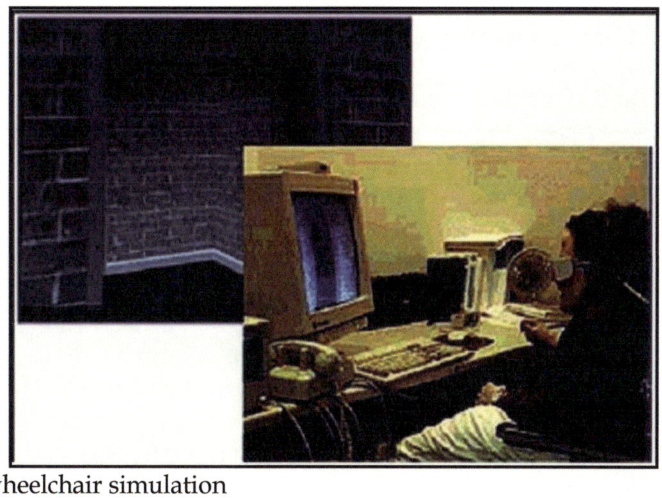
Fig. 4. Virtual wheelchair simulation

This project combines the expertise of Ohio State University researchers with Invacare Corporation competencies - a leader in assistive technologies design. This projet not only provides essential information on design and technology use for the best suited control, but it also constitutes a system to train users of such systems. Besides, the developped system can be used by architects, conceivers and builders so as to insure unlimited access to environments by the disabled person (public buildings, commercial centers, homes and offices, shops, etc.)

2.4 Wheelchair Training: Virtual Reality Laboratory, ORI (Oregon Research Institute), Oregon, USA

Oregon Research Institute (ORI) researchers have been among the first ones to use virtual reality to learn motor habilities to disabled children. They used this method to increase blind and deaf children mobility by training them to efficiently use their wheelchairs. In this project, the computer-generated environment simulates a street with people so that the child could have more experience as a wheelchair driver. Actually, researchers develop a set of education and rehabilitation tools for disabled persons. It is based on an experimental study to take into account specific consequences of different kinds of navigation learning on children using virtual reality.

This project also proposes the use of virtual reality to improve spatial perception through the movements (Inman & Loge, 1995), for instance by using "crash" sounns when collisions occur. The study can be based on three scenarii depending on motivation and personal rythm of evaluated patients. It aims at bringing secure navigation and experience to the user. This project proves the usefulness of virtual reality for children suffering from severe physical disabilities. With virtual reality, it is now possible to provide exploration and action on physical environment opportunities, in order to gain mobility competencies, i.e. cognitive strategies and perception capacities that can only be learnt from an initial experiment. In the simulator, a sound feedback has been also integrated for blind people (Inman et al., 2000).

2.5 H.M2.P.H.(Habitat Modulable et Mobile pour Personnes Handicapées): Tours University, France.

HM2PH project funded by Indre-et-Loire Conseil Général results from a collaboration between Computer Science Department Polytech'Tours Engineering School and HaNT (Handicap et Nouvelles Technologies) research team from Tours University LI Laboratory. It aims at bringing disabled (and elderly) people an autonomy gain thanks to adapted technical assistances. It also concerns the design of an adapted home :

- that can be inserted in a medical place (hospital, rehabilitation center)
- that can be brought at home

This project is related to a PhD work (Leloup, 2004) on « Software tools for assisted conception of adapted home ». Its aim is to specify and realize a software tool enabling to quickly conceive an adapted home for a disabled person. HM2PH 2D plans can be created and each room can be customized while respecting three kinds of constraints :

- constraints related to classic architecture and production costs (fixed HM2PH dimensions, water pipes optimizing…),
- architecture constraints for building where disabled people live (corridors with, rotation areas for wheelchair, rooms minimal dimensions…)
- preferences of the future resident (proximity of rooms, topography and specific distribution of each room, reduction or enlargement of some rooms…).

Difficulty is therefore to find different organization that satisfy these constraints the best way possible. Obtained results enable to generate a set of plans that take into account all

these constraints efficiently. This work is currently in progress in another PhD that started on same topics to have more precise results and to extend the nature of constraints by making the most of different modules.
In order to give a possibility to appreciate organization and equipments generated by the MoGAP (Moteur de Génération de Plans or Plan Generator Engine), it seemed interesting to make a 3D real time virtual visit from the 2D plan (Fig. 5). Simulation issues are important, since it must provide a maximal realism so that the user could have the feeling to be in his future home, and that he can appreciate aesthetics and functionalities. It is thus planned that the user can test different virtual equipments included in the scene. Moreover, visit includes collision detection that enable to test and validate home circulation conditions.
Conception assistive software deals with plan automated generation. Obtained result is used to build a 3D world for H2MPH virtual visit (Fig. 6). In order to satisfy the person in the concerned living space, it is necessary to put in place and activate services of life quality thanks to Technical Help Database.

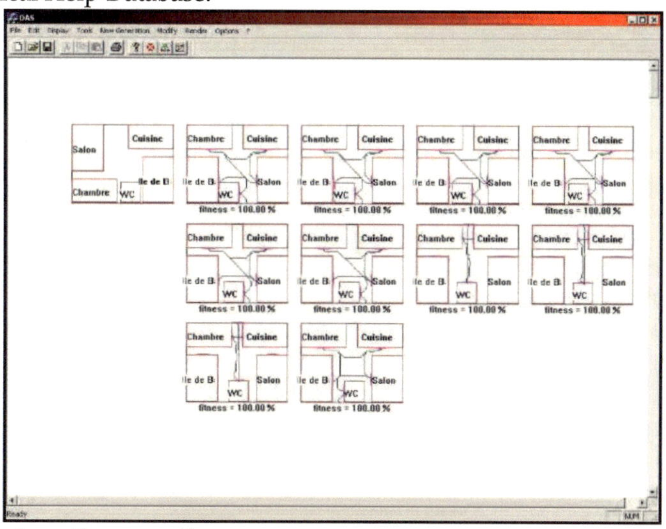

Fig. 5. 2D map automatic generation

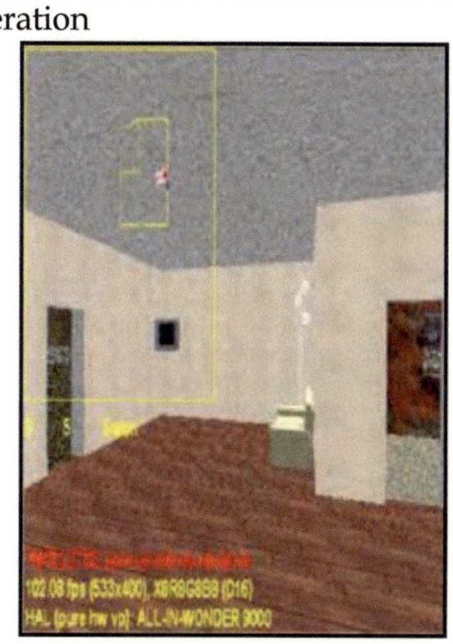

Fig. 6. 3D virtual visit

2.6 Assistive Technology Access Interfaces: Ohn School for the Physically Disabled, Tel Aviv, Israel.

This project is part of an investigating research field for developing and evaluating technological assistive interfaces so that people with severe physical incapacities to use more efficiently and with less fatigue their residual motor capacities. It aims at testing the capacity of a basic driving simulator software to evaluate and train disabled children to command an electric wheelchair (Hasdai et al., 1998).
An evaluation scale of considered capacities necessary to insure safety and command efficiency has been defined. Children with ages ranging from 7 to 22 suffering either from progressive muscular dystrophy or a cerebral palsy have been evaluated in their capacity of controlling an electric wheelchair. They have been divided in two groups : with or without previous experience.
After evaluation on a real wheelchair, unexperienced people have been trained on a computer game using a lever in which they evolved in labyrinths similar to the disposition of their own school environment. Another test labyrinth is then presented to them, before been evaluated on their wheelchair control capacities.
Research conductors noticed that althouigh being inferior to those of experienced drivers, unexperienced children have significantly increased their driving performances during the learning phase, hence their conclusion that a simulator software could help to develop and evaluate capacities required to drive an electric wheelchair.

2.7 Simulator of Powered Wheelchair: Research Institute, National Rehabilitation Center for the Disabled, Japan.

The context of study is the development and practical use of an outside environment simulator for an electric wheelchair. The simulator is made of two computer screens and a mobile platform (Fig. 7). Screens supply a virtual environment including road signs, railroads crossing, sidewalks, sloping ground… Walkers and bicycles can come in different ways and paths. Platform is connected to six actuators (i.e. electric servo-motors) producing accelerations and decelerations similar to those of a real electric wheelchair.

Fig. 7. Wheelchair driving simulator first version

15 persons tested the simulator. Results showed that users found many similarities between real and virtual drivings, despite having more difficulties when using the simulator. This has been caused by the lack of a lateral view and vestibular disease that has been felt by approximately half of the users. This point deserves a particular attention, particularly for first tests. Furthermore, researchers performed tests with a person having physical and mental disabilities to learn how to control a wheelchair with the simulator, but he lacked comprehension, concentration and motivation. A first attempt to actuate a sole switch to drive the simulator proved he could understand the system and focus his attention on the task to perform. Next attempt requiring two driving switches showed he could not stop when a collision risk occurred.

A head control has therefore been added, which showed in a last test that driving could be planned once safety problem was solved. This first simulation of the simulator has been considered efficient to determine if someone with important disabilities could drive an electric wheelchair or not, but decision is conditioned by a very good knowledge of the accompanying team in terms of technology. System evolutions drove to a new version (Niniss & Inoue, 2005) in which the most important modification was to replace the two screens display system with a hemispheric display proving a 110° field of vision (Fig. 8). Consequently, latest experiments drove to the definition of qualitative and quantitative evaluation criteria, which seem efficient when correctly combined.

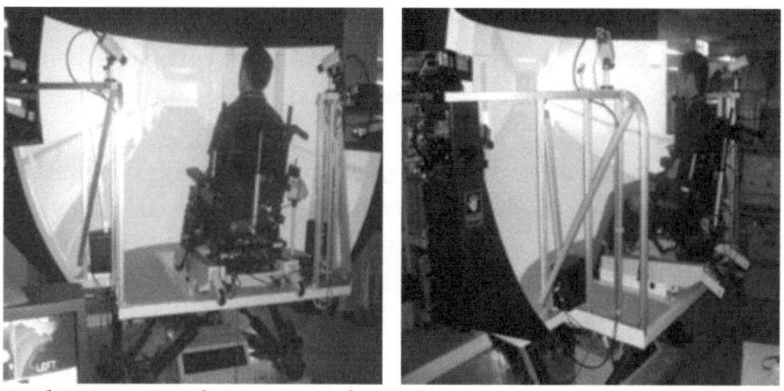

Fig. 8. Wheelchair driving simulator second version

2.8 Electric wheelchair Simulation, VAHM extension: LASC (Laboratoire d'Automatique des Systèmes Coopératifs), Metz University, France

This project is made of a simulation platform using virtual reality, where a wheelchair is placed (Fig. 9). It has three aims (Niniss et al., 2000a) :

- helping conceiving new mobility assistance functionnalities,
- helping choosing and prescribing electric wheelchairs,
- making driving learning easier

This work is an improvement of a previous project : VAHM (Véhicule Autonome pour Handicapés Moteurs) (Bourhis et al., 1998). It takes into account kinematics of an existing intelligent electric wheelchair, that has been conceived to ease driving learning (Niniss et al., 2000b). It also integrated the modeling of ultrasonic sensors in simulation and an intelligent system to help the driver controlling the motorized wheelchair.

Fig. 9. Simulation platform

Virtual reality system is here conceived to help the best suited assistance to locomotor assistance of any patient (Niniss & Nadif, 2003). The most important element of the systemis the simulator that controls the virtual world and its interactions with the real world. System includes a platform with a fixed base with a rigid structure to support the wheelchair. Two pairs of rolls enable motor wheels to turn freely while wheelchair frame remains fixed. In order to establish evaluation criteria, a test protocol has been defined. A test group is composed of good-health people, that have been divided in two sub-groups depending on their electric wheelchair knowledge.

Three wheelchairs have been used : Storm (Invacare), Moover95 (Flex) and Jet 2 (Pride Mobility). First, a global view of the environment and the virtual wheelchair has been presented to users, who had 5 minutes to freely explore it. Then, the research team analyzed quantitative and qualitative data obtained concerning users' capacities based on their evaluation in virtual reality. A second evaluation with disabled people is necessary before system conceivers approval. In this aim, a future foreseen evolution of this simulator will integrate the recording of several information and evaluation elements (duration and number of collisions and stops, environmental occupation…) (Dir et al., 2008; Niniss et al., 2004).

2.9 Evaluating and learning to drive an electric wheelchair: Royal Hospital for Neuro-disability & University of East London, Londres, UK.

This work results from the collaboration between the Department of Clinical Psychology (Astley Ainslie Hospital, Edinburgh), Department of Technology Clinic (Royal Hospital for Neuro-disability, West Hill) and the Psychology Department (East London University).

The first studies (Harrison et al., 2000) presented quantitative and qualitative data relative to conception and implementation of two non immersive virtual environments so as to evaluate and train adult electric wheelchair users having complex neurologic deficiencies. Manoeuvre and itinerary search capacities have been more particularly analyzed. Results showed that attendees consider the virtual environment to be realistic and representative, and that tasks are a good reflect of required capacities to drive an electric wheelchair but that the execution of manœuvrability tasks is more difficult in the simulator than in real life.

A second study then followed (Harrison et al., 2002) on system control efficiency so as to avoid touching obstacles manœuvrability and itinerary seeking capacity in a complex environment planification. Six unexperienced users participated to the project, either by avoiding an obstacle or planning a trajectory. Performance measures have been recorded in reality before and after learning phases and during virtual reality sessions. Questions about virtual environment aesthetics and simulator functional aspects have also been asked to users. Answers were globally positive. However, they found again manœuvrability tasks far more difficult in virtual reality than in reality, and put the light on some simulator command difficulties and various improvements to perform on both virtual and real tasks have been noted.

As a conclusion, authors said both chosen virtual environments showed potentially useful means for evaluating and training inexperienced users. However, they should be still improved so as to reinforce users motivation and improve their capacities.

2.10 Dynamic System Engineering Research Group, Portsmouth University, UK

This team worked on wheelchair obstacles avoidance, by using virtual reality for helping elderly people who have learning problems for manipulating a mobility system (Stott et al., 2000). The goal is to study different learning applications using virtual reality and their clinical involving for patients suffering from mobility problems. In this aim, automatized literature seeking has been performed using databases from MEDLINE, Cochrane, CIRRIE and REHABDATA. Three studies subcategories have been defined : five related to a training leading to qualifications, two "to physical exercises formation", and one to leisure activities. They proved interesting transfer of the learning from the virtual environment to the real use of mobility systems.

2.11 VEMS (Virtual Environment Mobility Simulator): Limerick University, Ireland

VEMS stands for Virtual Environment Mobility Simulator. Its development aims at providing simple and efficient solutions for learning and readaptation to drive an electric wheelchair (Adelola et al., 2002; Adelola et al., 2005a). Its main interest is analysis of manipulation effects of a real wheelchair to interact with an environment, since childhood development is strongly related to exploration and action with environment. This system provides a virtual and familiar environmental context on a computer screen and proposes simple tasks with various game elements to motivate the user (Fig. 10). The very diversified nature of the incapacities nevertheless requires system adaptability to each user specific needs. It uses a joint analysis technique to learn comportements, preferences and incapacites of the users to fulfill their conditions in order to have an efficient learning.

VEMS is still developed at Limerick University (Ireland) in order to obtain a virtual environment better adapted to both learning and patient disabilities, so that he could have better results in shorter time during context changes (Adelola et al.., 2003a). In fact, important waiting times between sessions can not reasonably occur with patients having an important fatigability.

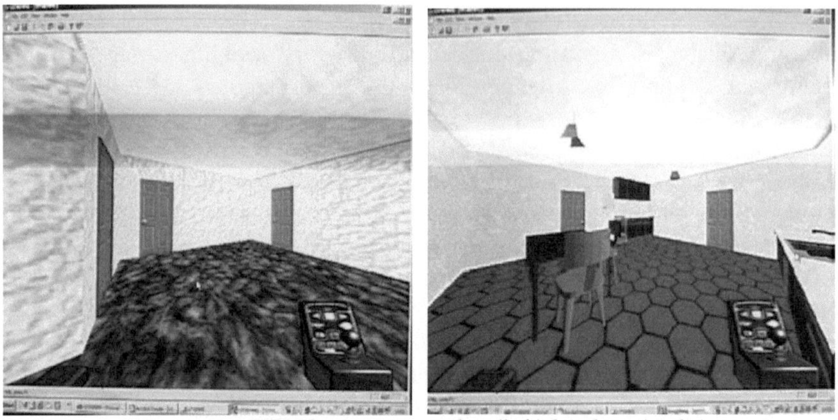

Fig. 10. Domestic environment VEMS view : corridor (left) kitchen (right)

VEMS is adaptive in its constitutive elements sizes (Adelola et al., 2003b) and enables to help children gaining some capacities by using a sure and safe virtual reality system thanks to :

- the study of an experimental and statistical method to measure qualifications required to drive a wheelchair in a virtual environment,
- the elaboration of an analytical procedure to obtain mathematical models of the behavior of a user in phase of learning to virtual environment modifications performance and user behaviour on optimal models evaluation and prediction

Evaluation by initial simulation uses joint analysis taking into account subjective answers, including patient reaction to demands done during wheelchair driving in various environments and circumstances. Authors conclude that virtual reality system is interesting for wheelchair driving learning. In fact, researches proved children with short attention times can be encouraged to use virtual reality, since virtual environments can easily captivate their attention. Virtual game conception is currently studied since it could enable children to use an electric wheelchair in several different virtual environments (Adelola *et al.*, 2005b).

One of the project most recent steps is the conception of a monitoring network in a Virtual Environment Simulator (VES) for motorized wheelchair drivers training (Dholkawala et al., 2005). Virtual environment enables therapists and welfare workers to watch and monitor the training process on a distant site, since some children are more at ease alone even if they require some supervision. The VES frame enables to provide a help to users if necessary. Improvements to its graphical interface are currently being studied.

2.12 OCAWUP (Obstacle Course Assessment of Wheelchair User Performance): Laval and Sherbrooke Universities, Québec, Canada.

It's a collaborative project between several Canadian rehabilitation centers and institutes, hospitals and universities working on assistive technologies. It deals with wheelchair navigation evaluation. In fact, authors think results obtained for clinically evaluate wheelchair users capacities and parameters which has influence on mobility are not complete enough. Initial work has therefore consisted in defining a conceptual frame that suggests a more global view for evaluating wheelchair driving capacity (Routhier et al.,

2003). In this aim, after selecting the factors that are suspected to influence wheelchair mobility, a conceptual frame has been defined to enable a precise examination of obtained results. Fine analysis of evaluation tools enabled to determine factors to consider in clinical evaluation of wheelchair mobility capacity, so as to develop a normalized method usable in hospitals.

Future improvements should concern additional factors conditionning mobility such as users profiles, environment complete characteristics, patients daily activities, social roles and received training. Analysis of results on a standardized route with normalized basic situation should enable rehabilitation therapists to evaluate patients for choosing a wheelchair and to note their progressions during the learning phase. Moreover, this tool should allow to predict results for daily activities.

Next steps for this study will also be on the development of a test to evaluate and record manual and motorized wheelchairs driving in a normalized context and in potentially difficult environmental situations. Authors particularly worked on the use of the validity process for obstacle choice and notation system. Specific obstacles include corridors, doors, sidewalks, smooth or damaged surfaces, slopes… Experts (ergotherapists, researchers, etc.) considered these choices to be representative of daily life situation. Authors hope OCAWUP will help clinicians in their interventions with patients to improve their participation in social activities (Routhier et al., 2004).

2.13 Accessibility haptic interface: Strathclyde University, Glasgow, Scotland, UK

Interest to life quality and respect of the law on disabled persons discrimination (DDA: Disability Discrimination Act (DDA, 1995) led to establish accessibility as a priority in recent constructions. In this context, a recent project from Strathclyde University concerned the development of a wheelchair platform (Grant et al., 2004) which is a virtual reality service that can be used to address buildings accessibility questions (Grant, 2003). This work is the result of collaboration between architects, biomechanicians and users, and consist in the study of materials related to conception and construction of the platform, test interfaces, and users evaluation. Research is oriented to a prototype development, extending existing simulation and exploration possibilities for a use in building conception (Fig. 11).

Project results led to the development a haptic interface so that wheelchair users could move into virtual buildings using their own wheelchair. This interface main point is that it provides the user a retroaction related to the required effort to drive wheelchair through surface changes and slopes This has been followed by the production of tools aiming at solving accessibility conception problems. Strathclyde University tried to combine advances graphics with a haptic interface in order to build a wheelchair platform able to simulation navigation in virtual buildings (Grant et al., 2005). This approach is one of the more sophisticated existing. Nevertheless, all problems encountered by a conceiver from access and environment interaction have not been solved yet.

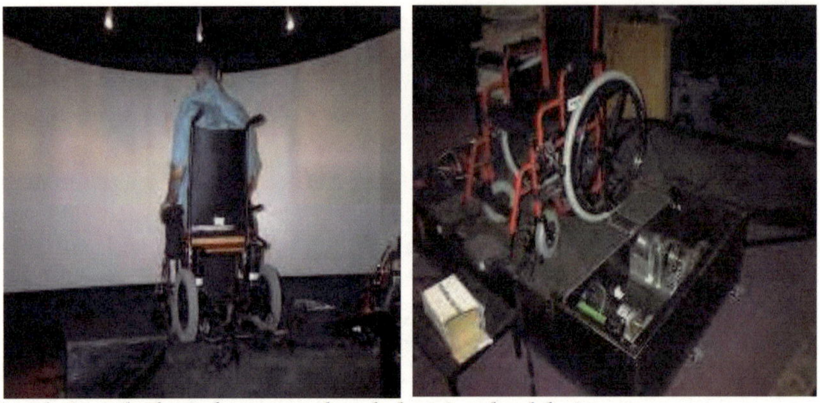

Fig. 11. Simulation platform for manual and electric wheelchair

2.14 VRWC (Virtual Reality Wheelchair): Clarkson University, New York, USA.

VRWC project deals with conception and realization of every parts of a manual wheelchair driving simulator (Baran et al., 2004). The price of an electric wheelchair can go up to $20.000,which is prohibitive for insure companies to help financing an acquisition to a child that has not showed yet driving skills. This simulator is used to train someone to efficiently control a wheelchair in a sure virtual environment, under supervising of a therapist. If this simulation is very precise, and that training proved to be efficient, the next step will be to make it accepted by insurance companies as a training for driving an electric wheelchair. Thus, families would obtain facilities for buying wheelchairs to their children.

This project uses a code that could transform a group of objects to detect collisions (for example). Collision detection and slope monitoring are possible thanks to a set of segments materializing a field contact (horizontal for obstacles, vertical for the ground) (Fig. 12). Interactions with virtual world are done with a lever, a video-tracker and a 6-DOF platform.

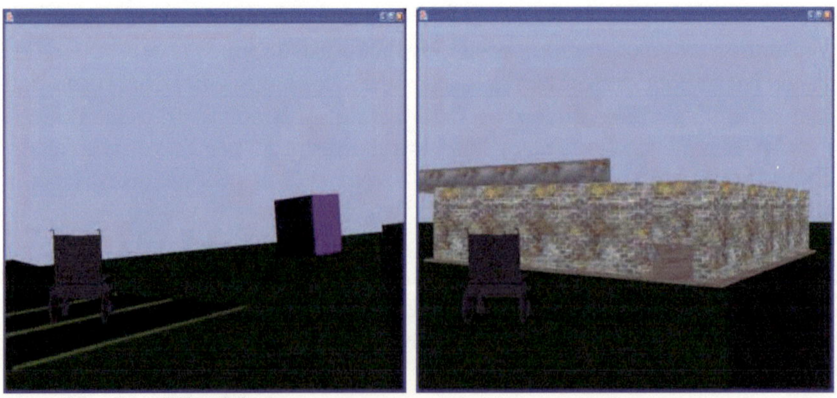

Fig. 12. Virtual Reality Wheelchair

For many people with physical and/or cognitive deficiencies, assistance engineering provided thanks to a powered wheelchair (PWC) solutions for mobilty in both personal and professional areas. However, in spite of independence provided by a PWC, indemnifying

organisms are often not disposed to finance its acquisition as long as the concerned person did not prove his capacity to command the wheelchair by himself.
Learning methods for manipulating electric wheelchairs are expensive and potentially unsafe. Virtal reality technology can therefore provide a learning tool, an efficient method and an information source on wheelchair driving capacities (Harrison et al., 2000), (Hasdai et al., 1998). Current VPW/C evolutions (Virtual Power WheelChair) aim at conceiving and realizing a power wheelchair providing precise visual, proprioceptive and vestibular information feedback in order to exactly simulate navigation for learning needs (Sonar et al., 2005).

2.15 Stroke Simulator : VRlab/HPC2N, Umeå University, Sweden

Initial goals were to study virtual environments capacities to influence social worker apathy (Maxhall et al., 2004). 9 persons from Norrland Universitary Hospital (NUS) perform three daily tasks such as reading a newspaper, filling a glass with water and putting toothpaste on a toothbrush. These tasks have been done in a virtual environment with both sane people and patiens suffering from tremors. The considered virtual environment was a normal flat that could be adapted to any perceptual disorder (Fig. 13).

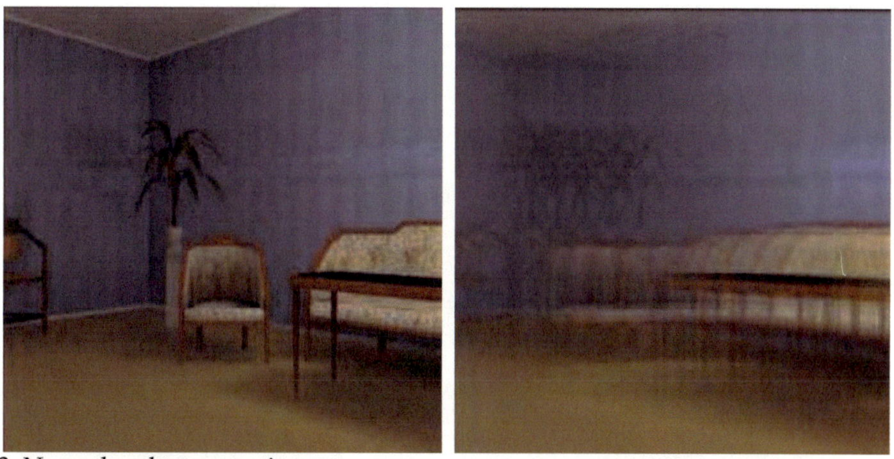

Fig. 13. Normal and tremors view

Observation and interview results showed that the simulator, despite usability problems, was efficient to influence social workers empathy. Since this time, the team is developing a real time 3D environment Colosseum3D (Backman, 2005) which architecture enable to easily create rich environments with rigid elements dynamics, 3D results using OpenGL Shaders, 3D sound and human avatars (Fig. 14).

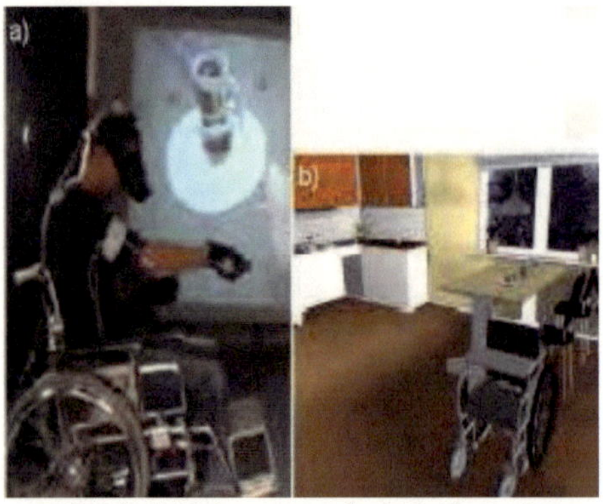

Fig. 14. Stroke Simulator : manipulating items and virtual environment view

2.16 Virtual Intelligent Wheelchair: CRVM (Centre de Réalité Virtuelle de la Méditerranée), Mediterranean University, Marseille, France

This team is currently developing a learning and evaluation simulator in the CRVM where a CAVE system has recently been installed so as to perform optimal immersion (Mestre et al., 2007). This interactive simulator will enable to optimize graphic and command interfaces, and human-machine cooperation by taking into account potential difficulties resulting from their sophistication (Mestre, 2004). It is also developed to become a learning system for commanding a virtual wheelchair, and an assistive tool for prescribing a wheelchair adapted to a specific user (Fig. 15).

Fig. 15. 3D intelligent wheelchair simulation with proximity sensors

More precisely, experiments are done by using a virtual wheelchair dynamic model using Virtools SKD (studies using real experimental intelligent wheelchair are planned) (Mestre et al., 2006). In previous works, wheelchair had a completely automated movement from its start to the end of the trajectory, while it is now necessary to consider an active patient.

2.17 Home electric wheelchair simulator: LISV (Laboratoire d'Ingénierie des Systèmes de Versailles), Versailles University, France

It is a simulation project of wheelchair navigation for a better building accessibility (Taychouri et al., 2008). In fact, adapting spaces of residential and industrial. In this complex and diversified context, space accessibility in wheelchair is a very interesting topic. The goal is there to make easier movings inside or outside homes. Proposed methodology consists in evaluating movings of wheelchair in a domestic place. Problem not only includes trajectories generation, but also their evaluation based on distance and time, trajectory curves, maneuverability and driving quality.

2.18 WheelSim © : LifeTool, Austria

This software is the only one dealing with commercial wheelchair simulation available on the market. WheelSim © is a product from LifeTool society (f. 1998), based in Linz, Austria. Software released version is now 1.2 and aims at adapting technology to human. This society unites competencies of diversified specialists, combining educative, psychological, social and realistic integrative of social organisms members with technical expertise of engineering researchers and specialists of disabilities and rehabilitation.

WheelSim software aims at making easier the learning of power wheelchair driving, and can also be used as a diagnosis and training software or even as a simple game.

Software aims at improving both personal mobility or circulation safety. The fact that learning system efficient command enable to move safely in circulation is however questionable, as several other factors have to be taken into account and rigourously integrated before taking decisions of moving capacities. Wheelchair can be commanded with a joystick, keyboard arrow keys or with contactors. Navigation levels can be very diversified in situation and complexity (Fig. 16).

Fig. 16. WheelSim navigation

WheelSim also includes fault notion, i.e. deviations from yellow lines showing security limits, navigation collisions, or circulation rules respect. Faults are quantified, indicated by visual feedbacks and can induce up to disqualification (navigation stop). Applications can concern rehabilitation, childhood, physical and/or mental disabililty, game of skills...

2.19 ISIDORE (Interface d'aide à la SImulation, à la DécisiOn et la REéducation), HandiBio, Toulon University , France

ISIDORE has been developped in partnership with Hôpital Renée Sabran (Giens/Hyères - Var) and Association des Paralysés de France (Var delegation). It is useful for learning and evaluating driving capacity for electric wheelchair prescription. It provides quantitative information to help therapist in decision making for user progressions and the instrumentation to be added for a safe driving. In this aim, it integrates reference comportemental information with a fuzzy approach (Randria et al, 2008a).

In order to obtain a reference trajectory, ISIDORE uses a genetic or determinist (A*) approach depending on precision required and patients fatigue (Randria et al., 2007a). It also uses behavioural, visual and sound information to have a better evaluation of the user and more pertinent information to therapists and prescribing doctors (Randria et al, 2007b). Simulator can be autonomously used (lever and computer screen) or integrated in a complete system using a virtual reality platform with a fixed wheelchair providing information of wheels speed (Fig. 17). It allows a very good cost-performance compromise and can be interfaced with 2D and 3D CAD software for environment conception. It can work with different controls (hand, foot, head, voice, blow…) and enable virtual tests before being in real conditions (Randria et al, 2008b).

Fig. 17. ISIDORE simulator

3. Summary

Table 1 below sums up all the particularities of each project detailed in section 2.

Projects	Year	Origin, Country	Characteristics	(1)	(2)	(3)	(4)	Info. feedback
PWMS	1993	New York, USA	Platform for manual wheelchair	No	Yes	No	No	No
Virtual Electric Power Wheelchair Driving	1993	Pittsburg Univ. / VA Healthcare System, PA, USA.	Speed control when perturbations Inside/outside environment, static and dynamic Evaluation : records position and speed of virtual wheelchair and pedestrians, number of collisions, execution time, path done.	Yes	Yes	Yes	Yes	Visual
Wheelchair User Proficiency Through Virtual Simulation	1994	OSC / Ohio Univ., OH, USA	Driving evaluation in a semi-structured world. Polarized stereovision glasses (non HMD) Spatial subdivision (Octree data structure) IR emitter DualScan stereo screen	No	No	Yes	No	Visual
Wheelchair Training	1994	Virt. Real. Lab., ORI, OR, USA	Platform « Crash » sound for collisions 3D spatialized sound Stereoscopic vision Different scenarii (user profile, environment complexity) Learning via Internet	No	Yes	Yes	No	Visual, stereoscopic, sonic, 3D
H.M2.P.H.	1998	Tours Univ, France	Automatic 2D map 3D virtual visit Databases	Yes	Yes	Yes	No	Visual
Assistive Technology Access Interfaces	1998	Ohn School for the Physically Disabled, Tel Aviv, Israel	Developing and evaluating technological assistive interfaces for people with severe physical disabilities Evaluating the ability of driving an electric wheelchair Defined evaluation scale	No	Yes	No	No	Visual
Simulator of Powered Wheelchair	1998	Research Institute, National Rehabilitation Center for the Disabled, Japon	Evaluating the ability of driving a wheelchair Mobile platform Flat screen to 110° screen Levers and behavior analysis Trajectory recording Different grounds and slopes Traffic signs	Yes	Yes	Yes	No	Visual

Projects	Year	Origin, Country	Characteristics	(1)	(2)	(3)	(4)	Info. feedback
VAHM extension for wheelchair driving simulation	2000	LASC, Université de Metz, France	Simulation platform using virtual reality Helping conception of new functionalities for mobility assistance Helping for choosing and prescribing wheelchairs Helping for learning of driving electric wheelchairs Exisiting wheelchair kinematics Sensor modelling Recording of behavior (execution time, collisions, stops in time and number…) Helping to fuzzy decision for functionalities activation	Yes	Yes	Yes	Yes	Visual
Evaluation and training for electric wheelchair driving	2000	Royal Hospital for Neuro-disability & University of East London, UK	Quantitaive and qualitative data for conception and implementation of two non-immersive environments Quantitaive and qualitative data for user evaluation Manoeuvrability and itinerary seeking capacities evaluation Collision detection	Yes	Yes	Yes	No	Visual
Automation and Robotics	2000	DSERG, Portsmouth Univ, UK	Virtual reality for elderly people	No	No	No	No	-
VEMS	2002	Limerick Univ., Ireland	Joint analysis of user performances when facing environment change and its interacting material Joint analysis of user profile and evaluation Navigation reexecution	No	Yes	No	No	Visual, haptic and sound
OCAWUP	2003	Laval and Sherbrooke Univ., QC Canada with rehab. centers	Navigation evaluation in a daily life context Standardized evaluation Determination of factors that have influence on driving capacities Modelizable environment	No	Yes	No	No	Visual
Haptic interface for accessibility	2003	Strathclyde Univ., Glasgow, UK	Wheelchair platform Virtual reality Buildings accessibility Haptic interface Surfaces changes and slopes	No	No	No	No	Haptic and visual
VRWC	2004	Clarkson	Transforming a group of	No	No	Yes	No	Visual,

Projects	Year	Origin, Country	Characteristics	(1)	(2)	(3)	(4)	Info. feedback
		Univ, New York, USA	objects Collision detection Following of slopes (set of segments) Interactions : lever, video tracker, 6-DOF platform					platform proprioceptive and vestibular
Stroke Simulator	2004	VRlab/HPC2N, Umeå Univ, Sweden	Simulator for tremors Data analysis (grounded Theory)	No	Yes	Yes	No	Visual and 3D sound
Virtual Intelligent Wheelchair	2007	CRVM, Mediterr. Univ., Marseille, France	Simulator for evaluation and training to intelligent wheelchair driving CAVE immersive interface Wheelchair dynamic model Automatic driving	Yes	No	Yes	Yes	Visual
Home electric wheelchair driving	2008	LISV, Versailles Univ., France	Wheelchair driving evaluation in a habitation Trajectory generations Criterions : distance, time, curves, manoeuvrability, driving quality	Yes	Yes	Yes	No	Numerical
WheelSim ©	2008	LifeTool, Autriche	Commercial software Coming from social, information technology, handicap and rehabilitation research Quantified driving evaluation	Yes	Yes	Yes	No	Visual and sound
ISIDORE	2008	Handibio, Toulon Univ., France and SIG, Polytech. Eng. School, Madagascar	Flatform Simulation interface Decision-making aid for therapists Rehabilitation tool Quantitative and qualitative evaluation of driving Navigation assistance through information feedback and collaborative control modes	Yes	Yes	Yes	Yes	Visual, sound and sensorial

Table 1. Summary of scientific works on wheelchair driving simulation / (1) : Numerical evaluation, (2) : Subjective evaluation (Questions), (3) : Obstacle detection, (4) : Navigation assistance

4. Conclusion

Numerous works are subjective and give priority to therapist, but lack quantified evaluation for driving capacity. Moreover, some projects are rather evolved while other are more simple but as interesting in terms of gain in conception and material costs. Thus, isometric levers use would almost have the same impact in navigation than the traditional lever of electric wheelchairs but they are most costly since very specific (Cooper et al., 2002). Other projects focus on the use of a 6-DOF platform so as to reduce inconvenience and to be more realistic (Niniss et al., 2005a). However, these gains in inconfort reduction and realism tend to be less interesting than expected (Mestre, 2004). Other use stereoscopic vision to have an improved 3D perception (Swan et al., 1994), but systems are costly and have not proved yet to bring important navigation improvement as their use quickly drive in establighing a visual discomfort or a persistant visual fatigue.

Nevertheless, conducting line of these works is visible : offering users (patients, prescribing and reahabilitation doctors, ergotherapists, ...) a performant evaluation and quantification tool, easy to use, open to interface with commercial tools and with a reasonable cost to be diffused on the wider possible scale.

5. References

ADA (1990). *Americans with Disabilities Act of 1990*, U.S. Department of Justice, U.S.A.

Adelola, I.; Cox, S. & Rahman, A. (2002). Adaptable virtual reality interface for powered wheelchair training of disabled children, *Proceedings of The Fourth International Conference on Disability, Virtual Reality and Associated Technologies*, pp. 173-179, ISBN 0704911434, Veszprém, Hungary.

Adelola, I.; Rahman, A. & Cox, S. (2003). Conjoint analysis in virtual reality based powered wheelchair rehabilitation of children with disabilities, *Proceedings of the California State University Northridge 2003 Conference*, Northridge, CA, U.S.A.

Adelola, I.; Cox, S. & Rahman, A. (2003). A framework for adapting wheelchair training in virtual reality, In: *Assistive Technology - Shaping the Future*, Craddock, G.M. ; McCormack, L.P.; Reilly, R.B. & Knops, H. (Ed.), pp. 122-126, IOS Press, ISBN 9781586033736, Amsterdam, The Netherlands.

Adelola, I.; Cox, S. & Rahman A. (2005). VEMS – Training wheelchair drivers. In: *Assistive Technology: From Virtuality to Reality*, Pruski, A. & Knops, H. (Ed.), pp. 757-761, IOS Press, ISBN 1-58603-543-6, Amsterdam, The Netherlands.

Adelola, I.; Rahman, A. & Cox, S. (2005). Motivation Elements in Virtual Reality for Disabled Children - The VEMS Experience. In: *Assistive Technology: From Virtuality to Reality*, Pruski, A. & Knops, H. (Ed.), pp. 762-766, IOS Press, ISBN 1-58603-543-6, Amsterdam, The Netherlands.

Baran, E.O & Clarvoe, R. (2007). *Virtual Reality Wheelchair Program Website*. kson University, Potsdam, NY, U.S.A., http://www.clarkson.edu/~vrlab/vrwc/index.html

Bourhis, G. & Agostini Y., (1998). The VAHM Robotized Wheelchair: System Architecture and Human-Machine Interaction. *Journal of Intelligent and Robotic Systems*, Vol. 22, No. 1, pp. 39-50.

Bresler, M. (1990). Turtle Trainer: A way to evaluate power mobility readiness, *Proceedings of the RESNA 13th Annual Conference*, pp. 399-400., Washington, DC, U.S.A.

Backman, A. (2005). Colosseum3D – Authoring framework for Virtual Environments, *Proceedings of the 11th Eurographics Workshop Virtual Environments*, pp. 225-226, Aalborg, Denmark, October 2005.

Carlson, W.; Swan, E.; Stredney, D. & Blostein, B. (1994). The application of virtual wheelchair simulation to the determination of environmental accessibility and standards compliance in architectural design, *Proceedings of the Symposium on Computers & Innovative Architectural Design, The 7th International Conference on Systems Research, Informatics & Cybernetics*, Baden-Baden, Germany, August 1994.

Cooper, R.; Spaeth, D.; Jones, D.; Boninger, M.; Fitzgerald, S. & Guo, S. (2002). Comparison of virtual and real electric powered wheelchair driving using a position sensing joystick and an isometric joystick. *Medical Engineering & Physics*, Vol. 24, No. 10, 2002, pp. 703-708.

Cooper, R.A.; Ding, D.; Simpson, R.; Fitzgerald, S.; Spaeth, D.; Guo, S.; Koontz, A.M; Cooper, R.; Kim, J. & Boninger, M (2005). Virtual reality and computer-enhanced training applied to wheeled mobility: an overview of work in Pittsburgh. *Assistive Technology*, Vol. 17, No. 2, 2005, pp. 159-170.

DDA. (1995) Disability Discrimination Act 1995 (c. 50). Office of Public Sector Information (OPSI), Norwich, UK.

Dholkawala, Z.; Cox, S. & Rahman, A. (2005). An architecture for "remote monitoring" in a networked virtual environment mobility simulator (VEMS). In: *Assistive Technology: From Virtuality to Reality*, Pruski, A. & Knops, H. (Ed.), pp. 224-228, IOS Press, ISBN 1-58603-543-6, Amsterdam, The Netherlands.

Ding, D.; Cooper, R.A.; Guo, S. & Corfman, T.A. (2003). Robust velocity control simulation of a power wheelchair, *Proceedings of the RESNA 26th International Annual Conference*, Atlanta, GA, U.S.A., June 2003.

Dir, S.; Habert, O. & Pruski, A. (2008). Critères pour la configuration dynamique d'un fauteuil électrique par la réalité virtuelle, *Proceedings of the HANDICAP Conference*, Paris, France, June 2008.

Grant, M. (2003). Space requirements for wheeled mobility, *Proceedings of the* Proce*Wheelchair Simulation in Virtual Reality Workshop*, Amherst, NY, U.S.A., October 2003.

Grant, P.; Harrison, C. & Conway, B. (2004). Wheelchair simulation, *Proceedings of Cambridge Workshop on Universal Access and Assistive Technology*, Cambridge, UK, March 2004.

Grant, P.; Harrison, C. & Conway, B. (2004). The development of a wheelchair simulator to determine qualitative and quantitative performance metrics. *Journal of Assistive Technology*, Vol. 17, ISSN 1040-0435.

Harrison, A.; Derwent, G.; Enticknap, A.; Rose, F. & Attree, E. (2000). Application of virtual reality technology to the assessment and training of powered wheelchair users, *Proceedings of the 3rd International Conference Disability, Virtual Reality and Associated Technologies*, Alghero, Italy, September 2000.

Harrison, A.; Derwent, G.; Enticknap, A.; Rose, F. & Attree, E. (2002). The role of virtual reality technology in the assessment and training of inexperienced powered wheelchair users. *Disability & Rehabilitation*, Vol. 24, No. 11-12, pp. 599-606.

Hasdai, A.; Jessel, A. & Weiss, P. (1998). Use of a computer simulator for training children with disabilities in the operation of a powered wheelchair. *American Journal Occupational Therapy*, Vol. 52, No. 3, pp. 215-220.

Inman, D.P. & Loge, K. (1995). Teaching motorized wheelchair operation in virtual reality, *Proceedings of the CSUN Virtual Reality Conference*, Northridge, CA, U.S.A.

Inman, D.P.; Loge, K. & Cram, A. (2000). Teaching orientation and mobility skills to blind children using computer generated 3-D sound environments, *Proceedings of the ICAD International Conference on Auditory Display,* Atlanta, GA, U.S.A., April 2000.

Inoue, T.; Hirose, H.; Yasuo, S. & Kazunori, H. (1998). Development of a simulator of powered wheelchair, *Proceedings of the RESNA'98 Conference,* pp. 182-184, Minneapolis, MN, U.S.A., June 1998.

Inoue, T.; Sudoh, Y.; Hirose, H. & Nagumo, N. (2002). Development of a simulator for a powered wheelchair and practical use for a person with physical and mental disabilities. *Biomechanisms,* Vol. 16, pp. 183-194.

Kirby, R.L.; Swuste, J.; Dupuis, D.J.; MacLeod, D.A. & Monroe, R. (2002). Wheelchair skills test: pilot study of a new outcome measure. *Archives of Physical Medicine and Rehabilitation,* Vol. 83, pp. 10-18.

Leloup, J. (2004). *Le projet HM2PH, habitat modulaire et mobile pour personnes handicapées : spécifications d'un espace de vie adapté pour personnes en déficit d'autonomie.* PhD Thesis, Tours University, France.

Marchuk, N.D.; Ding, D. & Gaukrodger, S. (2007). Development of a virtual platform for assessment and training of power wheelchair driving, *Proceedings of the Rehabilitation Engineering & Assistive Technology Society of North America (RESNA) Conference,* Phoenix, AZ, U.S.A., 2007.

Maxhall, M.; Backman, A.; Holmlund, K.; Hedman, L.; Sondell, B. & Bucht, G. (2004). Caregiver responses to a stroke training simulator, *Proceedings of ICDVRAT2004 Conference,* Oxford, U.K., September 2004.

Mestre, D. (2004). Activités sensori-motrices : apports de la réalité virtuelle à la psychologie ergonomique, In: *Psychologie ergonomique: tendances actuelles,* Hoc, J.M.; Darses, F. (Ed.), Presses Universitaires de France, Paris, France.

Mestre, D.; Pergandi, J. & Mallet, P. (2004). Virtual reality as a tool for the development of a smart wheelchair, *Proceedings of the Virtual Reality International Conference,* Laval, France, April 2004.

Mestre, D.; Pergandi, J. & Mallet, P. (2007). Designing a navigation aid for a smart wheelchair. *AMSE Journals, Modelling C,* Vol. 67, pp. 86-95.

Niniss, H.; Nadif, A. & Bourhis, G. (2000). Système de simulation pour fauteuil roulant électrique : premiers résultats. *Journal Européen des Systèmes Automatisés,* Vol. 34, No. 6-7, pp. 799-815.

Niniss, H. & Nadif, A. (2000). Simulation of the behaviour of a powered wheelchair using virtual reality, *Proceedings of the 3rd International Conference on Disabilities, Virtual Reality and Associated Technologies,* pp. 9-14, Alghero, Italy, September 2000.

Niniss, H. & Nadif, A. (2000). Simulation System for Powered Wheelchairs : Evaluation of Driving Skills Using Virtual Reality, In: *Assistive Technology - Shaping the Future,* Craddock, G.M. ; McCormack, L.P.; Reilly, R.B. & Knops, H. (Ed.), pp. 112-116, IOS Press, ISBN 9781586033736, Amsterdam, The Netherlands.

Niniss, H. & Nadif, A. (2004). Système de simulation pour fauteuils roulants électriques: Evaluation de la commande à l'aide de la Réalité Virtuelle, *Proceedings of the HANDICAP Conference,* pp. 163-69, Paris, France, June 2004.

Niniss, H. & Inoue, T. (2006). Electric wheelchair simulator for rehabilitation of persons with motor disability, *Simpósio Brasileiro de Realidade Virtual,* Belém, Brazil.

Niniss, H. & Inoue, T. (2006). Assessment of driving skills using Virtual Reality: comparative survey on experts and unskilled users of electric wheelchairs. *Journal Technology and Disability*, Vol. 18, No. 4, pp. 217-226.

Pan, Z.; Cheok, A.D.; Yang; H.; Zhu, J. & Shi, J. (2006). Virtual reality and mixed reality for virtual learning environments. *Computers & Graphics*, No. 30, pp. 20-28.

Pronk, C.; de Klerk, P.; Schouten, A.; Grashuis, J.; Niesing, R. & Bangma, B. (1980). Electric wheelchair simulator as a man-machine system. *Scandinavian Journal of Rehabilitation Medicine*, Vol. 12, No. 3, pp. 129-135.

Randria, I.; Abellard, P.; Abellard, A. & Ben Khelifa, M. (2008). ISIDORE : Interface d'assistance à la SImulation, à la DécisiOn et à la Reéducation, *Proceedings of the Handicap Conference*, Paris, France, June 2008.

Randria, I.; Ben Khelifa, M.; Bouchouicha M. & Abellard, P. (2007). A comparative study of six basic approaches for path planning towards an autonomous navigation, *Proceedings of the 33rd Annual IEEE IECON Industrial Electronics Conference*, Taipei, Taiwan, November 2007.

Randria, I.; Ben Khelifa, M.; Abellard, A.; Abellard, P.; Gorce, P. & Bouchouicha M. (2007). A virtual reality application to disabled transportation simulation, *Proceedings of the Virtual Reality International Conference*, Laval, France, May 2007.

Randria, I.; Abellard, A.; Ben Khelifa, M.; Abellard, P. & Ramananstizehena P. (2008). Evaluation of trajectory applied to collaborative rehabilitation for a wheelchair driving simulator, *Proceedings of the 4th European Congress for Medical and Biomedical Engineering*, Antwerp, Belgium, November 2008.

Roussel, P. (2002). Une estimation de la diffusion des aides techniques à partir de l'enquête HID de l'INSEE. *Handicap*, No. 96, pp. 47-54.

Routhier, F.; Vincent, C.; Desrosiers, J. & Nadeau, S. (2003). Mobility of wheelchair users: a proposed performance assessment framework. *Journal Disability & Rehabilitation*, Vol. 25, No. 1, pp. 19-34.

Routhier, F.; Vincent, C.; Desrosiers, J.; Nadeau, S. & Guerette, C. (2004). Development of an obstacle course assessment of wheelchair user performance (OCAWUP): a content validity study. *Technology and Disability*, Vol. 16, No. 1, pp. 19-31.

Schmeler, M.R. (1993). *Performance Validation of a Powered Wheelchair Mobility Simulator*. Thesis (M.S.), State University of New York at Buffalo, U.S.A.

Schmeler, M.R.; Johnson, D. & Granic, J. (1993). Powered wheelchair mobility simulator, *Proceedings of the RESNA 16th Annual Conference*, pp. 357-359, Las Vegas, NV, U.S.A.

Sonar, A.; Burdick, K.; Begin, R.; Resch, E.; Thompson, E.; Thacher, E.; Searleman, J.; Fulk, G. & Carroll, J.J. (2005). Development of a virtual reality-based power wheel chair simulator, *Proceedings of the IEEE International Conference on Mechatronics and Automation*, pp. 222-229, Niagara Falls, ON, Canada, July-August 2005.

Stott, I. & Sanders, D. (2000). The use of virtual reality to train powered wheelchair users and test new wheelchair systems. *The International Journal Of Rehabilitation Research*, Vol. 23, No. 4, pp. 321-326.

Swan, E.; Stredney, D. & Carlson, W. (1994). The determination of wheelchair user proficiency and environmental accessibility through virtual simulation, *Proceedings of the Second Annual International Conference on Virtual Reality and Persons with Disabilities*, pp. 156-161, California State University, Northridge, CA, U.S.A., June 1994.

Taychouri, F.; Monacelli, E.; Hamam, Y. & Chebbo, N. (2008). Méthodologie d'analyse de l'aménagement d'un habitat pour une déambulation en fauteuil roulant, *Proceedings of the Handicap Conference*, Paris, France, June 2008.

12

Magneto-rheological technology for human-machine interaction

Jose Lozada[1], Samuel Roselier[1], Florian Periquet\Xavier Boutillon[2] and Moustapha Hafez[1]
[1]CEA-LIST, Sensory Interfaces Laboratory, 18 route du panorama, 92265 Fontenay-aux-Roses Cedex, France [2]

1. Introduction

Human Machine Interfaces aim to give feedback to the user according to a given mechanical model or behavior. These systems are in constant interaction with the users senses. Haptic interfaces are mechatronic systems designed for providing force and tactile feedback to the operator while immersed into a virtual environment. The user inputs a motion or force into the system that reacts to this mechanical energy according to the virtual environment dynamical behavior.

The mechanical behavior needed to give correct feedback is then directly related to the mechanical impedance of human being. The system should be capable to display free motion as well as a rigid contact (for virtual reality applications). Traditional actuators, DC motors for example, are limited due to stability issues for high forces and usually their weight and size do not allow free motion.

Magneto-rheological fluids can be used to design semi-active systems (controlled damping or braking forces) that dissipate the mechanical energy applied by the user, this ensures their stability over all the working range. Moreover, the fluid exhibits a ~1 ms response time that permits real time control in the human interaction bandwidth.

Two basic operational modes have been reported for MR fluids: the "valve mode" and the "direct shear mode" Jolly et al. [1999]. Active dampers usually operate in the valve mode in order to develop high stiffness and damping. The potential drawback of systems based on this mode of operation is the high level of frictional forces induced by the piston configuration, even in the absence of magnetic field.

Current limitations for the magneto-rheological devices are often due to the design which is either in a brake configuration or a piston design. These limitations are mainly high uncontrolled forces due to friction and high inertia because of heavy moving parts. We propose in this work, a novel operating mode that reduces uncontrolled forces as well as the inertia of moving parts. This operating mode is based on shearing the fluid by a thin element (plate or cylinder), rather than shearing with the magnetic poles, increasing the performances and reducing the overall size.

We present the modeling and experimental characterization of the novel operating mode through two haptic interfaces.

The first example concerns the design of a haptic interface for musical keyboards. The presented system aim at reproducing the behavior of a traditional piano to give high quality feedback to a musician using a numerical keyboard. The system is then capable of precise force feedback controlled in real time for fast actuation.
The second illustrating example is a novel Human Machine Interface for automotive cockpits. The behavior of the system is given by prerecorded behaviors. The system should be user friendly, easy to use, stable and compact while generating moderate torque.
Both systems use the novel actuation principle, in a linear configuration for the first one and in a rotational configuration for the second.

2. Magneto-rheological (MR) Fluids Behaviour

"Rheological fluids" are liquids with viscosity properties that can be modified by an external stimulus. This stimulus can be a magnetic or electric field. In both cases, the external field acts on the micron sized particles in suspension into the carrier fluid. These particles form chains along the magnetic or electric flux lines thereby the apparent viscosity of the fluid increases as the external field intensity increases. The rheological modification can be exploited to dissipate the mechanical energy applied by the user. This semi-active behavior makes these fluids particularly interesting for real-time controlled systems because of their intrinsic stability.
We can cite two different technologies, magneto-rheological (MR) and electro-rheological (ER) fluids according to the external field nature. Their properties are similar but their integration on a controlled user friendly system is different. MR fluids use low voltage (~10V) and current up to 2A input whereas ER fluids use high voltage (~2kV) which limit the application, and low current (~10mA). The electrical power consumption is then more or less equivalent. Magnetic field generation needs coils and ferromagnetic parts which makes the MR-fluid-based systems more bulky than ER-based systems.
The resistant force available is a function of the yield stress under the maximum field intensity (given by saturation for magnetic field and by dielectric insulation for electric field). This maximum yield stress is of about 100 kPa for MR-fluids and 5 kPa for ER-fluids. The maximum resistant force is then 20 times higher for a MR-based system at equal active surface. The table 1.1 shows a comparison of typical values for both fluids.
Low voltage and higher yield stress make MR-fluids better for mass-market applications and for interfaces which are in direct contact with the operator.

2.1 Working Principle

Magneto-rheological fluids are suspensions of ferrous particles of a few microns size (typically 1 to 10 /xm) in a carrying fluid such as water or oil (synthetic or mineral oil). In the absence of a magnetic field, the fluid

Property	ER fluid	MR fluid
Yield strength (Field)	2–5 kPa (3–5 kV/mm) Field limited by breakdown	50–100 kPa (150–250 kA/m) Field limited by saturation
Viscosity (No field)	0.2–0.3 Pas at 25°C	0.2–0.3 Pas at 25°C
Operating temperature	−10–+90°C (ionic, DC) -25–+125°C (non-ionic, AC)	-40–+150°C (Limited by the carrier fluid)
Current density	2–15 mA/cm² (4kV/mm, 25°C) (x 10 100 at 90°C)	Can energise with permanent magnets
Specific gravity	1–2.5	3–4
Ancillary materials	Any (conductive surfaces)	Iron/steel
Colour	Any, opaque or transparent	Brown, black, grey/opaque

Table 1. Typical values for MR and ER fluids [Carlson et al., 1996]

behaves like a Newtonian[1] fluid. It's viscosity depends on that of the carrying fluid and on the concentration of particles.
Usually the particles represent 80 to 85% of the fluid weight which corresponds to 20 to 40 % of the volume if the carrying fluid is a mineral oil. MR-fluids can achieve shear stress of 100 kPa under a magnetic field of 200 kA/m.
In the presence of a magnetic field, the particles align themselves and form chains along the magnetic flux lines. These cohesive chains resist to the fluid flow. The amplitude of the magnetic field controls the apparent viscosity.
According to [LORD, Dec. 1999], the Bingham rheological model - Eq (1.1) - is used to describe the fluid behavior under direct shear.

$$\tau = \eta\dot{\gamma} + \tau_y(H)\widehat{\dot{\gamma}} \tag{1}$$

where the operator $\hat{z}$ is defined by

$$\hat{z}\begin{cases} = Sign(z) & \text{for} |z| > 0 \\ \in [-1, 1] & \text{for} z = 0 \end{cases} \tag{2}$$

In (1), τ is the shear stress, γ the shear strain, η the fluid viscosity, and τ_y a magnetic field dependent yield stress. Figure 1 shows the theoretical behavior of MR-fluids under direct

[1] A Newtonian fluid is characterized by a shear stress lineary dependent from the shear strain rate

shear. If the magnetic field is controlled as a function of the shear rate, we obtain a group of operating points. The apparent viscosity of the fluid can then be controlled.
The particles organization under a magnetic field follow five steps (see figure 2):

- when there is no magnetic field, the particles are distributed homogeneously in the carrier fluid
- a magnetic field produces a local magnetization of the ferrous particles

'A Newtonian fluid is characterized by a shear stress linearly dependent from the shear strain rate

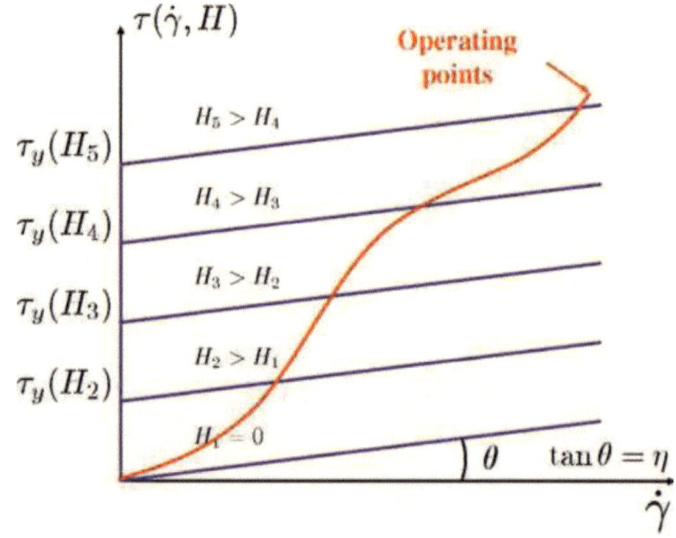

Fig. 1. Theoretical rheogram of a MR-fluid

- the ferrous particles act as magnetic dipoles that attract themselves
- under these attracting forces the particles form chains along the magnetic flux lines
- if the magnetic field is canceled, the thermal agitation brakes the chains and disperse the particles

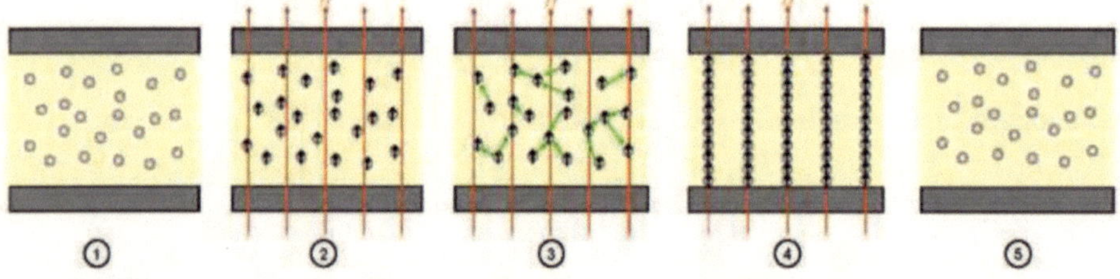

Fig. 2. Magneto-rheological effect

Since the magnetic field can only increase the apparent viscosity, systems based on MR fluids are intrinsically dissipative. The induced stability makes them appealing for usage in haptic interfaces.

2.2 Operating Modes

Two basic operational modes have been reported for MR fluids: the "valve mode" and the "direct shear mode" [Jolly et al., 1999]. Valve mode (see figure 3) is driven by a pressure drop between two cavities which creates a fluid flow pass an active zone. A magnetic field applied across the active zone increases the apparent viscosity of the fluid into the active zone thereby increasing the pressure difference.

Active dampers usually operate in the valve mode in order to develop high stiffness and damping. The potential drawback of systems based on this mode of operation is the high level of frictional forces induced by the piston configuration, even in the absence of magnetic field.

In the direct shear mode(see figure 4), the fluid is sheared by the relative motion between the magnetic poles, as in rotating machines such as brakes or clutches that do not need to produce large forces [Ahmadian,2002].

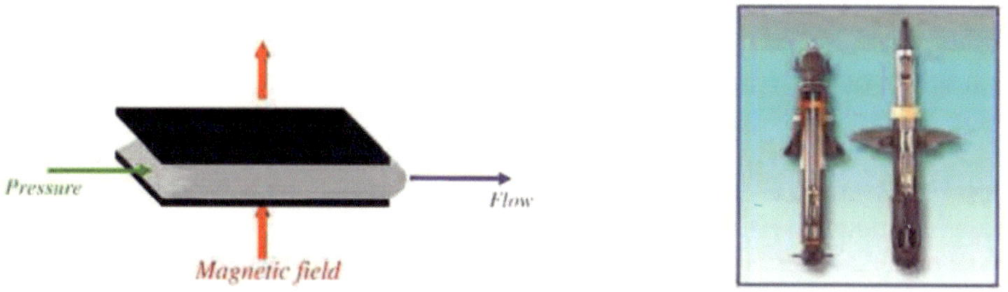

Fig. 3. Valve mode and application example: automotive damper [Delphi, 2005]

Fig. 4. Valve mode and application example: semi-active haptic interface [Reed and Book, 2004]

The relative motion of magnetic poles shears the particle chains which mechanical resistance is a function of the magnetic field, the shear stress increases as the magnetic field increases. In most cases, the rotor is
magnetic. However, the drawback of this operational mode is a large inertia of the moving parts. In order to lower the inertia, we introduce a novelty to the latter mode [Lozada et al., 2007]. Poles are held fixed and the gap between them is filled with MR fluid. A thin plate moving within the gap shears the fluid (see figure 5). This operating mode is used on both applications presented further. The haptic interface for musical keyboards uses it in a linear way and the MR-drive uses it in a rotational configuration (see figure 20).

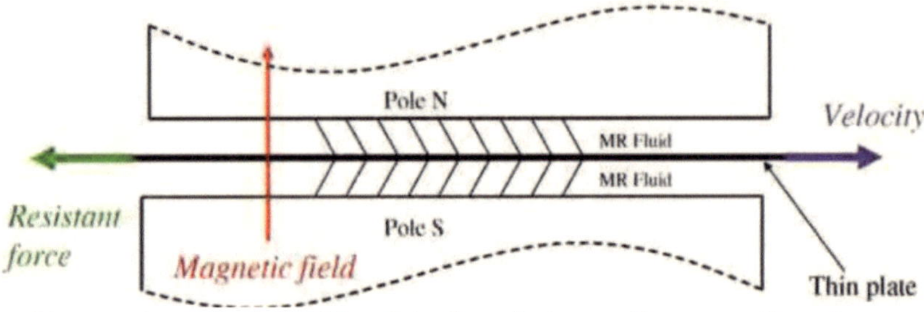

Fig. 5. Novel interaction principle: thin plate shear between the magnetic poles

2.3 Experimental Characterization and Modeling for the Novel Actuation Principle

The Bingham rheological model is not entirely satisfactory for the use of MR fluid in the context of the haptic interfaces presented further.

Usually in rotating machines under operating conditions, the relative velocity is usually high and the shear stress exceeds the yield stress: the behavior below the threshold can be neglected (see eq. (1)). In the application of a haptic interface, the velocity of the moving device of the interface changes continuously, the dynamics of the system is controlled in real-time by means of the control parameter H. Since a permanent regime is seldom reached, the behavior below the yield stress must be considered.

The function fluid behavior when sheared by a thin magnetic plate has been characterized by the experimental procedure described below. The device represented in Fig. 6 is composed of two magnetic poles (1), two coils (2), a plate support (3), a fluid container supporting also the poles and the coils (4), a frame (5), and a force/torque sensor (6).

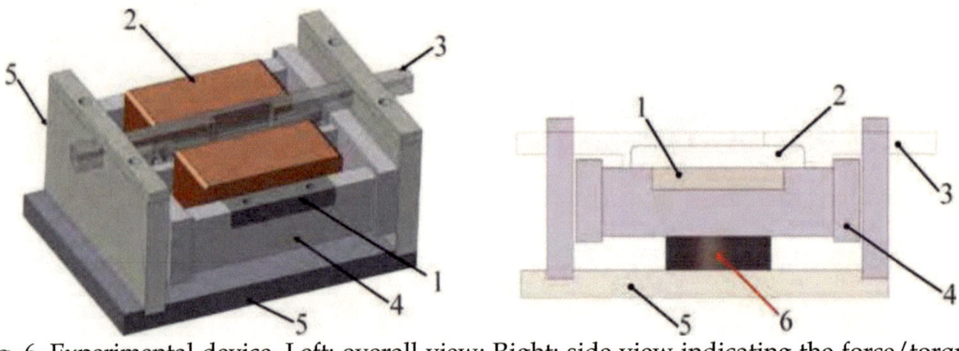

Fig. 6. Experimental device. Left: overall view; Right: side view indicating the force/torque sensor (6) position

The MR fluid fills the gap between the magnetic poles in which a 0.15 mm thick plate made of magnetic material moves linearly, shearing the fluid. Electrical current in coils induce a controlled magnetic field across the gap.

The plate support (Fig. 7.) is made of aluminum, two nuts, and two screws. The plate is attached to the nuts by metal pins. Screws control the tensile stress in the plate, thus ensuring flatness. This whole slider is constrained to move linearly by the frame (5).

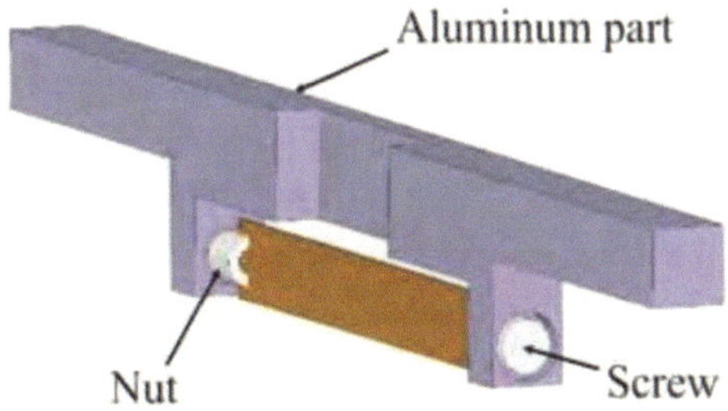

Fig. 7. Slider: plate and support

A test consists in applying an arbitrary motion $\xi(t)$ to the slider for a given current I in the coils.
The motion ξ of the slider is measured by a non-contact displacement sensor (Keyence, LB-72) and gives access to the shear strain. In order to measure the interaction forces between the slider and the MR fluid, the fluid container has been isolated mechanically from the frame and attached to a strain gauge 6-axis force sensor (ATI mini 40). This sensor (6) is placed between the fluid container (4) and the supporting frame (5) so that reaction forces and torques are fully transmitted to the supporting frame through the force/torque sensor (Fig. 6). The system guiding the slider is part of the frame which ensures that the corresponding friction is not applied to the force sensor. A careful calibration yields the relationships that transform the measurements of the sensor into the force F_s exerted by the slider on the MR fluid.
In this experiment, the fluid is sheared on each side of the plate, each facing a magnetic pole. The device has been designed so that the gaps and the shear areas associated to the north and the south pole are the same. It follows that the shear strain and shear stress are identical on each side of the plate. With the gap g = 0.5 mm, the area associated to each pole A = 400 mm^2, and £ the displacement of the plate.

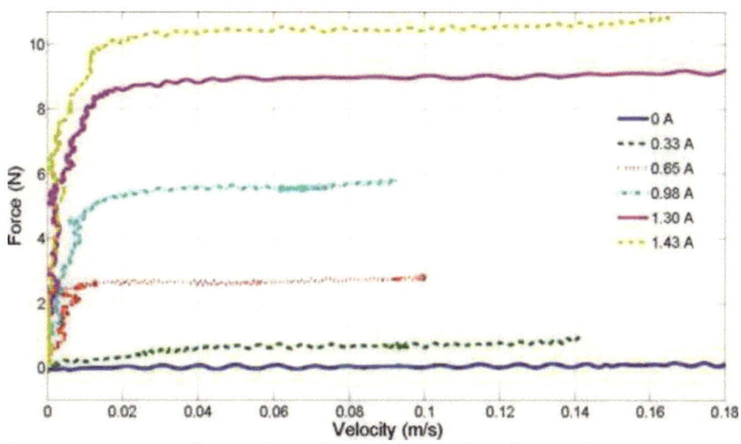

Fig. 8. Shearing force exerted by the MR fluid on the slider. The curves correspond to different levels of current in the coils, proportional to the the magnetic field.

The curves of figure 1.8 indicates two different regimes:

- the Bingham behavior is observed above a certain level of the force F_s with the same linear relationship between the increase in force and the increase in velocity, as described by (1),
- the transition regime where the force increases from 0 to the force threshold.

In the Bingham regime, the yield force increases with the magnetic field. Since the reluctance of the magnetic circuit is not altered by the motion of the slider, the magnetic field H is assumed to be proportional to the value I of the current. This relationship has not been quantitatively measured.
The Bingham regime is analyzed in Fig. 9. the parameters η and τ_y are obtained by fitting the experimental data points of the Bigham region (here: for a velocity above 0.0136 m/s) to the prediction of Eq. (1). The viscosity is found to be 2.8 Pa.s and the yield stress τ_y 12.4 kPa (force threshold 9.92 N).
When the applied shear stress τ is lower than the yield stress τ_y (H), we have observed that the behavior of the fluid depends also on the shear strain γ. We propose the following modified rheological model with three state-parameters γ, $\dot{\gamma}$, and H :

$$\tau = \eta\dot{\gamma} + \widehat{\dot{\gamma}}\min\{G(\gamma), \tau_y\} \tag{3}$$

where the value of τ_y and the function $G(\gamma)$ depend on the magnetic field H. Since, the carrying fluid is not expected to react significantly to a magnetic field, the viscosity η is assumed to be constant but depends on the particle content of the fluid.

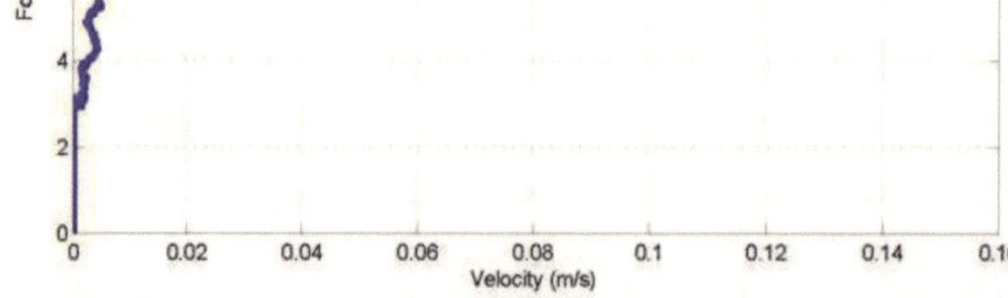

Fig. 9. Experimental results and predicted behavior according to the Bingham model of the MR fluid. Since one of the state variable of the fluid is not represented, the model is not valid in the transient region below the Bingham regime.

The local rheological model given by equation (3) is integrated as:

$$F_t = 2A\left|G\left(\frac{\xi}{g}\right)\right|$$
$$F_s = \eta\frac{2A}{g}\dot{\xi} + \hat{\dot{\xi}}\min\{F_t, 2A\tau_y\} \quad (4)$$

Where F_t is the magnitude of the transition force between the fluid the slider.

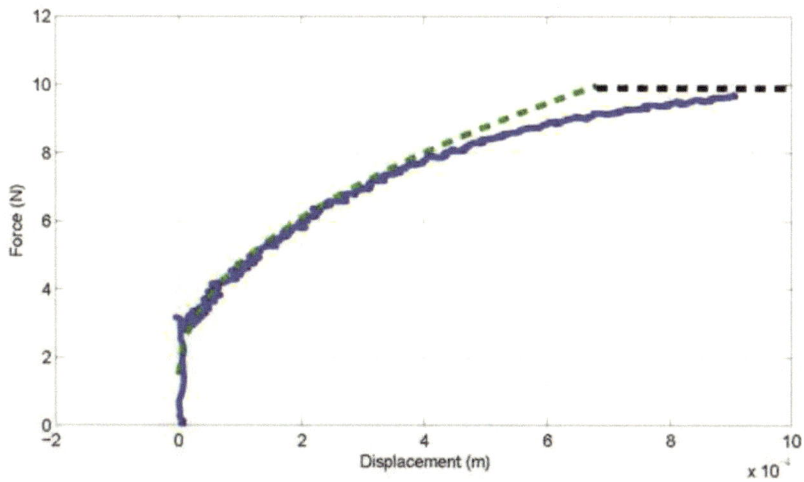

Fig. 10. Experimental results and fitted analytical model for the transient force F_t below the Bingham regime. The horizontal dashed line corresponds to the force threshold of the Bingham model.

The transition regime is analyzed in Fig. 10 the plotted force is F_t as a function of the displacement ξ of the plate. The force F_t is obtained by subtracting $2A\eta\dot{\xi}/g$ from the measurement of F_s with the numerical value of η given by the previous optimization. It appears that F_t is well fitted by the fllowing expression:

$$F_t = F_0 + \phi\sqrt{\xi} \quad (5)$$

with $F_0 = 1.5\text{N}$ and $\phi = 325\text{uSI}$. A log-log plot (not shown here) confirms the value 1/2 of the exponent. One observes also a transition between the behavior described by Eq. 5 and the force threshold of the Bingham model. At this point, we have not found a physical ground for this analytical expression. We consider that the
mechanical behavior of the fluid is given by the expression:

$$\tau = \eta\dot{\gamma} + \hat{\dot{\gamma}}\min\{\tau_o + \varphi\sqrt{\gamma}, \tau_y\} \quad (6)$$

with $\tau_0 = 1.875\text{kPa}$ and $\varphi = 401\text{kPa}$.
Finally, the yield stress is a linear function of the magnetic field in a MR fluid until saturation occurs [LORD, 2005]. In the absence of physical understanding for the rheological behavior

below the Bingham threshold, we assume that the parameters τ_0 and φ also vary linearly with the magnetic field H:

$$\tau_y = K_y H \quad \tau_0 = K_0 H \quad \varphi = K_\varphi H \tag{7}$$

The proportionality constants are derived numerically from the results of the test described above and by a determination of H based on the analysis of the magnetic circuit.
The fluid behavior described above is used to simulate and control the haptic interfaces presented further. The experimental characterization gives also a way to precisely estimate the maximum resistant forces given by the interfaces according to its design.

3. Haptic Interface for Musical Keyboards

The traditional acoustic piano action mechanism (Fig. 11.) is composed of many different parts of wood, wool felt, leather, metal, and steel-springs. These parts form a multi-degree-of-freedom system that transmits energy from the player to the hammer. In return, the action mechanism generates a specific tactile rendering that is felt by pianists during playing. The haptic feedback is essential for a precise control of timing and loudness [Repp, 1999].

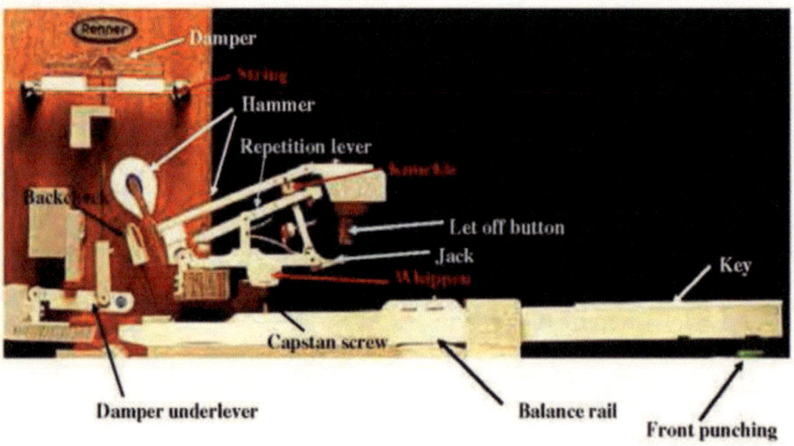

Fig. 11. Traditional piano action mechanism

The action mechanisms used in numerical pianos are much simpler but provide a poor haptic feedback. In the last few years, many developments have been carried out by keyboard manufacturers in order to improve the touch feeling of their products. Most of these systems are not actively controlled and are based on simplified models of the dynamical behavior of traditional pianos. According to users, improvements are still required in terms of size, performance and realism of the device.
Active systems capable of reproducing the traditional piano dynamics have been addressed by [Gillespie, 1996] [Cadoz et al., 1990] and commercial products [Meisel, Nov. 2003]. They are based on simplified models or pre-recorded dynamics that do not satisfactorily match the

dynamical behavior of the traditional piano key. Moreover, the size of these systems based on electromagnetic actuators is not suitable for an industrial keyboard implementation.
The resistant force provided by the traditional piano action mechanism varies from 0.5 N (the minimum force that initiates key motion), to 15N (at *fortissimo* nuance, according to [D. Parlitz, 1998]). Extensive measurements of the kinematics of a grand piano action mechanism [Askenfelt and Jansson, 1990] [Askenfelt and Jansson, 1991] indicate that the duration of the key motion varies from 20 to 250 ms depending on the nuance whereas the key velocity varies from 0.1 $m.s^{-1}$ to 0.6 $m.s^{-1}$.
We present here a system aimed at reproducing the dynamical behavior of the piano action. The traditional mechanism is passive and it can generate a opposite force to the finger's motion during the whole duration of the note stroke. Therefore, we have chosen to develop a system controlling a *resistant* force in real-time, dissipating the mechanical energy given by the pianist by means of a magneto-rheological fluid, sheared by a slider pushed by a simple lever.

3.1 Interface Description

The haptic piano interface shown in Fig. 12. is composed of two main parts: a mechanical structure with various sensors (not shown in figure) and an active system composed of a slider and electromagnetic elements with MR fluid, similar to those described in the previous section.

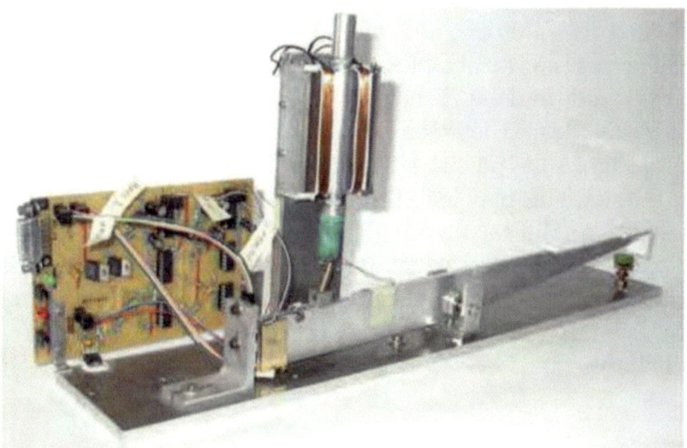

Fig. 12. Piano Interface with Active Feedback

In this prototype, the aluminum key has the same inertia and the same geometry as those of a traditional wooden white key. The pivot between the key and the keybed is ensured by high quality bearings. The contact between the front-end of the key and the keybed is ensured by the same felt pad as in traditional pianos.
A connecting rod is pinned between the key and the slider of the active system in order to transform the rotational motion of the key into a translational motion for the slider.
The motion of the key is measured by *MEMS* accelerometers (ADXL 103 and ADXL 321 for two different measurement ranges 0... 10 $m.s^{-2}$ and 0... 150 $m.s^{-2}$); this instrumentation would be sufficient for the interface operational usage.

The active system (Fig. 13.) is composed a magnetic circuit (1), moving-parts (2), sealing fixtures (3) encapsulating the MR fluid, and guiding parts (4). The whole system is attached to a fixed frame (5).

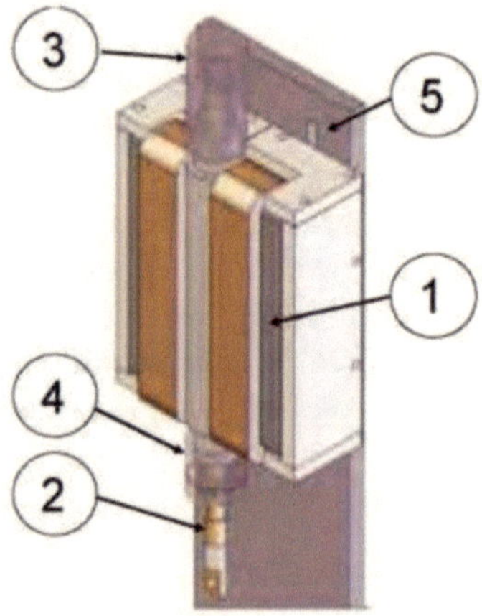

Fig.13. View of the active system

The magnetic circuit (Fig. 14.) includes two I "-shape ferromagnetic cores, each of them supporting a 100-turn coil. This circuit is closed by the 1 mm gap containing the MR fluid. The magnetic field across the gap activates the MR fluid. The moving part consists of a slider formed by three brass rods and a 0.15 mm thick iron plate (left part of Fig. 14.).
The slider moves inside the gap between the poles filled by the MR fluid. A central part with a slot (right part of Fig. 14.) is positioned around the poles so that the magnetic poles face each other across the slot. This part seals the vertical openings of the gap. The top and bottom of the gap are sealed by a solid cap on top ((3) in Fig. 13.), a tube attached to the bottom of the central part, and a nitrile membrane glued to this tube. The two bottom rods of the slider are screwed together through a hole at the center of the membrane as to pinch it and ensure proofness. This design is an alternative to the usual O-ring joint arrangement which generates more friction.

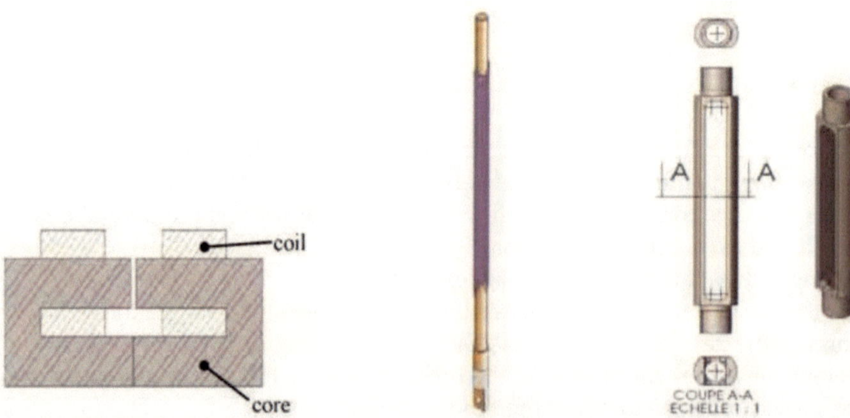

Fig. 14. Cross-section view of the magnetic circuit (left), Slider (center) and slot frame (right)

A guiding mechanism, enclosed in the tube attached to the bottom of the central part, is required to maintain the lateral position of the slider at the center of the gap and to avoid contacts between the plate and the poles.
With a current of 1 A in the coils, this active system can produce a force of 50 N opposing the motion, nearly independent of the velocity. This force becomes 25 N at the key front end, much higher than the 15 N required by specifications . A further optimization could help reducing size and power consumption.

3.2 Mechatronic Model of the Interface

The behavior of the haptic keyboard interface results from mechanical, magnetic, and electrical interactions. Since they are independent up to a certain extent, they are presented separately here: mechanical modeling, electromagnetic behavior, and coupling due to the MR fluid; finally, a box-diagram presents synthetically how the whole system operates and how its behavior can be computed.

Mechanical modeling

The mechanical model of the interface is based on the following approximations:

- all parts are rigid bodies
- friction is negligible in rotational links
- the mass and the inertia of the connecting rod are negligible
- the angle θ describing the motion of the key remains very small: $\cos\theta \approx 1, \sin\theta \approx \theta$.

A schematic view of the interface is presented in Fig. 15. The points of interest for the model are represented either those subjected to a particular force or centers of rotation, as summarized in Table 2.

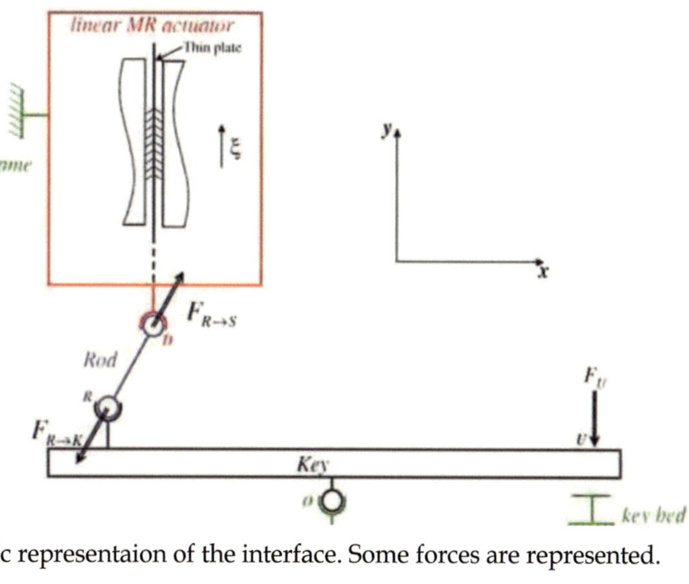

Fig. 15. Shematic representaion of the interface. Some forces are represented.

Point	x	y	Forces	Description
U	x_U	y_U	$\overrightarrow{F_U}$	Force applied by the user (musician)
T	x_T	y_T	$\overrightarrow{F_T}$	Front end keybed contact
G	x_G	y_G	$m_K \overrightarrow{g}$	Key weight
R	x_R	$y_R \approx y_{R_0} + x_R\theta$	$\overrightarrow{F_{R\to K}}$	Force exerted by the rod on the key
D	x_D	$y_D = \xi$	$\overrightarrow{F_{S\to R}}$ $F_{g,y}\overrightarrow{y}$	Force exerted by the slider on the rod Friction exerted by the guiding system

Table 2. Geometrical parameters and associated dynamical variables

Thee coordinates listed in the second and third columns of Table 2 belong to the reference frame $(O, \overrightarrow{x}, \overrightarrow{y})$. The axis $\overrightarrow{Ox}$ corresponds to the horizontal.
It is now described how constraints on the three solid bodies - key, rod, and slider - reduce the number of mechanical degrees of freedom of the system to one. The key is constrained to rotate with an angle θ around O. The motion of the slider is constrained to be a translation in the y directorion by ξ. In other words, the abscissa x_D of the top of the rod or bottom of the slider is kept constant. Let us define x_R as the abscissa of the bottom of the rod; since this point is attached to the key, x_R is also constant within the approximation given by $\cos\theta = 1$. The projection of the rod on Ox is therefore the constant L_x. Finally, the fixed length L of the rod gives ξ as a function of θ:

$$\begin{aligned} & (y_D - y_R)^2 + (x_D - x_R)^2 = L^2 \\ \Rightarrow \quad & \xi^2 - 2y_{R_0}\xi + 2x_R y_{R_0}\theta(1-\xi) + y_{R_0}^2 + L_x^2 - L^2 = 0 \end{aligned} \tag{8}$$

to the first order in θ, with intuitive notations.

In normal use, the key is subjected to forces applied by the rod $\overrightarrow{F_{R\to K'}}$ the user $\overrightarrow{F_U}$, the contact with the key-bed at the front $\overrightarrow{F_T}$, and to its own weight. To zero-order in θ, the angular momentum balance yields:

$$\begin{aligned} J_K\ddot{\theta}\overrightarrow{z} \quad &= \quad m_K\overrightarrow{OG}\times\overrightarrow{g} + \overrightarrow{OR}\times\overrightarrow{F_{R\to K}} \\ & \quad + \overrightarrow{OU}\times\overrightarrow{F_U} + \overrightarrow{OT}\times\overrightarrow{F_T} \\ \Rightarrow \quad F_U x_U \quad & J_K\ddot{\theta} + m_K g x_G - F_T x_T - x_R F_{R\to K,y} \end{aligned} \tag{9}$$

where J_K is inertia momentum of the key in O, and m_K its mass.
In Eq. (9), the first term is derived from the measurement of $\ddot{\theta}$, the second term is constant, the force F_T exerted by the keybed at the end of the key motion depends on the

characteristics of the felt (not studied in this paper), and the last term is given by the following analysis.
Since the mass of the rod is neglected, forces on the rod are balanced:

$$F_{R\to K,x} = -F_{R\to S,x} \tag{10}$$
$$F_{R\to K,y} = -F_{R\to S,y} \tag{11}$$

The moments of forces are also balanced and may be derived at any point, for example in D:

$$F_{R\to K,y} = F_{R\to K,x}\tan\theta_R \tag{12}$$

where θ_R is a geometrical data given by $\cos\theta_R = \frac{L_x}{L}$
Coupling between the rod and the slider involves some friction due to the guiding system. It is assumed that friction is localized in D and follows the Coulomb model. With the geometrical setting represented in Fig. 15., the force $F_{g,y}$ exerted by the guiding system on the slider in the y direction is derived as follows:

$$F_{g,y} = -\mu F_{R\to S_x}\widehat{\dot{\xi}} \tag{13}$$
$$\text{Eq.(10)} \Rightarrow \quad F_{g,y} = +\mu F_{R\to K,x}\widehat{\dot{\xi}} \tag{14}$$
$$\text{Eq.(12)} \Rightarrow \quad F_{g,y} = \frac{\mu}{\tan\theta_R}F_{R\to K,y}\widehat{\dot{\xi}} \tag{15}$$

where μ is the friction coefficient of the guiding system.
Forces acting on the slider in its direction of motion are:
- the slider weight
- the friction force due to the guiding system $F_{g,y}$,
- the restoring force due to air compression in the fluid cavity, $F_{sp} = -k\xi + F_{sp0}$ where k is the stiffness of the spring equivalent to the compression of air in the cavity,
- the force applied by the rod, equal to $-F_{R\to K,y}$ according to the dynamics of the rod,
- the force F_f exerted by the MR fluid which can be calculated with (6).

Applying Newton's law to the slider yields:

$$m_S\ddot{\xi} = -m_s g + F_{g,y} + F_{sp} + F_{R\to S_y} + F_f \tag{16}$$

Finally:

$$F_{R\to K,y}\left(1 - \frac{\mu}{\tan\theta_R}\widehat{\dot{\xi}}\right) = -m_S\ddot{\xi} - m_s g - k\xi + F_{sp0}$$
$$-\frac{\eta A_S}{g_f}\dot{\xi} - \widehat{\dot{\xi}}A_S\min\,\tau_0 + \varphi\sqrt{\frac{\xi}{g_f}},y \tag{17}$$

where m_s is the mass of the slider, A_S the area of sheared flu id, and g_S the width of sheared fluid.

Electromagnetic modeling

The electromagnetic behavior of the coils and the magnetic circuit determines the magnetic field H in the MR fluid as a function of the voltage V applied to the coils.
Electrically, the system behaves as a RL circuit with:

$$L = \frac{N^2}{\mathcal{R}} \tag{18}$$

$$R_\rho \frac{4L_w}{\pi d_w^2} \tag{19}$$

where N is the number of turns, $\mathcal{R}$ is the reluctance of the magnetic circuit, ρ, L_w, and d_w are respectively the resistivity, the length, and the diameter of the wire.

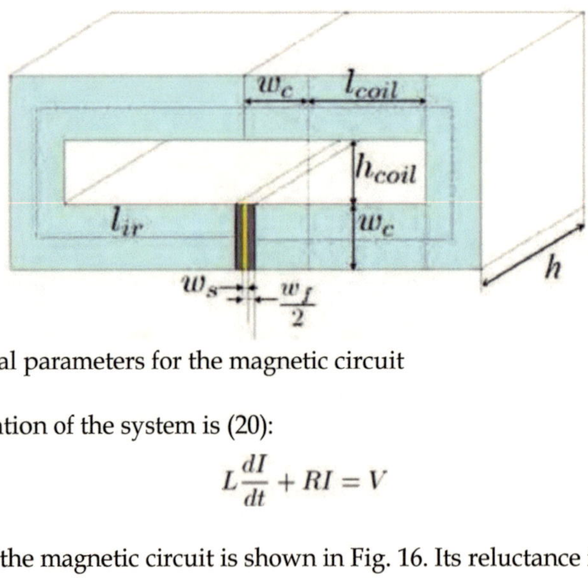

Fig. 16. Geometrical parameters for the magnetic circuit

The electrical equation of the system is (20):

$$L\frac{dI}{dt} + RI = V \tag{20}$$

A cross-section of the magnetic circuit is shown in Fig. 16. Its reluctance is:

$$\mathcal{R} = \mathcal{R}_s + \frac{g_f}{\mu_f A_f} \tag{21}$$

where $\mathcal{R}_s$ is the reluctance of the solid part of the magnetic circuit (ineluding the slider), g_f and A_f the total width and area of the fluid gap, and μ_f the permeability of the MR fluid. With μ_i and μ_p the magnetic permeabilities of the iron (used in the magnetic circuit) and the moving plate, l_i and e_i the average lengths of the magnetic lines in the iron magnetic circuit and in the slider, and A_i and A_p the cross section areas, the reluctance $\mathcal{R}_s$ is given by:

$$\mathcal{R}_s = \frac{l_i}{\mu_i A_i} + \frac{e_p}{\mu_p A_p} \tag{22}$$

The magneto-rheological effect in the fluid changes somewhat the permeability of the fluid; if this effect remains small, the total reluctance of the circuit can be considered constant and independent of the mechanical evolutions of the system: the magnetic flux is solely determined by the tension applied to the coils.
The Ampere's theorem yields (1.23):

$$NI = \mathcal{R}\Phi \tag{23}$$

where Φ is the magnetic flux in the circuit. Considering that the magnetic induction $\vec{B}$ is constant across the gap between the poles, the flux, the magnetic field H in the MR fluid is given by:

$$H = \frac{\Phi}{A_f \mu_f} \tag{24}$$

$$= \frac{NI}{\mathcal{R} A_f \mu_f} \tag{25}$$

In order to maximize the magnetic field, $\mathcal{R}$ must be kept as small as possible. This implies compactness of the circuit and a minimal width of the fluid gap (that improves also the validity of the approximation stated above). The former is limited by the size of the coils and the latter by mechanical constrains: rigidity of the slider plate and tolerance for smooth motion.

Block diagram of the haptic interface

All the models developed in the previous sections can be summarized biny the block-diagram of Fig. 1.17 which shows how the output force F_U felt ba the user is determined by the applied voltage V on the coils and by the angular motion θ.

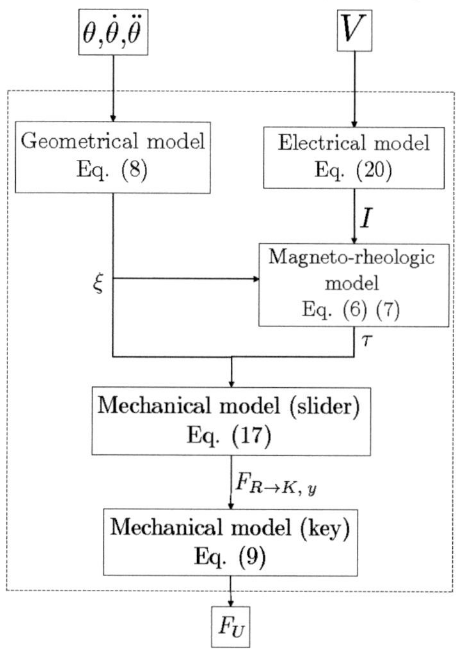

Fig. 17. Haptic piano interface model

At this point, it should be noticed that the user does not *impose* a force or a motion to the interface but follows its own {force-motion} law; consequenty, the choice of θ and F_U as input and output variables respectively is rather arbitrary. It has been guided by the availability, price, and ease of use of motion sensors rather than force sensors.
The system presented in this section exhibits all the required properties for displaying the behavior of a traditional piano, especially the force range (0.5N...50N) and the electrical response time (0.6 ms) The interface model presented above can be used for parameter identification and for the control strategy. Future work will implement a model-based real time control strategy that permits to link the interface to the dynamical model of a virtual piano. The virtual piano will give the needed force in real time and the control strategy should ensure that the force felt by the user corresponds to the desired force.
This interface is a linear application of the novel operating mode presented in section 1.2. The system performances are increased in regard to systems described on the literature of linear MR-actuators. The force ratio (controllable over passive forces) is of about 100. By applying this operating mode to a rotational interface, we can observe the same kind of advantages over existing systems as shown in section 1.4.

4. MR-Drive: A new 1-DOF MR Fluid Based Haptic Interface

GPS navigation, communication, climate and audio systems are widespread functions available in recent cars. The resulting control panels of such systems usually look like an array of controls. Moreover, some "all-in-one" button-based solutions have been developed and despite their ease of use, visual feedback is still necessary to activate the functions, even while driving [BMW, AUDI]. These developments have underlined the interest for haptics in cars. When using a force feedback button, different haptic patterns assigned to specific functions or areas within the menus of the HMI (Human Machine Interface) of be displayed. Haptic interaction decreases the visual load of the driver. Several button designs based on new actuation technologies are presented in [Melli-Huber et al., March 2003] and [Badescu et al., March 2002].

4.1 System design

The MR-Drive system (see figure 1.18) is composed of two magnetic poles facing each other (1), a coil around the central pole (2), a rare earth permanent magnet (3), all these parts are fixed to a frame (5). This architecture is defined as a permanent-electromagnet system. A thin cylinder (4) and a cap (6) are connected to the shaft (8) and guided by two bearings. A nitrile joint is used to prevent fluid leakage. The rotation angle is measured by an optical encoder (7). The thin cylinder rotates in the gap between the magnetic poles. This gap is filled with MR fluid.

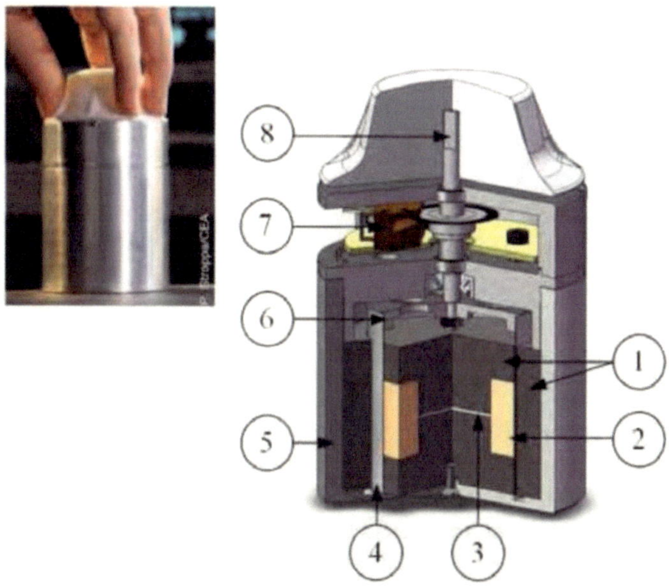

Fig. 18. Overall view of MR-drive

Figure 19 shows the different parts of the magnetic circuit. The thickness of the cylinder (center of figure 1.19) is 0.2 mm with an average diameter d_c of 44 mm. The height h of the magnetic circuit is 40 mm. The coil has an inner diameter of 27 mm and a thickness of 29 mm with 770 turns, its resistance is 24 ohms. Central and outer poles are made of pure iron Armco Telar 57 and the cylinder is made of conventional magnetic steel.

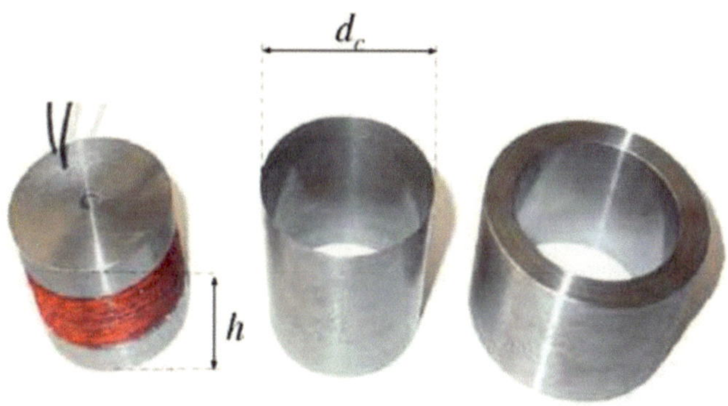

Fig. 19. Parts of the magnetic circuit. Coil arround the central pole (left), moving cilynder (center) and external pole (right)

Figure 20 shows the working principle of the MR-drive system. The shaft, that rotates when the system is actuated by the user, is linked to the cylinder. The cylinder rotates in the gap g between the two poles. The gap is 1 mm thick and it is filled with 4.5 cm^3 of MR fluid. The cylinder rotation shears the fluid into the gap which produces a resistant force tangent to the cylinder surface.

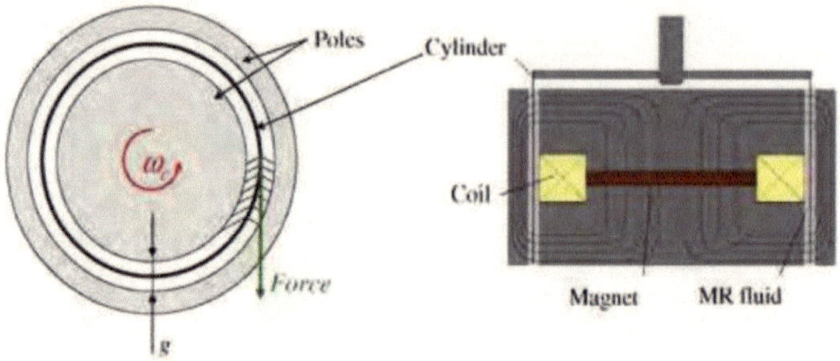

Fig. 20. Working principle

The operating mode of the system is a variation of the thin plate shear mode presented in section 1.2. When the user turns the button, the cylinder rotates around its axis and shears the MR fluid into the gap, the magnetic poles remain static. Thereby the inertia of the moving parts is highly reduced. A current applied to the coil generates a magnetic field across the gap. The value and the current direction set in the coil can either cancel or reinforce the magnetic field established by the magnet. Both the inner and outer surfaces of the thin cylinder shear the MR fluid, therefore the active surface is larger which increases the maximum torque.

4.2 Magnetic Simulation and Torque Estimation

The fluid is sheared at a small rate by a thin cylinder. The MR fluid model presented in section 1.2 can be applied to describe this device operation.
Lets dS be an elementary area of the cylinder defined by dS = dz x d0 where z is the vertical coordinate of the cylinder and 0 the circular coordinate. The magnetic circuit produces a magnetic field H assumed perpendicular to the surface dS. Under these conditions, the elementary torque is expressed as follows:

$$\frac{dT_z}{dS} = \frac{d_c}{2}\tau_y(H)dS \tag{26}$$

with d_c the inner diameter of the cylinder.
According to the model presented in 1.2.3, the yield stress τy is considered as a linear function of the magnetic field *H*. The magnetic circuit was designed to maintain the magnetic induction B under the fluid saturation which validates the linear approximation for the relationship between the magnetic field and the magnetic flux density. We define the two constants K_y and μ_f as follows:

$$\tau_y = K_y H \quad \text{and} \quad H = \frac{1}{\mu_f}B \tag{27}$$

The normal magnetic flux density B_s varies continuously along the two sides of the surface of the cylinder and it is a function of the vertical coordinate z. h is the height of the magnetic circuit and the thickness e of the cylinder is neglected. The field dependent torque T_z is given by:

$$T_z = \iint_S \frac{d_c}{2} \frac{K_y}{\mu_f} |B_s(z)| \, dS \tag{28}$$

$$T_z = \frac{d_c}{2} \frac{K_y}{\mu_f} \int_0^{2\pi} \frac{d_c}{2} d\theta \int_0^h |B_s(z)| \, dz \tag{29}$$

$$T_z = \frac{d_c^2 \pi}{2} \frac{K_y}{\mu_f} \int_0^h |B_s(z)| \, dz \tag{30}$$

Simulations are carried out with FEMM[2] software . The positive values of the current through the coil reinforce the magnetic field of the magnet whereas the negative ones cancel it. The normal magnetic flux density has been simulated for currents in the coil ranging from -1.2 A to 2 A. The permanent magnet placed in the central pole is 1.6 mm height and has a diameter of 25.4 mm.

Figure 21-left presents the results of the simulation by Finite Elements Method of the magnetic circuit behavior with a current of 1.75 A in the coil. $|B|$ and the flux lines are represented. The central part of the magnetic circuit is near the saturation limit ; the value of $|B|$ is over 2 T. Normal flux density B_s along the gap between the central pole and the thin cylinder is extracted from this simulation. Figure 1.21-right presents its topology at different current levels. The minimum of $|B_s|$ (5 mT) is obtained with a current of -0.44 A. At 0 A, the value of $|B_s|$ is 0.3 T due to the presence of the magnet.

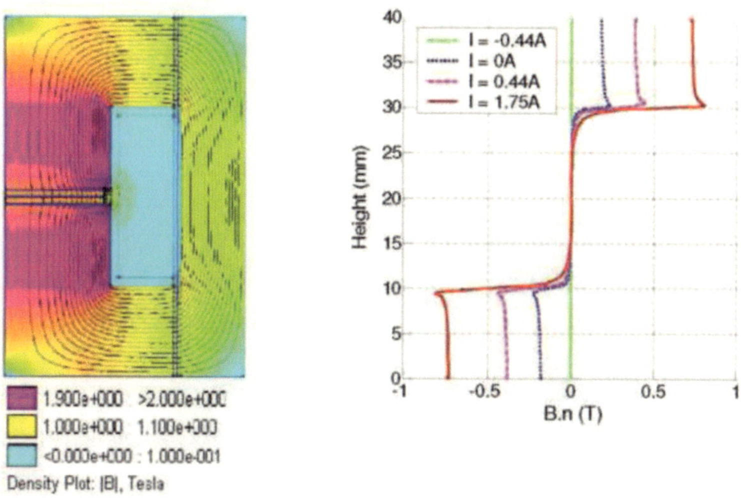

Fig. 21. Finite elements simulation of the magnetic field for a current of 1.75 A. Magnetic circuit (left) and normal magnetic flux density in the gap (right)

[2] http://femmfoster-miller.net

According to Eq. (30), the magnetic field density B_s has been integrated over the surface of the cylinder to simulate the maximum torque available. Results are reported in figure 22.

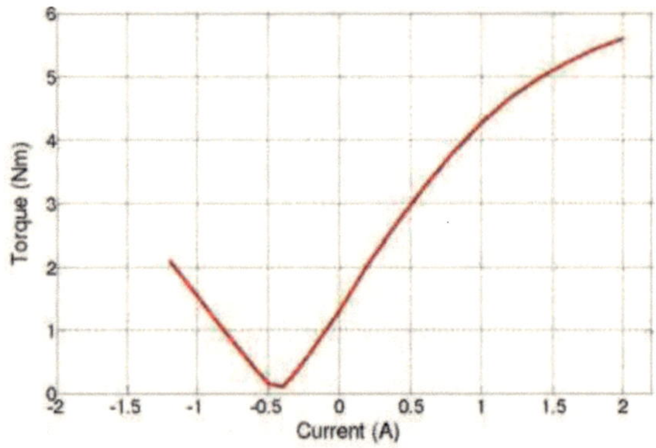

Fig. 22. Maximum torque estimation as a function of the current set in the coil

The minimum torque is obtained for a current of -0.44 A. The torque due to the presence of the magnet alone is 1.25 Nm and the maximum torque is expected over 5 Nm. Magnetic saturation of the material occurs for currents up to 1 A. Under -0.44 A, the magnetic field is inverted and the torque increases.

4.3 Experimental Results

An experimental setup (Figure 23) has been mounted to validate the simulations. A torque is applied to the shaft of the interface. The frame is mounted on a 6-degree of freedom force/torque sensor ATI mini40. The angular displacement of the shaft is measured with an optical encoder HEDS 5500. Data are recorded with Lab View software.

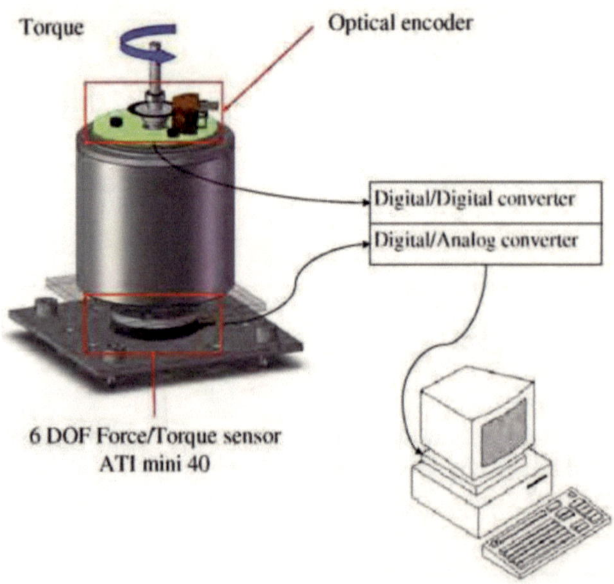

Fig. 23. Experimetal septup

Measurements have been carried out with current intensities through the coil between -1 A and 1.75 A. Results are reported in figure 1.24 and compared to the simulation.

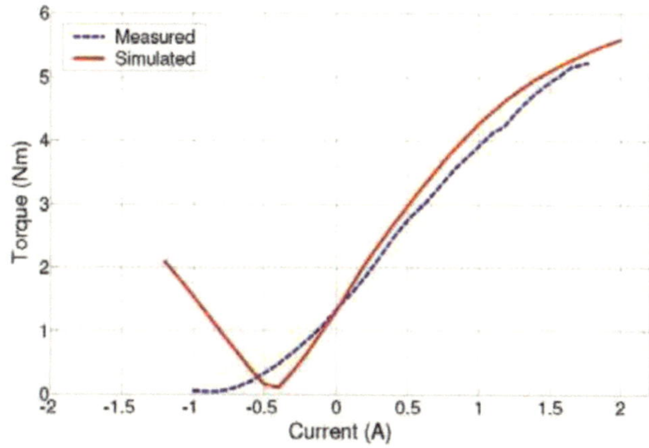

Fig. 24. Experimental results and comparison with numerical simulation

Experimental results and simulation are close for positive values of the current. The maximum deviation does not exceed 8%. Discrepancies can be observed for negative values of the current. The current intensity required to reach the minimum torque is -0.8 A which is twice more than the results of the simulations. This deviation is probably due to high values of the magnetic flux density at the surface of the magnet that might not be supported by the electromagnetic simulation. Simulating the non linear behavior of materials at high magnetic field densities should explicitly be mentioned [LORD, 2002].
It is worth noting that the minimum $T_{(min)}$ and the maximum $T_{(max)}$ torques measured are respectively 0.03Nm and 5.25Nm. The ratio between T_{max} and T_{min} is 175.

5. Conclusion and future work

Magneto-rheological fluids exhibit an apparent viscosity that varies as a function of an external magnetic field. This property can be exploited to the dissipate mechanical energy applied by the operator. Thereby, MR fluids based systems have a semi-active behavior which makes them intrinsically stable. Moreover, their low response time make them appealing for human machine interaction. Current limitations concern mainly the moving parts inertia or friction needed for proofness.
In order to reduce these limitations, a novel operating mode for magneto-rheological based systems is presented. This operating mode is based in shearing the MR fluid with a thin element (plate or cylinder). The thin element permits low inertia moving parts to increase the mechanical bandwidth while increasing the active surface thereby increasing the maximum force.
Two application design using this novel approach are presented. The piano haptic interface which uses the operating mode in a linear configuration. The design allows high force range and quick response time needed to reproduce the traditional piano dynamical behavior. A novel haptic interface for musical keyboards has been presented in this paper, based on controlled damping by a magneto-rheological fluid in its direct shear mode:

• the standard shear mode principle with relative motion between poles was replaced by a slider (made of a thin plate) moving between the static poles in order to reduce overall size and to allow the implementation on a real keyboard,
• the fluid behavior in this shear mode was explored experimentally and a new model has been derived for the behavior of the fluid below the Bingham threshold.
The analytical predictive model of the overall system was presented and implemented. The parameters influence on the global behavior has been studied and the critical parameters have been identified.
Future work will address the physical nature of the rheological model below the Bingham threshold, the improvement of the measurement of the key motion in order to improve the accuracy of the predicted behavior of the interface. A real-time control scheme will be implemented. In order to avoid the use of a force sensor on each key, we will explore an open loop control approach, using a traditional piano dynamical model and the interface model.
The Mr-drive system is a HMI for automotive applications. It exhibits high torque/volume ratio in comparison with traditional actuators and intrinsic stability. The design of a rotary interface based on magneto-rheological fluids has been presented. A thin cylinder that shears the fluid in a gap between two static poles minimizes the overall inertia. In addition, a permanent-electromagnet has been introduced to widen the output torque range. The ratio between the minimum and the maximum torque is 175. This ratio is far better than all what has been reported so far in the literature on MR brakes. Finite Element simulation and experimental measurements have been compared. The first results are promising and the application of the actuation principle to other haptic applications will be studied.

6. References

M. Ahmadian. On the application of magneto-rheological fluid technology for improving rail vehicle dynamics. In *International Mechanical Engineers Conference.* New Orleans, 2002.

A. Askenfelt and E.V. Jansson. From touch to string vibrations .1. timing in the grand piano action. *Journal of the Acoustical Society of America,* 88(l):52-63, 1990.

A. Askenfelt and E.V. Jansson. From touch to string vibrations .2. the motion of the key and hammer. *Journal of the Acoustical Society of America,* 90(5):2383^93, 1991.

AUDI. Mmi controller. URL: http://www.audi.com.

M. Badescu, C. Wampler, and C. Mavroidis. Rotary haptic knob for vehicular instrument controls. In *Proceedings of the Tenth Symposium on Haptic Interfaces for Virtual Environment and Teleoperator Systems.* Orlando, FL, March 2002.

BMW. idrive controller. URL: http://www.bmw.com.

C. Cadoz, L. Lisowski, and J. L. Florens. A modular feedback keyboard design. *Computer Music Journal,* 14 (2):47-51, 1990.

J. D. Carlson, D. M. Catanzarite, and K. A. StClair. Commercial magneto-rheological fluid devices. *International Journal of Modem Physics B,* 10(23-24):2857^2865, 1996.

E. Altenmuller D. Parlitz, T. Peschel. Assessment of dynamic finger forces in pianists: effects of training and expertise. *Journal of Biomechanics,* 31(11):1063^67, 1998.

Delphi. Hydrocarbon-based mr fluid mrf-132dg. Product Bulletin, 2005.

B. Gillespie. *Haptic Display of Systems with Changing Kinematic Constraints : The Virtual Piano Action.* PhD thesis, Stanford University, 1996.

M. R. Jolly, J. W. Bender, and J. D. Carlson. Properties and applications of commercial magnetorheological fluids. *Journal of Intelligent Material Systems and Structures,* 10(1):5-13, 1999.

LORD. *Permanent-Electromagnet Systems.* The Lord Corporation, Thomas Lord Research Center, Cary, NC (USA), 2002.

LORD. *Hydrocarbon-Based MR Fluid MRF132-DG Product Bulletin.* The Lord Corporation, Thomas Lord Research Center, Cary, NC (USA), 2005.

LORD. Designing with mr fluids. In *Lord Corporation Engineering note,* Thomas Lord Research Center, Cary, NC (USA), Dec. 1999.

J. Lozada, M. Hafez, and X. Boutillon. A novel haptic interface for musical keyboards. In *Advanced Intelligent Mechatronics.* Zurich, 2007.

D. Meisel. Key actuation systems for keyboard instruments, Nov. 2003.

J. Melli-Huber, B. Weinberg, A. Fisch, J. Nikitczuk, C. Mavroidis, and C. Wampler. Electro-rheological fluidic actuators for haptic, vehicular instrument controls. In *11th Symposium on Haptic Interfaces for Virtual Environment and Teleoperator Systems (HAPTICS'03).* Los Angeles, California, U.S.A, March 2003.

M. R. Reed and W. J. Book. Modeling and control of an improved dissipative passive haptic display. In *2004 Ieee International Conference on Robotics and Automation, Vols 1- 5, Proceedings,* Ieee International Conference on Robotics and Automation, pages 311-318. 2004.

B.H. Repp. Effects of auditory feedback deprivation on expressive piano performance. *Music Perception,* 16(4): 409-38, 1999.

13

Impedance Control of Two D.O.F. CPM Device for Elbow Joint

Shota Miyaguchi, Nobutomo Matsunaga and Shigeyasu Kawaji
Graduate School of Science and Technology, Kumamoto University
Japan

1. Introduction

Recently, new technology that supports human will be an emergent issue as population ages and birthrate declines. As human having a lot of joints utilizes many degrees of freedom (d.o.f.) for living and working, reduction or loss of d.o.f. will deteriorate the quality of life. So, it will be important to recover joint functions immediately after injury or operation. Accordingly, many studies have been performed on the development and application of orthopedic treatment or rehabilitation training devices (Furusho et al., 2005; Duong et al., 2005; Mavroidis et al., 2005; Hogan et al., 2006). In the tendency, continuous passive motion (CPM) was proposed as an orthopaedic treatment and a physiotherapy method that promotes recovery from the injuries after surgery of joints (Salter et al., 1960). The CPM is intended to accelerate the regeneration of periarticular tissues, to prevent contracture, and to correct range of motion (ROM) (Salter, 1993), and is more effective than conventional treatment method with casts (Salter, 1993; O'Driscoll et al., 2000).

There are many works on the CPM devices for the lower limbs. Particularly, the CPM devices for the knee joint have been widely used. It was reported that these devices induce quick recovery and improve the joint ROM more effectively (Tanaka, 1998; Sakaki et al., 2000; Akdogan et al., 2006). The knee joint has a simple hinged joint, which is the same structure as the CPM devices for the knee joint, so that the CPM devices can perform effective CPM by easy setting. On the other hand, there are few researches about CPM devices for the upper limbs because of the musculoskeletal complexity. Most conventional CPM devices for the elbow joint have only one d.o.f., and perform the flex-/extension with fixed forearm to avoid the complex setting of the device (Kawaji, 2006).

The human's elbow joint, which consists of a radius, ulna, humerus and biological tissues, has complex structure, and is structurally different from the conventional CPM devices for the elbow. This difference causes some problems such as fluctuation of rotation center of the flex-/extension, excessive reaction force due to the inappropriate pose of the forearm, etc. For the former, Usui et. al. have proposed an algorithm that compensates the fluctuation of the rotation center while the rehabilitation(Usui et al., 2004). But, for the latter there are few researches about the variation of the reaction force due to the forearm position i.e., the pro-/supination.

On the other hand, high safety is required for the CPM device in practical use. It is well known that the stiffness of joints increases after surgery (O'Driscoll, 2000; Kim, 2005). So, the reaction force near the end of the ROM increases excessively due to the large stiffness, and this excessive reaction force may aggravate joint injury. Because of the excessive reaction force due to the inappropriate pose of the forearm, the conventional CPM devices for the elbow joint have lower therapeutic effect than those for the knee joint, so practical use of the CPM device for the elbow joint is fewer than that for the knee joint.

For the problem that the reaction force increases excessively, the method with pneumatic actuators (Noritsugu et al., 1997; Tsgarakis et al., 2003) and impedance control for decrease of load to joint (Moughamir, 2005) were proposed. Although these actuators and control method can realize the suppression of the reaction force, but narrow the range of the flex-/extension because of soft control. Narrowing the range of the flex-/extension decreases the effectiveness to correct the joint ROM. Therefore, to correct the ROM with CPM devices, it is needed to suppress the excessive reaction force without narrowing the range of the flex-/extension. In the upper limbs, the reaction force increases excessively due to the inappropriate pose of forearm. Therefore, control of the pro-/supination is effective to decrease the reaction force.

In this paper we propose a new impedance control scheme of two d.o.f. CPM device for the elbow joint, which performs not only the flex-/extension but also the pro-/supination in order to suppress the excessive reaction force without narrowing the range of the flex-/extension. First, the relation between the pro-/supination and flex-/extension is confirmed, and it is clarified that adequate reference trajectory of the pro-/supination is required to suppress the excessive reaction force. Next, a new impedance control is designed for the reference trajectory measured from normal subject. Finally, some experimental results with simulated patient are shown, from which the effectiveness of the proposed scheme is evaluated.

2. Issues of Conventional CPM Devices

The upper limb, which consists of a radius, ulna, humerus and biological tissues, has the complex viscoelastic property. As a result of the complex viscoelasticity, the forearm is naturally pro-/supinated as the elbow is flexed/extended. Also, the reaction force of the elbow changes due to the pro-/supination. For the upper limb with the complex viscoelasticity, CPM devices with one d.o.f., which perform the flex-/extension with fixed forearm, have been used in clinical practice. As an example, a CPM device in Fig.1(a) which was developed in our laboratory has fixed gripper that constrains the pro-/supination. The constraint of the pro-/supination involves the problem that the reaction force increases excessively near the end of the ROM because the natural pro-/supination is prevented.

In order to cope with this problem, a CPM device with a new gripper was developed in our laboratory as shown in Fig.1(b) (Kawaji, 2006). The gripper of this CPM device can rotate around pro-/supination axis freely, and consequently realize the natural pro-/supination due to the flex-/extension in normal subject. But, in the case of the patient with contracture, the natural pro-/supination is prevented. The prevention of the natural pro-/supination causes the excessive reaction force. Thus, it will be difficult to expand the ROM effectively. For this difficulty, it will be effective to control the pro-/supination motion actively. Simultaneously, it is required for the CPM devices to take safety strategy to the mechanism and the control algorithm, because the elbow might be injured at the end of ROM in the case that forearm is moved forcibly.

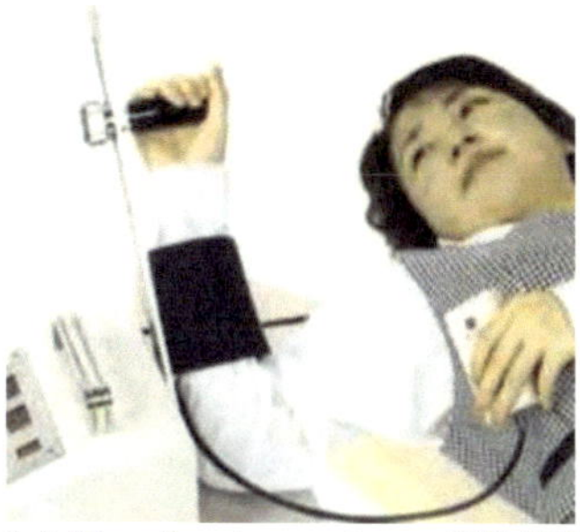

(a) Fixed gripper type (KM702)

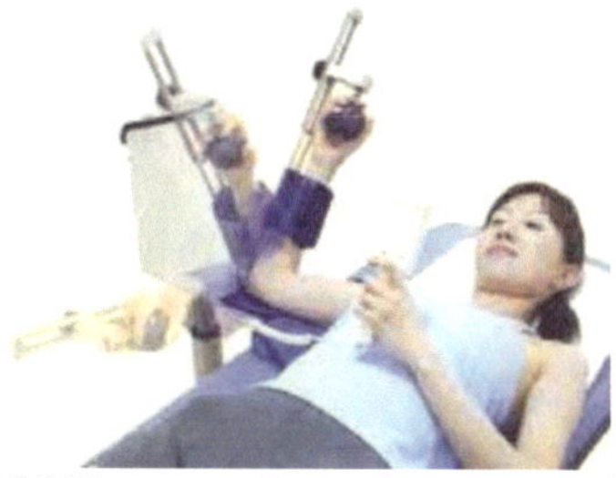

(b) Free rotation gripper type (AH706)

Fig. 1. Developed CPM devices

3. Analysis of Pro-/Supination Angle in CPM

In order to investigate the characteristic of the pro-/supination, the pro-/supination angle and the reaction force due to the flex-/extension are measured using the free rotation gripper. As we aim to clarify the natural pro-/supination due to the flex-/extension, the subject for the measurement is a normal subject.

3.1 Analysis of pro-/supination with free rotation gripper

At first, we measure in the case of the free rotation gripper. A CPM device for the measurement is shown in Fig.2. The device is usable as the CPM device with both the free and fixed pro-/supination by replacement of the mechanical parts. The wrist is constrained by the belt attached to the CPM device in order to suppress the motion of the wrist. A force sensor and encoders are attached in order to measure the norm of the reaction force F [N],

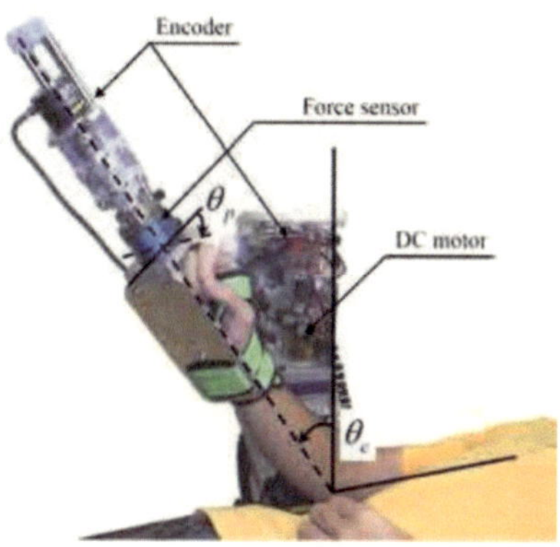

Fig. 2. CPM device for measurement

the flex-/extension angle θ_e [deg] and the pro-/supination angle θ_p [deg].

In the measurement experiment, the initial angle of the upper limb was set as follows: the elbow is flexed to a right angle, the angle between the upper arm and the trunk is 10[deg]

and the forearm is at intermediate position between the pronated position and the supinated position. The norm of force sensor is set as 0[N] at the initial position. The velocity of the CPM device's arm is controlled with constant velocity 0.75[rpm]. The measurement time was 5 minutes.
The relationship between pro-/supination angle θ_p and flex-/extension angle θ_e are shown in Fig.3. And, the relationship between θ_e, θ_p and reaction force F is shown in Fig.4.

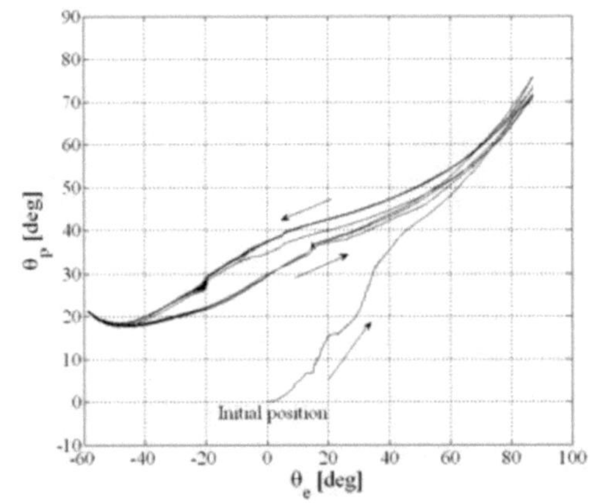

Fig. 3. Trajectory of pronation angle with free rotation gripper

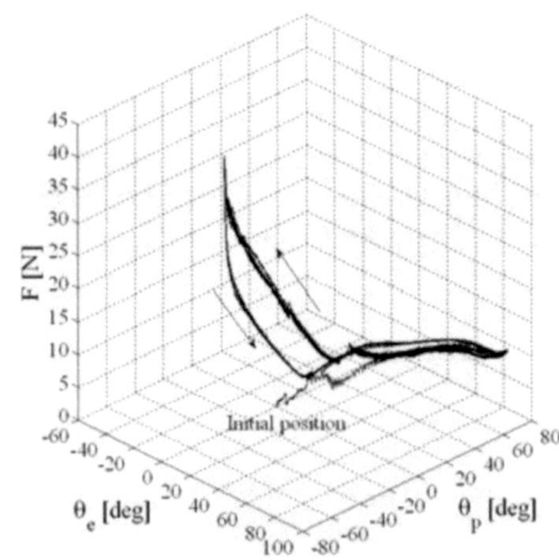

Fig. 4. Reaction force with free rotation gripper

In Fig.3, It is confirmed that the forearm is naturally pronated and supinated as the elbow is extended and flexed. Also, it is observed that the trajectory of the pro-/supination is converged to a certain trajectory. In Fig.4, the reaction force F varies due to the movement of the upper limb, and increases excessively near the end of the ROM where the elbow joint is flexed.

3.2 Analysis of pro-/supination with fixed gripper

Secondly, the reaction force F in the case of the fixed pro-/supination is measured to evaluate the type of the CPM device with fixed gripper as shown in Fig.1(a). In the measurement, θ_p is fixed at every 10[deg] between −80[deg] to +80[deg]. The measuremental result is shown in Fig.5.

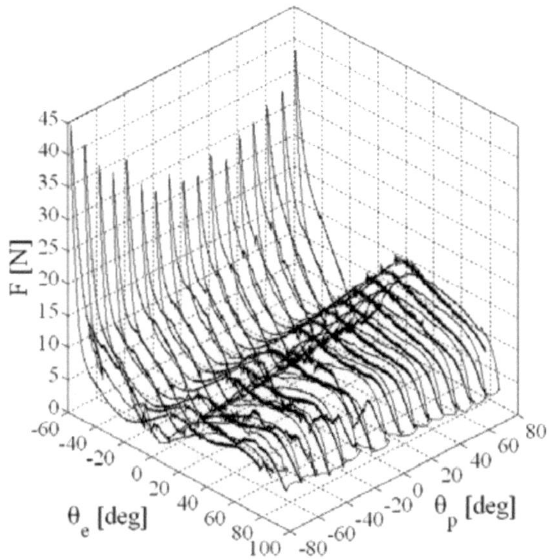

Fig. 5. Reaction force with fixed gripper

From the figure, it is observed that F increases at the flexed position ($\theta_e = -60$[deg]), and the reaction force at the flexed position is excessively enlarged if the pro-/supination angle is not set appropriately. Such the excessive reaction force might injure the elbow joint. And θ_p that minimizes the reaction force at the flexed position in Fig.5 is approximately same as θ_p at the flexed position in Fig.4. It is clarified that the pro-/supination of the normal subject in the case of free pro-/supination suppresses the reaction force at the flexed position.

3.2 Approximation of pro-/supination trajectory

As it is difficult for patient to pro-/supinate as normal subject, the pro-/supination trajectory of the patient using the free rotation gripper deviates from that of the normal subject. This causes that the excessive reaction force is applied to the patient. So, it is expected that an ideal CPM that suppresses the reaction force without narrowing the range of flex-/extension can be realized when an adequate reference pro-/supination angle trajectory is given. Since θ_p of the normal subject using the free rotation gripper suppresses reaction force at flexed position, in the following we regard the pro-/supination trajectory shown in Fig.7 as an adequate trajectory to suppress the reaction force.

Because the normal subject's trajectory is a certain trajectory with respect to the flex-/extension, we express the reference trajectory of the pro-/supination as the function of θ_e. The non-linear trajectory in Fig.3 is approximated with following polynomials,

$$\theta_r(\theta_e) = \sum_{i=0}^{N} \alpha_i \theta_e^i \tag{1}$$

where α_i is the coefficients determined by least squares method and we set $N = 7$ so as to minimize the error sufficiently. The approximated trajectory θ_r is shown in Fig.6. From the figure, it is confirmed that the trajectory of normal subject can be approximated well by eq.(1).

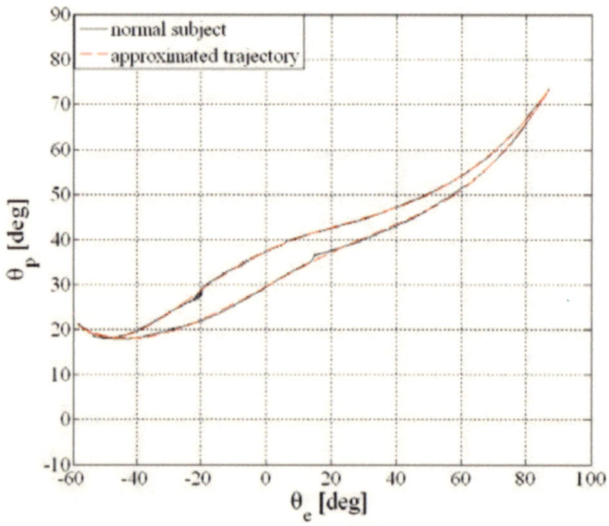

Fig. 6. Approximated pronation trajectory

4. Two d.o.f. CPM device

For the control of the pro-/supination and flex-/extension, we developed the two d.o.f. CPM device for the elbow joint as shown in Fig.7. This CPM device controls and measures θ_p and θ_e directly by DC motors with rotary encoder. A force sensor is attached at the end of the arm which measures the reaction force F.

The velocity of the arm for the flex-/extension is controlled with slow constant velocity 0.75[rpm]. The range of the flex-/extension of the CPM device is controlled by switch box operation of patient according to medical condition. It realizes treatment or rehabilitation for the flex-/extension suited to the patient (Kawaji et al., 2006).

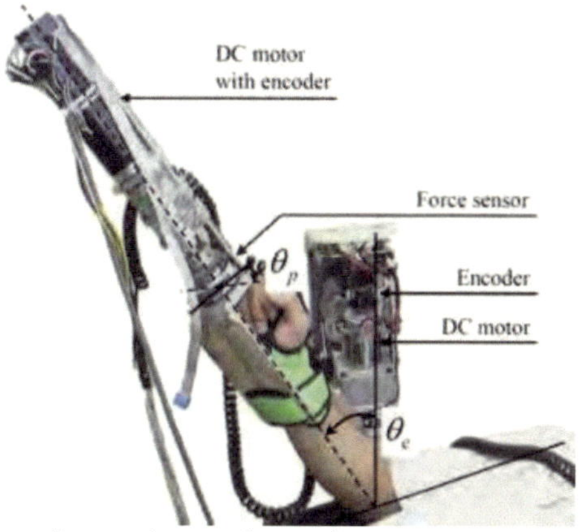

Fig. 7. CPM device with flex-/extension and pro-/supination

5. Impedance Control Scheme

It is obvious that the pro-/supination trajectory of the patient who has contracture is different from that of normal subject because the patient's pro-/supination that consists of the movement of the ulna and radius is restrained. Thereby, with the CPM device used in clinical treatment, automatic control of pro-/supination function will be required to follow the reference trajectory approximated from the normal subject. To control the pro-/supination angle during the flex-/extension, we will propose the control scheme shown in Fig. 8.

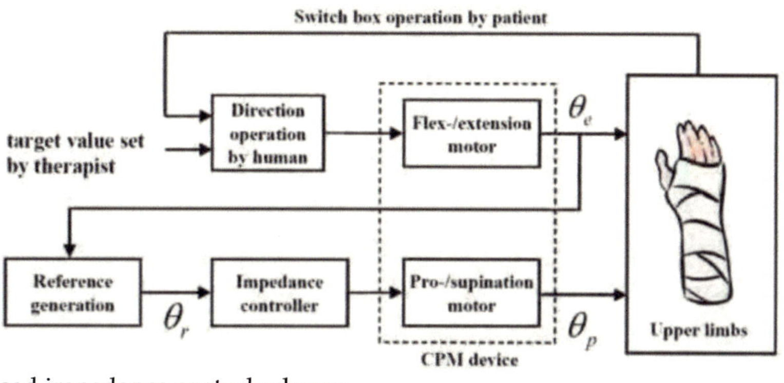

Fig. 8. Proposed impedance control scheme

In this scheme, using the extension angle, the reference angle of the pro-/supination is generated by eq.(1). To follow the reference trajectory of the pro-/supination while CPM movement, next two points should be considered.

A) The CPM device is controlled to follow the reference trajectory smoothly so as not to generate the large force and velocity.
B) The assistance force to follow the reference must be small near the end of the ROM.

To realize the flexible control of CPM device, active control methods such as impedance control or force control etc. have been studied (Gorinevsky, 1997). However, the force control is not suited for medical use since the arm of CPM device make sudden movements in the case that the human's arm comes free from the arm of CPM device. For this, we introduce an impedance control in this paper.
When the pro-/supination angle of the patient is close to that of the normal subject, the patient's condition is similar to the normal subject's one. In this case, the generated assistance torque of the CPM device should be small. When the pro-/supination angle of the patient deviates from that of the normal subject, it will be effective to make the assistance torque large.
For the conditions A) and B), let the reference impedance at the wrist be given as

$$\tau_p = \mathrm{M_I}\ddot{e}_p + \alpha \mathrm{D_I}\dot{e}_p + \alpha K_I e_p \tag{2}$$

$$e_p = \theta_r(\theta_e) - \theta_p \tag{3}$$

$$: \alpha = \frac{|e_p|}{\hat{e}_{p\max}} \tag{4}$$

where θ_r is the reference of normal subject in eq.(1), τ_p is the pronation torque, $\hat{e}_{p\max}$ is the maximum error and $\mathrm{M_I}$, $\mathrm{D_I}$ and K_I are the inertia, damping and stiffness factor, respectively. The stiffness factor is determined as follow so that the assist torque becomes the maximum value $\tau_{p\max}$ at $\alpha = 1$.

$$K_I = \frac{\tau_{p\max}}{\hat{e}_{p\max}} \tag{5}$$

For the condition A), impedance control is applied within the area $|e_p| \le \hat{e}_{p\max}$. In eq.(2) viscoelasticity terms are nonlinear with respect to $\dot{e}_p$ and e_p. So, $\alpha \mathrm{D_I}$ and αK_I become small in the case that θ_p gets close to θ_r, and the equation (2) becomes inertia system when $e_p = \dot{e}_p = 0$. When $|e_p|$ is large, the CPM device moves to reference slowly since the viscosity term is sufficiently large.
For the condition B), viscoelasticity terms should be changed due to the flex-/extension angle. So let maximum error area $\hat{e}_{p\max}$ be modified by θ_e based on the following equation.

$$\hat{e}_{p\max} = \exp(-\mathrm{W_F}\theta_e) + \exp(\mathrm{W_E}\theta_e) \tag{6}$$

where $\mathrm{W_F}$ and $\mathrm{W_E}$ are setting parameters to design the maximum error area at the flexed and extended position, respectively. From the viewpoint of control system design, the impedance characteristics can be adjusted with $\mathrm{W_F}$, $\mathrm{W_E}$ in clinical use. Fig.9 shows the reference trajectory θ_r and $\hat{e}_{p\max}$ when $(\mathrm{W_F}, \mathrm{W_E}) = (0.04, 0.03)$. The width of hatching area indicates the maximum error area $\hat{e}_{p\max}$. From the figure, $\hat{e}_{p\max}$ is enlarged near the end of the ROM. It realizes small assistance torque because K_I becomes small with eq.(5). So, the condition B) will be satisfied.

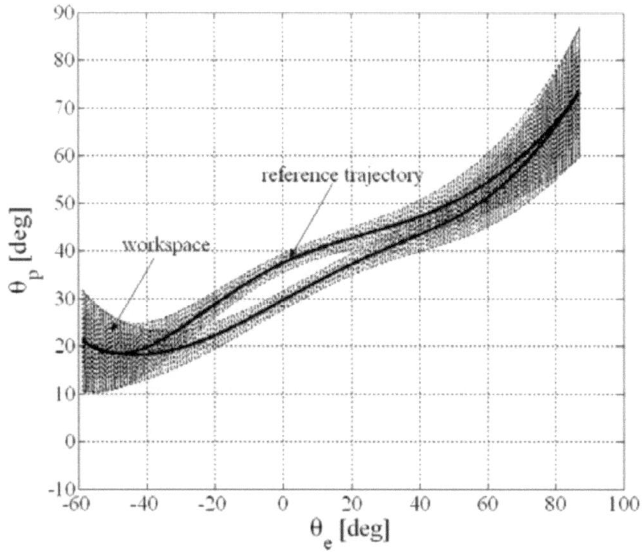

Fig. 9. Error space in pro-/supination plane

From eq.(2), (3) and (6), τ_u can be expressed as follows,

$$\tau_u = (\mathrm{M_c} + \mathrm{M_I})\ddot{e}_p + (\mathrm{D_c} + \alpha \mathrm{D_I})\dot{e}_p + \alpha K_I e_p - \mathrm{M_I}\ddot{\theta}_r - \alpha \mathrm{D_I}\theta_r - \alpha K_I \theta_r \tag{7}$$

Also, the parameters $\mathrm{M_I}$, $\mathrm{D_I}$ are adjusted by trial and error to prevent oscillatory response.

6. Experimental Result

We experiment in order to ascertain whether the proposed control scheme suppresses the reaction force without narrowing. In the experiments, the reaction force of subject whose pro-/supination is controlled is compared to that of the subject whose pro-/supination is free in the same range of the flex-/extension.

The subject for the experiments wears the bracing device (Armbrace, made by Bledsoe) as shown in Fig.10. The working range of the bracing device can be set mechanically. If the working range of the bracing device is set smaller than one of the CPM device, the bracing device deforms near the end of the ROM, and emulates the large stiffness of the contracture. The pro-/supination is constrained by the belt of the bracing device. The bracing device can emulate restriction of the elbow joint due to the contracture although the subject is normal. Thus, we regard the subject with the bracing device as a simulated patient whose condition is similar to arthrogryposis.

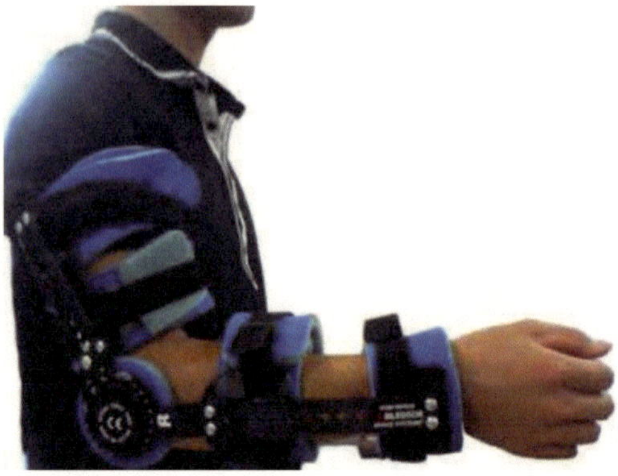

Fig. 10. Bracing device

In the experiments, the range of the flex-/extension of the CPM device was determined as $-60[\text{deg}] \leq \theta_e \leq 90[\text{deg}]$. The working area of the bracing device was set as $-45[\text{deg}] \leq \theta_e \leq 90[\text{deg}]$ so as to be smaller than one of the CPM device. $\tau_{p\max}$ was set as 3.0[Nm] based on the maximum torque examined by normal subject, and the impedance parameters were set as $(M_I, D_I) = (0.004, 0.8)$. The parameters of the maximum error area were adjusted as $(W_F, W_E) = (0.04, 0.03)$.

Experimental results are shown in Fig.11 and Fig.12. In Fig.11, the reference trajectory of the normal subject, the trajectory of the patient with the free rotation gripper and the experimental result using the proposed impedance control are represented by solid line, dotted line and heavy line, respectively. From the figure, the variation of the prosupination angle of the simulated patient is smaller than that of the normal subject because the arm is restricted by the bracing device. On the other hand, using the proposed impedance controller, the experimental trajectory is close to that of the normal subject.

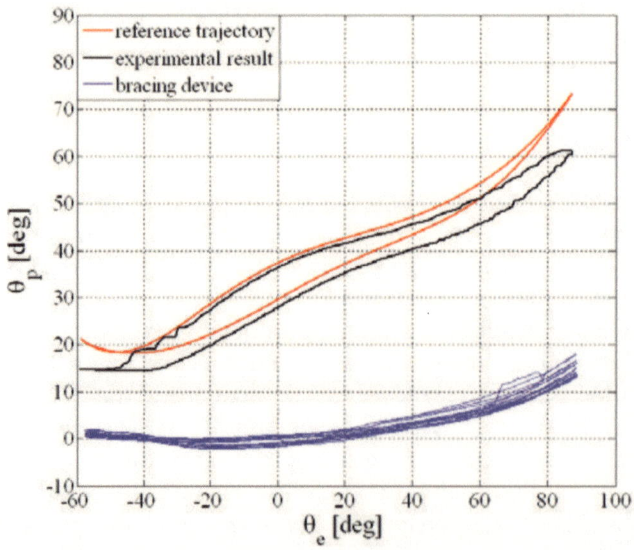

Fig. 11. Trajectory of pronation angle in CPM using the impedance control

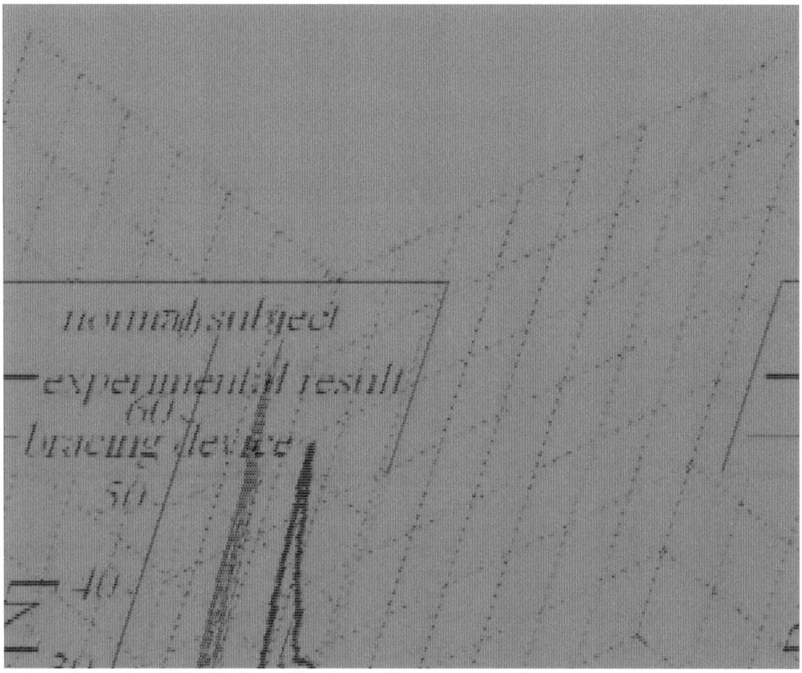

Fig. 12. Reaction force in CPM using the impedance control

Fig.12 shows the reaction force. The reaction force of the simulated patient increases excessively comparing with that of the normal subject. On the contrary, the reaction force of the simulated patient with the proposed control scheme comes close to that of the normal subjects and is suppressed near the flexed position.
From these experimental results, it is clarified that the proposed scheme is effective to realize the ideal CPM that the excessive reaction force is suppressed without narrowing the range of the flex-/extension.

7. Conclusion

In this paper, we have proposed the novel impedance control scheme of the CPM device for the elbow joint which controls the pro-/supination angles to follow the reference of the normal subject. The effectiveness was validated with the simulated patient wearing the bracing device.
From the experimental results for the subject with bracing device, it was clarified that the proposed scheme is effective to suppress the excessive reaction force. This will realize the CPM which does not burden the patient. Furthermore, the characteristics of the proposed control system can be adjusted intuitively by W_F, W_E.
In the proposed scheme, the trajectory of normal subject is applied as the reference trajectory. However, the parameters such as arm length, shape and dynamic characteristics of the joint etc. are different individually. So, the reference trajectory generation will be an important problem in practice application. Also, impedance control for flex-/extension, motion

analysis and evaluation of the proposed scheme by doctors or physical therapists remain as future investigations.

8. Acknowledgment

This work was supported by Grants-in-Aid for Scientific Research (No. 16360213) from the Ministry of Education, Science, Sports and Culture of Japan.

9. References

Akdogan, E., Tacgin, E., & Adli, M.A. (2006). Intelligent Control of a robot Manipulator for Knee Rehabilitation, *Proceedings of the 5th International Symposium on Intelligent Manufacturing Systems,* pp.695-703, Turkey, May 2006

Duong, M.D.; Terashima, K.; Imamura, T.; Master-Slave System with Teleoperation for Rehabilitation, *Proceedings of the 16th IFAC World Congress,* Th-A01-TP/9, ISBN: 008045108X, Prague, July 2005, Elsevier

Furusho, J.; Koyanagi, K.; Imada, Y.; Fujii, Y.; Nakanishi, K.; Domen, K.; Miyakoshi, K.; Ryu, U.; Takenaka, S. & Inoue, A.; A 3-D Rehabilitation System for Upper limbs Developed 5-Year NEDO Project and its clinical Testing, *Proceedings of the Rehabilitation Robotics,* pp.53-56, June, 2005, IEEE Press, Chicago

Gorinevsky, D.M.; Formalsky, A.M.; Schneider, A.Y. (1997) *Force Control of Robotics Systems,* CRC Press, ISBN: 978-0849326714

Hogan, N.; Krebs, H.I.; Rohrer, B.; Palazzolo, J.J.; Dipietro, L.; Fasoli, S.E.; Stein, J.; Hughes, R.; Frontera, W.R.; Lynch, D. & Volpe, B.T. (2006). Motions or Muscle? Some Behavioral Factors Underlying Robotic Assistance of Motor Recovery, *Journal of Rehabilitation Research and Development,* Vol.43, No.5, (Sep., 2006), pp.605-618, ISSN: 0748-7711

Kawaji, S. & Sakurai, I. (2006). R&D on User-friendly Passive Motion Device for Upper Limbs, *KINZOKU Materials Science and Technology,* Vol.76, No.1, 2006, pp.37-40, ISSN: 0368-6337

Kim, P.D., Grafe, M.W., Rosenwasser, M.P. (2005). Elbow Stiffness: Etiology, Treatment, and Results, *Journal of the American Society for Surgery of the Hand,* Vol.5, No.4, (Nov., 2005), pp.209-216, ISSN: 1531-0914

Mavroidis, C.; Nikitczuk, J.; Weinberg, B.; Danaher, G.; Jensen, K.; Pelletier, P.; Prugnarola, J.; Stuart, R.; Arango, R.; Leahey, M.; Pavone, R.; Provo, A.; & Yasevac, D. (2005). Smart Portable rehabilitation device, *Journal of Neuro-Engineering and Rehabilitation,* Vol.2, No.18, (July, 2005), pp.1-5, ISSN: 1743-0003

Moughamir, S.; Deneve, A.; Zaytoon, J.; Afilal, L. (2005). Hybrid Force/Impedance Control for The Robotized Rehabilitation of The Upper Limbs *Proceedings of 16th IFAC World Congress,* pp.Fr-M19-TO/6, ISBN: 0-08-045108-X, Czech Republic, (July, 2005), Elsevier, Praha

Noritsugu, T.; Tanaka, T. (1997). Application of Rubber Artificial Muscle Manipulator as a Rehabilitation Robot, *IEEE/ASME Transaction on Mechatronics,* Vol.2, No.4, (Dec, 1997), pp.259 - 267, ISSN: 1083-4435

O'Driscoll, S.W., Giori, N.J. (2000) Continuous Passive Motion (CPM): Theory and Principles of Clinical Application, *Journal of Rehabilitation Research and Development*, Vol.37, No.2, (April, 2000) pp.179-188, ISSN: 0748-7711

Sakaki, T.; (1999). TEM: Therapeutic Exercise Machine for Recovering Walking Functions of Stroke Patients, *Industrial Robot: An International Journal*, Vol.26, No.6, (1999), pp.183-186, ISNN: 0143-991X

Salter, R.B. & Field, P. (1960) The effects of continuous compression on living articular cartilage : An Experimental Investigation. *The journal of bone and joint surgery*, Vol.42-A, No.1, (Jan. 1960) pp.31-90, 1960, ISSN: 00219355

Salter, R.B. (1993). *Continuous Passive Motion(CPM)*, Williams & Wilkins, ISBN: 0683074970

Tanaka, N.; Okajima, Y.; Taki, M.; Uchida, S.; Tomita, Y.; Horiuchi, T.; Sakaki, T. & Kimura, A. (1998) Effect of Continuous Range of Motion Exercise on Passive Resistive Joint Torque, *The Japanese Journal of Rehabilitation Medicine*, Vol.35, No.7, (May, 1998), pp.491-495, ISSN: 0034-351X

Tsgarakis, N.G.; Cardwell, D.G. (2003). Development and Control of a Soft Actuated' Exoskeleton for Use in Physiotherapy and Training, *Autonomus Robotics*, Vol.15, No.1, pp.21-33, ISSN: 0929-5593

Usui, T.; Kato, N.; Nomura, Y.; & Matsui, H. (2004). Study on Assistive Robot Manipulator for Arm Kinesitherapeutic Exercise, *Transaction of Japan Society of Mechanical Engineers, Part C*, Vol.70, No.694, (June, 2004), pp.1781-1794, ISSN: 0387-5024

14

A Far Sign Recognition by Applying Super-Resolution to Extracted Regions from Successive Frames

Hitoshi Yamauchi[1], Atsuhiro Kojima[2] and Takao Miyamoto[2]
[1]*Okayama Prefectural University*
[2] *Osaka Prefecture University*
Japan

1. Introduction

Many researches are proposed for realizing Intelligent Transport Systems (ITS). As a driving assistant system, a part of ITS, many techniques are proposed that extracting and recognizing road signs placed at the front of a car. At present, road-to-vehicle communication systems using dedicated short range communications (DSRC), currently used for electronic toll collection system (ETC) as one of applications, are in experimental phase. By using this communication, road sign information are able to be notified to drivers. Then, informing way of road signs may change from by watching into by such communication systems. Changing informing way by equipping infrastructure may be achieved quickly in urban area. However, long period is needed for completing infrastructure equipment spreading into rural area. Moreover, it is hardly expected that road signs placed at roadside will be abolished and removed after completing infrastructure equipment, while cars are controlled by human being.

For realizing road sign recognition and alerting systems, two techniques are needed. One is a technique for extracting regions including road signs from image sequences. The other is for identifying the type of those signs. In previous techniques, road signs are extracted using genetic algorithms (Uchiyama et al., 1998), active networks (Yabuki et al., 1998), and so on. Since these techniques require too much time to process, it is not always applicable to in-vehicle systems in which real-time processing is necessary. To overcome this problem, fast detection techniques are proposed using specific colors of road signs (Matsuura et al., 2002; Ohara et al., 2002; Zin & Hama, 2005).

As recognition techniques, template matching based technique (Gavrila, 1998) is proposed. However, template matching requires large amount of resize and rotation processing, then processing time will be long. As another approach, feature extraction based on frequency analysis, wavelet transform for example, are difficult to fast processing. To achieve fast processing, a contour vector matching technique is proposed (Yamauchi & Takahashi, 2003). It extracts outlines of the extracted road sign as vectors, and matches the vectors with

reference signs' vectors. By using outline vectors, the technique shows robustness against variation of extraction size and slight rotations.
Target images for these extraction and recognition techniques are still images in a scene. Then, these techniques are applied to frame by frame independently. This means that recognition results may be failed at the frame image which has illumination variances or occlusions. These failed results may cause drivers to be thrown into confusion. Against still images, successive frames in a scene have some correlations each other. By using these correlations, more effective processing can be done. Moreover, this can also remove occasional occlusion.
By the way, a technique of generating pseudo high resolution image by analyzing multiple still images of a same target is proposed (Park et al., 2003). This technique is called as super-resolution. This is used for capturing precise images of objects for digital museum or medical imaging. These targets may require high resolution images over the camera capacity.
In case of road sign detections, a targeted sign will appear in successive frames. And then, the size of the sign region will become larges and high resolution in latter frames. By treating extracted sign images as low resolution images in super-resolution technique, a pseudo high resolution image can be generated. Though an extraction image from a single frame does not have enough quality, early recognition can be done by combining number of successive frames.
In this chapter, a robust road sign recognition technique by applying super-resolution to extracted regions from successive frames is proposed. This technique consists of three steps: 1) searching and tracking road sign, 2) applying super-resolution technique to extracted sign regions, and 3) identifying type of the sign.
In the following, proposed technique is explained in Section 2. Next, the experimental results are shown in Section 3, and the considerations from the experimental results are in Section 4. Finally, the conclusion is in Section 5.

2. Recognition by Applying Super-Resolution to Extracted Regions from Successive Frames

There are many techniques for extracting and recognizing road signs from a frame image. Their target image is single frame, namely, a still image. This means that it is difficult to recognize because of low resolution when the distance to target sign is long yet. Thus, recognition process cannot be applied until the distance becomes short and the resolution is enough for recognition.
In case of road signs extracted from images taken by an in-vehicle camera, the extracted sign is placed continuously on the successive frames. Thus, plural images of a sign can be obtained by applying tracking process. These plural images may generate a pseudo high resolution image by using super-resolution processing, and the generated high resolution image may lead into high accuracy recognition.
The proposed technique is organized from 4 steps, 1) initial extraction of a road sign, 2) tracking the road sign, 3) super-resolution process, and 4) recognizing the super-resolution image. Details of each process are explained in the following.

2.1 Initial Extraction of a Sign Region

To track a sign, it is needed to extract an initial region. In this proposition, extraction of a faraway road sign is needed because early recognition is required. In this initial extraction, high resolution image is not yet necessary because latter super-resolution process may improve the resolution. At this moment, it is enough to judge by only sign shape and colors. For example, in case of restriction signs in Japan, their shapes must be circular and they have red ring at the outer edge (refer to Fig. 1 (a)). For the other example, in case of caution signs, their shapes must be rectangular and their inner area is almost filled by yellow (refer to Fig. 1 (b)). An extraction technique of such kind is presented in (Matsuura et al., 2002).

(a) An example of restriction signs (speed limitation to 50 km/h).

(b) An example of caution signs (existence of left curb ahead).

Fig. 1. Examples of road signs in Japan.

2.2 Tracking of the Sign Region

The proposition needs plural images that include a same road sign. These images are obtained by tracking from the initial extracted road sign region. For successive frames of a video image, many object tracking methods are proposed, CONDENSATION (Isard & Blake, 1998) is one of them. CONDENSATION selects plural candidates which are distributed just before tracking. Then, new candidates are generated from previous ones, obeying normal distribution with plausibility weights. Finally, a candidate which has highest fitness is selected from these new candidates as the tracking result at current frame.

To apply CONDENSATION, an evaluation function calculating fitness has to be designed properly. For road sign tracking, a function based on sign shape, color composition may work properly. For a red circular sign, for example, the parameters to be estimated by CONDENSATION can be selected as three-tuple, the center coordinates (x, y) and the radius r of the circle. In the following, let k as the number of parameter sets generated by CONDENSATION, and let (x_k, y_k, r_k) as parameter values of k -th candidate.

Then, the sign shape will be circular or slight ellipse, and red belt is placed at the outer edge of the circle. In addition, difference of the parameters between tracking results of next-by-next frame will be in small in most cases. Thus, these parameters are able to be employed for the evaluation function.

Let $p_R(x, y)$ as red component value of a pixel at (x, y), N as the number of division on circumference, W_{edge} as the weight for the edge strength. On this condition, the evaluation value E_k^{circle} is denoted as follows.

$$
\begin{aligned}
E_k^{circle} &= \sum_{i=0}^{N-1} \left\{ W_{edge} p_R\left(x_i^{in}, y_i^{in}\right) - p_R\left(x_i^{out}, y_i^{out}\right) \right\}, \\
x_i^{in} &= r_k \cos(2\pi i / N) + x_k, \\
y_i^{in} &= r_k \sin(2\pi i / N) + y_k, \\
x_i^{out} &= (r_k + 1)\cos(2\pi i / N) + x_k, \\
y_i^{out} &= (r_k + 1)\sin(2\pi i / N) + y_k
\end{aligned}
\tag{1}
$$

By this equation, a parameter is fit to the outer edge of a red circle when E_k^{circle} is large value. The second evaluator E_k^{color} denoted as eq. 2 represents similarity of consistent color distribution between previous extraction result and current extraction. Where, let D_k and D^* as circular regions of k -th parameter of current frame and previous frame, respectively. In addition, let $|D|$ as the number of pixels in region D, and $P_Y(x,y)$, $P_B(x,y)$ as yellow, blue component values at coordinate (x,y), respectively.

$$
\begin{aligned}
E_k^{color} = & \sum_{(x,y)\in D_k} \frac{p_R(x,y)}{|D_k|} - \sum_{(x,y)\in D^*} \frac{p_R(x,y)}{|D^*|} \\
& + \sum_{(x,y)\in D_k} \frac{p_Y(x,y)}{|D_k|} - \sum_{(x,y)\in D^*} \frac{p_Y(x,y)}{|D^*|} \\
& + \sum_{(x,y)\in D_k} \frac{p_B(x,y)}{|D_k|} - \sum_{(x,y)\in D^*} \frac{p_B(x,y)}{|D^*|}
\end{aligned}
\tag{2}
$$

The third evaluator $E_k^{distance}$ denoted as eq. 3 represents difference between current parameter and parameter (x^*, y^*, r^*) which is previous tracking results.

$$
E_k^{distance} = W_{pos}\sqrt{(x^* - x_k)^2 + (y^* - y_k)^2} + W_{rad}\sqrt{(r^* - r_k)^2} \tag{3}
$$

Where, let W_{pos} and W_{rad} are weight constants for position variance and radius variance, respectively.

These three evaluation values are weighted properly, and totally evaluated. Then, the parameter of the highest evaluation is selected as a tracking result for each frame. For the other signs, such as blue circular, yellow rectangular signs, tracking process may be achieved by defining similar evaluators.

2.3 Application of Super-Resolution to Tracked Regions

A pixel image is achieved from a receptor of an image sensor by gathering rays through a lens. Then, the observed ray on each receptor becomes a mixed ray from acceptable area at the ratio specified by ray gathering characteristics. On the assumption that the target is a high resolution image, each pixel of the observed image is got by convolution of the high resolution image and a filter function which is represented as ray gathering characteristics. The super-resolution process is a technique to recover the original high resolution image from some observed low resolution images. Some super-resolution techniques are proposed, a local iterative calculation method is presented in (Irani & Peleg, 1991), which is one of the standard techniques of super-resolution. In the following, the outline of this method is described.

Let $f(x,y)$ as the true image to be calculated, and $c_i(u,v)$ as the observed low resolution images. And let $h_c(x,y)$ as a point spread function (PSF) which affects observed image. The super-resolution technique is done by iteration of 4 steps described as follows.

1. Let $f^{t=0}(x,y)$ as a first solution of an estimated high resolution image. In this paper, the latest observed image $c_{last}(x,y)$ is normalized to the regular size and adopted as solution at this moment.
2. Calculate simulated observed image $c_i^t(u,v)$ by applying PSF $h_c(x,y)$ to an estimated high resolution image $f^t(x,y)$ which is a last solution (refer to eq. 4).

$$c_i^t(u,v) = \sum_{(x,y)} f^t(x,y) h_c(x_u - x, y_v - y) \tag{4}$$

   Where, coordinates (x_u, y_v) on an estimated high resolution image is corresponding to coordinates (u,v) on a simulated observed image.
3. Calculate a sum of squared error between a simulated observed image $c_i^t(u,v)$ and a real observed image $c_i(u,v)$ (refer to eq. 5).

$$E^t = \sqrt{\sum_i \sum_{(u_i,v_i)} \left\{c_i(u_i,v_i) - c_i^t(u_i,v_i)\right\}^2} \tag{5}$$

   If E^t is under a threshold T_e, quit this iteration process.
4. Modify an estimated high resolution image $f^t(x,y)$ so that the error E^t may become small. This modification result $f^{t+1}(x,y)$ is a last estimated high resolution image. Then, return to step 2.
   In this step, a modification result $f^{t+1}(x,y)$ is calculated as eq. 6, 7, 8. Where, let coordinates $(u_{i,x}, v_{i,y})$ on a real observed image $c_i(u,v)$ as pixels observing coordinates (x,y) on the estimated high resolution image $f^t(x,y)$.

$$\Delta e_i^t(u,v) = c_i(u_{i,x}, v_{i,y}) - c_i^t(u_{i,x}, v_{i,y}) \tag{6}$$

$$\Delta E_i^t = \sum_{(u_{i,x}, v_{i,y})} \Delta e_i^t(u_{i,x}, v_{i,y}) \frac{h_c(x - x_{u_i}, y - y_{v_i})^2}{C} \tag{7}$$

$$f^{t+1}(x,y) = f^{t}(x,y) + \sum_{i} \Delta E_i^t \tag{8}$$

In the process of super-resolution, each observed image must be registered in order to make correspondence between a pixel on a real observed image (u_x, v_y) and a pixel on a high resolution image (x, y). In this chapter, an ordinal template matching technique is used after resizing extracted images from observed images to a same size.
Note that extracted images by tracking are varying size related with the distance between the camera and the sign. This means that resolution of the extracted image grows. Thus, larger extraction size may mean that the accuracy is higher.
In applying super-resolution process to road sign recognition techniques, each extraction size is varyingly according to relative distance. This leads addition in eq. 8 may be varyingly. Thus, the accuracy varying may reflect the equation results.

2.4 Recognition of Super-Resolution Image

As the final process, a recognition process is applied to a generated super-resolution image. For road sign recognition process, it can employ an already proposed technique, such as template matching which is widely used in the image processing area, and so on.

3. Confirming Experiments

Some experiments are done for confirmation of the proposed technique.

3.1 Experimental Conditions and Preliminary Experiment

To confirm effectiveness of the proposed technique, some experiments were done. As input images, a camera SONY DFW-VL500 with 1/3 inch CCD sensor and vary-angle lens 5.5 to 64mm is used. The camera was placed on dashboard of a car. Images are taken at 15 fps in fine or cloudy daytime. The image sizes of all images are 640×480 pixels. Some of these images include occasionally lighting varying as affection by high reflection of daylight or occasionally clouding.
Tracking process was done from the frame that an extraction road sign's radius is more than 7 pixels. This extraction size is corresponding that the relative distance between the camera and the target sign is about 125 m. For tracking and extraction process from frames following the first extraction frame, CONDENSATION was employed which was described in 2.2. The parameters for fitness calculation were $W_{edge} = 1.0$, $W_{pos} = 1.25$, $W_{rad} = 1.0$. As the recognition technique, a template matching algorithm which is generally used was employed. In the extraction, tracking and recognition process, the technique proposed in (Yamauchi & Takahashi, 2004) is used for specific color detection.
Prior to verification experiments, a measurement experiment was done. In this experiment, relation of relative distance between the camera and the target sign and size of the extracted sign was obtained. The measurement was done by taking image at distance one by one meters from 100 m to 11 m, in which the target sign becomes frame-out. As a result of the measurement, a relation between relative distance and extraction radius was achieved. By

using regression analysis, the regression equation can be represented as eq. 9, where r is radius of extracted sign image in pixels, and d is relative distance in meters.

$$d = \frac{1348.8}{r^{1.2284}} \tag{9}$$

The R^2 value, which is used for an index of suitableness, of this equation is 0.9914, which represents that eq. 9 is a suitable equation. In the following, relative distance is used for evaluations, which is more important index in practice case, by using eq. 9.

3.2 Experimental Result

As an experimental results, Fig. 2 is expressing the result targeted a yellow precaution sign, left curbing caution (refer to Fig. 1 (b)). These experimental observed images were taken in a fine day of December at around 14:15. In this figure, results of 1) extracted images from each single frame, 2) superposed images of the extracted images as comparison targets, and 3) proposed super-resolution images, are shown. The superposing technique is proposed in (Yamauchi et al., 2007). These superposed images are generated by weighted averaging of extracted images according to sizes of extraction images.

In Fig. 2, extracted images from single frames that are picked up from a scene every 5 frames, and processing result images of these extracted images are shown. This picking up is from frames which is taken at relative distances range in 100 to 50 m. The columns are relative distances calculated by eq. 9 and extracted sizes, extracted images from single frames, superposed images, and super-resolution results, respectively. Moreover, correct recognition ratios on the images are shown under each image. These ratios are average value of 10 frames which consists of the target frame, formerly 5 frames and following 4 frames. In these relative distances, sizes of extracted images are lower than 30×30 pixels. Most of former road sign recognition techniques are hard to recognize correctly for such sizes. Then, it is relative difficult to recognizing by human observation. In this condition, not only extractions from a single frame but also superposed images are not clear because of blur affection. In spite of such condition, super-resolution results are clearly than these comparison images.

From the figure, the recognition results on extracted images from single frames are almost all failed in the relative distance range. The main reason of this may be low resolution resulted from long relative distance. On another comparison technique, superposition technique proposed in (Yamauchi et al., 2007), high recognition ratio is shown in range near than 60 m. However, results at more far distance are almost all failed, similar to results on extracted images from single frames. This means that blur affection resulted from low resolution may not be greatly improvable by using superposition.

As comparison with these results, results of the proposed technique using super-resolution are almost full correct recognitions. This due to resolution estimation from plural low resolution images which have less resolution information themselves. Thus, the estimated high resolution images may be properly estimation, and this leads correct recognitions.

Note that the correct recognition ratios on super-resolution images shown as (c8) and (c9) are low. At such relative distance, extracted images are in higher resolution themselves. These failures may be caused by position shifts because of low position detection accuracy in CONDENSATION of the experiments. Thus, these ranges of position shifts are exceeded search range of the registration process of super-resolution process. Then, super-resolution process was done using incomplete registration images.

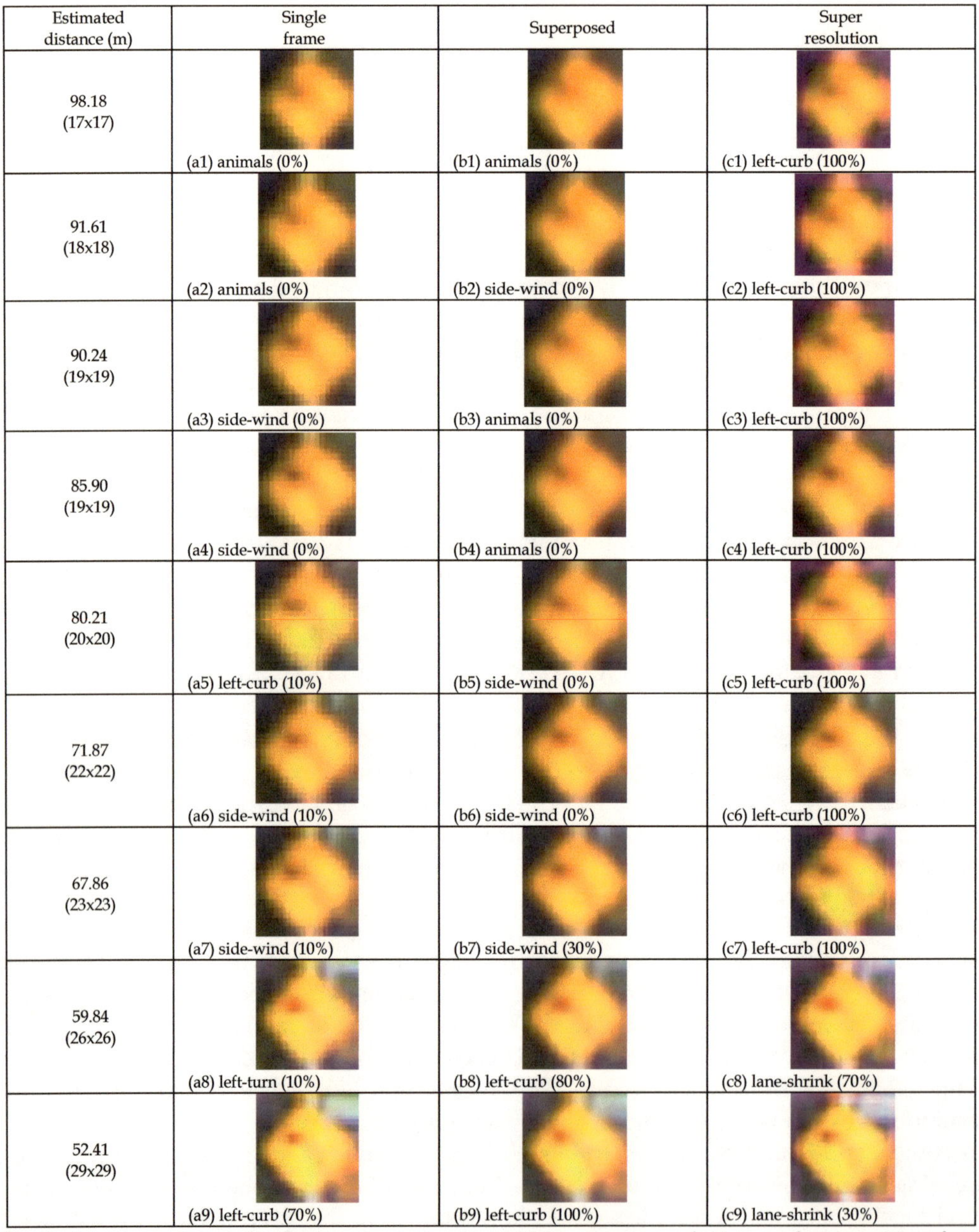

Estimated distance (m)	Single frame	Superposed	Super resolution
98.18 (17x17)	(a1) animals (0%)	(b1) animals (0%)	(c1) left-curb (100%)
91.61 (18x18)	(a2) animals (0%)	(b2) side-wind (0%)	(c2) left-curb (100%)
90.24 (19x19)	(a3) side-wind (0%)	(b3) animals (0%)	(c3) left-curb (100%)
85.90 (19x19)	(a4) side-wind (0%)	(b4) animals (0%)	(c4) left-curb (100%)
80.21 (20x20)	(a5) left-curb (10%)	(b5) side-wind (0%)	(c5) left-curb (100%)
71.87 (22x22)	(a6) side-wind (10%)	(b6) side-wind (0%)	(c6) left-curb (100%)
67.86 (23x23)	(a7) side-wind (10%)	(b7) side-wind (30%)	(c7) left-curb (100%)
59.84 (26x26)	(a8) left-turn (10%)	(b8) left-curb (80%)	(c8) lane-shrink (70%)
52.41 (29x29)	(a9) left-curb (70%)	(b9) left-curb (100%)	(c9) lane-shrink (30%)

Fig. 2. An example result of extraction, superposition, super-resolution and recognition for a yellow rectangle sign.

4. Consideration

The characteristics of the proposed technique are described as follows.

1. Road sign images are extracted from successive frames by tracking. Using these extracted images, a relatively high resolution image is estimated by employing super-resolution technique. The experimental results show that this technique achieves improvement of recognition ratio.
2. In real driving scene, correct recognition of road sings at far relative distance is more useful and meaningful. It is shown that the proposed technique improves quite improvement at more far range 50 to 100 m than ordinary road sign recognition techniques' target ranges.
3. In the experiments, it was not in consideration that extraction images of a targeted road sign may be affected by angle changes. In such cases, estimating affine parameters is useful for more accurate recognition. But increasing processing time by adding this process is in trade-off relation against low recognition accuracy for real-time processing applications such as road sign recognitions. To overcome this increasing processing time, employing some object matching techniques, such as Scale Invariant Feature Transform (SIFT) (Lowe, 2004) based technique, may be usefulness.

5. Conclusion

In this paper, a road sign recognition technique applying super-resolution to extracted images from successive frames is proposed to achieve high recognition accuracy. The super-resolution process is applied on low resolution images extracted from successive frames by tracking using CONDENSATION.

As results of comparison to recognition ratio of extracted images from single frames and superposition images, high accurate recognition ratio is achieved for signs at far position by the proposed technique.

Verification of weights in super-resolution process, development of eliminating technique cased by occlusions or shadowing, are future works.

6. References

Gavrila, D.M (1998). Multi-feature Hierarchical Template Matching Using Distance Transforms, *Proceedings of IEEE International Conference on Pattern Recognition*, pp. 439-444, 0-8186-8512-3, Brisbane, Australia, September 1998.

Irani, M. & Peleg, S. (1999). Improving Resolution by Image Registration, CVGIP:Graphical Models and. Image Processing, Vol. 53, No. 3, May 1991, pp. 231-239, 1049-9652.

Isard, M. & Blake, A. (1998). CONDENSATION -- Conditional Density Propagation for Visual Tracking, *International Journal of Computer Vision*, Vol. 29, No. 1, 1998, pp. 5-28, 0920-5691.

Lowe, D.G. (2004) Distinctive image features from scale-invariant keypoints, *International Journal of Computer Vision*, Vol. 60, No. 2, 2004, pp. 91-110.

Matsuura, D.; Yamauchi, H. & Takahashi, H. (2002). Extracting Circular Road Signs Using Specific Color Distinction and Region Limitation, *IEICE Transactions on Information and Systems, Part 1 (Japanese Edition)*, Vol. J85-D-II, No. 6, June 2002, pp. 1075-1083.

Ohara, H.; Miki, S.; Yabuki, N.; Fukui, Y. & Sumi, Y. (2002). Detection and Recognition of Road Signs Using Neural Network, *Proceedings of the 15th Workshop on Circuits and Systems in Karuizawa*, pp. 505-510, Karuizawa, Japan, Apr. 2002 (in Japanese).

Park, S.C.; Park, M.K. & Kang, M.G. (2003). Super-Resolution Image Reconstruction: A Technical Overview, *IEEE Signal Processing Magazine*, Vol. 26, No. 3, May 2003, pp. 21-36, 1053-5888.

Uchimura, K.; Kimura, H. & Wakiyama, S. (1998). Extraction and Recognition of Circular Road Signs Using Road Scene Color Images, *IEICE Transactions on Fundamentals of Electronics, Communications and Computer Sciences (Japanese Edition)*, Vol. J81-A, No. 4, April 1998, pp. 546-553.

Yabuki, N.; Matsuda, Y.; Kimura, H.; Fukui, Y. & Miki, S. (1998). Extraction of Special Color Region Using Feature Energy of Color Image, *Proceedings of the 11th Workshop on Circuits and Systems in Karuizawa*, pp. 101-106, Karuizawa, Japan, Apr. 1998 (in Japanese).

Yamauchi, H. & Takahashi, H. (2003). A Road Sign Recognition Technique by Tracing Outline Vectors, *Journal of The Institute of Image Information and Television Engineers*, Vol. 57, No. 7, July 2003, pp. 847-853 (in Japanese).

Yamauchi, H. & Takahashi, H. (2004). A Specific Color Detection Technique for Road Sign Recognition Using Outline Vectors, *Technical Report of IEICE*, No. IE2004-88, November 2004, pp. 11-16 (in Japanese).

Yamauchi, H.; Kojima, A.; Miyamoto, T.; Takahashi, H. & Fukunaga, K. (2007). A Robust Road Sign Recognition by Superposing Extracted Regions from Successive Frames, *IEICE Transactions on Information and Systems (Japanese Edition)*, Vol. J90-D, No. 2, February 2007, pp. 494-502 (in Japanese).

Zin, T.T. & Hama, H. (2005). A Robust Road Sign Recognition Using Segmentation with Morphology and Relative Color, *Journal of the Institute of Image Information and Television Engineers*, Vol. 59, No. 9, September 2005, pp. 1333-1342.

15

Mechatronics Design of an Unmanned Ground Vehicle for Military Applications

Pekka Appelqvist, Jere Knuuttila and Juhana Ahtiainen
Helsinki University of Technology
Finland

1. Introduction

In this chapter the development of an Unmanned Ground Vehicle (UGV) for task-oriented military applications is described. The instrumentation and software architecture of the vehicle platform, communication links, remote control station, and human-machine interface are presented along with observations from various field tests. Communication delay and usability tests are addressed as well. The research was conducted as a part of the FinUVS (Finnish Unmanned Vehicle Systems) technology program for the Finnish Defense Forces. The consortium responsible for this project included both industry and research institutions. Since the primary interest of the project was in the tactical utilization scenarios of the UGV and the available resources were relatively limited, the implementation of the vehicle control and navigation systems was tried to be kept as light and cost-effective as possible. The resources were focused on developing systems level solutions.

The UGV was developed as a technology demonstrator using mainly affordable COTS components to perform as a platform for the conceptual testing of the UGV system with various types of payloads and missions. Therefore, the research methodology was rather experimental by nature, combining complex systems engineering and mechatronics design, trying to find feasible solutions. The ultimate goal of the research and development work was set to achieve high level of autonomy, i.e., to be able to realistically demonstrate capability potential of the system with the given mission. In this case, the test tasks were related to tactical reconnaissance and surveillance. The targeted end-result, task-oriented autonomous mission capability, means that the UGV is able to cope with the predefined tasks as autonomously as possible, requiring operator support only in demanding decision making and unexpected problem situations.

The contribution of this research relies in the systems level solution that combines autonomous motion and task execution capabilities of the UGV, as well as scalable level of autonomy to support the UGV operator. Additional contribution consists of systems integration and information flow between the higher level command & control site and the UGV remote control station. So far, there have been relatively few attempts to combine all these ambitious features in realistic field environment tests, especially in a cost-effective approach. Along with the presented UGV demonstrator case (see Fig. 1), the chapter provides for an overlook to the essential research problems and presents solutions used in

the domain of autonomous field robots. The research challenges are generic by nature in similar robotic systems.
The emphasis of this chapter is rather on practical observations and feasible solutions instead of describing algorithms in detail. Section 2 reviews similar UGV systems and their development, while generic design considerations of various subsystems are discussed in Section 3. Next, in Section 4, the implementation and design solutions of the UGV demonstrator systems are explained in detail. This is followed by the results and observations based on several demonstrations and field testing in Section 5. Finally, the achievements are concluded with the analysis of developed solutions and observations from the field tests.

Fig. 1. The fully instrumented UGV demonstrator at the primary test site.

2. Related Work

2.1 Historical perspective

The term Unmanned Ground Vehicle or UGV only defines that the operator controlling the vehicle, if one exists, is not on the vehicle itself. It does not specify the kind of control that an operator might have over the vehicle at any given time. Sheridan (1992) divides control modes into manual control, supervisory control and fully automatic control. Manual and supervisory modes can be further divided depending on the degree of computer involvement in the control loop. In this chapter we consider vehicles that can be used in any one of these modes, possibly in addition to manned driving. However, more emphasis is on the supervisory and autonomous modes since that is where most research is currently focused on. Sheridan does not give preference to any mode over another, but it is evident that systems with more autonomy also have to be technically more sophisticated. However, after attaining full autonomy, adding high-level communication between human and the vehicle can be seen as an even higher level of sophistication as suggested by SPAWAR (2007a). In that classification, full autonomy is followed by semi-proximal autonomy and proximal autonomy. Semi-proximal autonomy signifies a mode where a robot can act as a guide to a human in unknown environment. Proximal autonomy is a mode where the human and the robot can cooperate on human terms. Also note that the term UGV does not rule out the possibility of human passengers on the vehicle. Such situations might arise within mass transport systems and military evacuation missions.

So far, the development and use of remotely controlled vehicles has been dominated by the military. During their existence, these vehicles have been a relatively high-tech and as such, quite expensive. The military however has had the need and the resources to develop and deploy systems that are able to successfully perform specific tasks that might be otherwise too costly. One of the earliest, if not the first, were German wire-guided electric boats during the First World War. These boats had an explosive charge on them and were steered into big, slow warships. (Lightoller, 1935; Neumann, 1921) In the Second World War, all the major parties had remote controlled systems. There were Soviet "TT-26" tanks (Armchair General, 2004), German "Goliath" tracked mines (Kaleun, n.d.) and U.S. "TDR" aircraft (SPAWAR, 2007b). Despite their technical similarity, the boats and tracked mines just mentioned and also guided torpedoes and cruise missiles - to name a few - are however not considered unmanned vehicles as they are destroyed when used.

The space race saw the deployment of Soviet planetary Lunokhod rovers on the Moon in the 1970's (Wade, 2008). Three U.S. rovers have successfully landed on Mars: Pathfinder in 1997, Spirit and Opportunity in 2003. The latter two are still operational (NASA 2009). While the Lunokhods were teleoperated, the same cannot be said for the Mars robots because the delay caused by the distance between Mars and the Earth makes it impractical. As a result, the Mars robots are programmable and more autonomous. Unmanned aerial vehicles are currently in use by armed forces all over the developed countries. The challenges on the ground - obstacles to moving, sensing and communicating - have proved much more difficult than in the air or under water.

Autonomous driving on roads and highways has been under development since the late 1960's when the available computing power allowed the first ones to be built. The early systems used only computer vision to track the road and lane markings. For a more detailed history on the development of unmanned ground vehicles, see "UGV history 101" (Cage, 1995). The VaMP ja VITA-2 vehicles developed by Universität der Bundeswehr München and Mercedes-Benz became a significant advance in unmanned ground vehicle technology (Schmidhuber, 2005) when they drove over 1000 km on highways in 1994. They reached speeds of 130 km/h and drove in heavy traffic, but required some human intervention (UniBwM, n.d.).

The state-of-the-art UGV projects around the world have proved that current autonomous UGV systems are capable of coping even in dynamic environments with high speeds. In the DARPA Urban Challenge (UC) competition (DARPA, n.d.), the vehicles had to navigate in an urban environment with other vehicles and perform a simulated logistics task. A digital map of the road network was provided, but some of the roads were blocked and the vehicles were required to use their sensors to detect the blocks and replan their route on the fly. Since then DARPA has not announced new competitions in the UGV domain. The UC competition showed that the Velodyne laser scanner (Velodyne, n.d.) is the best readily

available sensor for environment modeling at the moment. All the major teams that completed the UC used this sensor. It provides a very accurate model about the environment using a column of 64 vertically aligned laser beams that rotate at 5-15 Hz frequency.
In the Military ELROB event arranged in 2006 and 2008 (ELROB, n.d.), a different approach was chosen. It emphasized completing real military missions - such as detection of explosives and tactical awareness in urban reconnaissance - but the rules did not require fully autonomous vehicles. Practically all participating vehicles were purely teleoperated. In 2007, a civilian ELROB was arranged where the tasks were related to civilian scenarios such as improving situation awareness in an incident with improvised explosive devices (IED). Military and civilian versions alternate the annual ELROB event.

2.2 Current military systems

There are a number of small man-portable robots in military and police use. These are typically used to defuse bombs and other explosive devices and are remotely operated. The first two fully deployed vehicle-size autonomous unmanned systems are the Israel Aerospace Industries' Guardium Unmanned Security Vehicle (IAI, n.d.) and the MDARS developed by General Dynamics Robotic Systems (GDRS, n.d.). The Guardium is designed for guarding and surveillance tasks and it can both drive autonomously and be teleoperated. MDARS is developed to provide an automated intrusion detection and inventory assessment capability for use in warehouses and storage sites. In 2008, 24 MDARS systems were ordered and expected to be delivered in January 2009 (Washington Technology, 2008). A number of other vehicle-size systems exist, but they are more or less limited to direct teleoperation. Several implementations of remotely operated de-mining vehicles exist, but so do some more versatile vehicles such as the Accident Response Mobile Manipulator System (ARMMS) by Sandia (Sandia, 2008). The ARMMS is built on a HMMWV and is equipped with two manipulators in the front. Its main purpose is to act as a remotely operated vehicle in radioactive or otherwise hazardous environments.
Several research projects aim at developing more autonomous ground vehicles for military use. The most notable ones are being developed in the United States. R-Gator is a robotic version of the John Deere M-Gator, a multipurpose military transport vehicle. The R-gator seems likely to be deployed in the near future. It is able to follow a pre-defined path or a human and can also be controlled with a remote controller up to a distance of 300 m (John Deere, 2008). The MULE, part of the U.S. Army Future Combat System, is a 2,5 ton wheeled vehicle that is planned to have different versions for transport, countermine and assault (U.S. Army, 2008. Research projects have taken place around other similar vehicle platforms such as the TAGS-CX by Foster-Miller and Spinner, Crusher and Autonomous Platform Demonstrator by the National Robotics Engineering Center (NREC), part of the Carnegie-Mellon University Robotics Institute. NREC has also been developing the Gladiator, an armed vehicle-size unmanned vehicle for the U.S. Marines (NREC, n.d.).

2.3 Commercial civilian systems and autonomous features in cars

The term unmanned vehicle usually implies military applications. The same technology in civilian applications is often called driverless, automatic, autonomous or robotic. A number of such mass transport systems are in use. Most of these are trains, but also road vehicle systems such as the ParkShuttle and Phileas (Lohmann, 2007). Industrial solutions such as the Sandvik Automine (Sandvik, 2008) or Brisbane harbour automatic container terminal (Tekes, 2006) also make use of unmanned vehicles in material handling. All current systems work in structured environments and are typically restricted to moving in an area with no humans present due to safety reasons.

Teleoperation kits or systems to convert a traditional vehicle to a remotely controlled one also exist. Some of these systems such as the Pronto4 by Kairos Autonomi state that they can be installed in a matter of hours on any wheel operated vehicle (Kairos Autonomi, n.d.).

Vehicle automation features are becoming increasingly common in private cars. A modern car typically has several processors taking care of engine and transmission control, anti-lock brakes and vehicle stability to name a few. These are not any more the "smart features" that get special attention from consumers. More advanced systems include sensors that can sense the environment outside the vehicle. The currently new vehicle automation features can control the speed or position of the vehicle or help the driver to do it.

Adaptive cruise control (also active cruise control and intelligent cruise control) typically uses a radar or laser scanner to measure distances to the obstacles in front of the car. The cruise control system can then adjust vehicle speed to match the speed of the vehicle driving in front and keep sufficient distance to it. Some of these systems can also apply the brake to avoid collisions or lessen the damage of an unavoidable collision.

Intelligent speed adaptation is a system where the vehicle can access a database of speed limits to determine the currently active one for the vehicle. This information can be used to limit the speed that the vehicle can be driven at, or it can use force-feedback on the accelerator pedal so that the driver feels it in the pedal when he is speeding.

Lane departure warning systems use cameras to monitor lane markings. If the driver is about to exit the lane without signaling, the system alerts by vibrating the steering wheel. Some systems include lane keeping assistance by applying the brake on the opposite side or by steering to keep the vehicle in the lane. The manufacturers stress that the driver still has to be alert and driving.

Parking automation systems are also becoming more common. These systems can parallel park between obstacles on the side of the road. The system takes care of steering and the driver typically has to control the vehicle speed.

Automated highway systems are systems where the vehicles can communicate and cooperate to drive automatically in queues. The aim is on the other hand to improve safety and on the other, lower the consumption of the vehicles. When the vehicles are able to communicate, they can brake simultaneously. This allows shorter distanced between cars and thus lower consumption. No systems exist, but a demonstration of such a system has been held in 1997 (PATH, n.d.).

3. Design Considerations

The development of the UGV demonstrator was divided into three phases in the project plan. In the first phase the vehicle was instrumented for computer controlled operation. The UGV consists of subsystems that were: locomotion system, energy system, positioning system, sensor system, navigation system, motion control system and communication system. Naturally, there is also a need for an interface for human-machine interaction (HMI). In case the UGV is used for actual missions, there is also a need for mission control system and if there is some mission specific payload there has to be a control system for the payload. Due to the limited resources and focusing in the UGV project, the actual implementation of the UGV demonstrator was a so called minimum realization. This means that COTS components were used whenever possible and the number of sensors was kept as low as possible. Resources were aimed for developing software rather than hardware.

Overall architecture of the UGV demonstrator was designed according to the figure 2 where most of the main components of the UGV demonstrator are shown. The remote control station is shown in gray and the white boxes represent the actual vehicle. Next subchapters summarize the design considerations related to each subsystem of the UGV demonstrator and explain the figure 2 in detail.

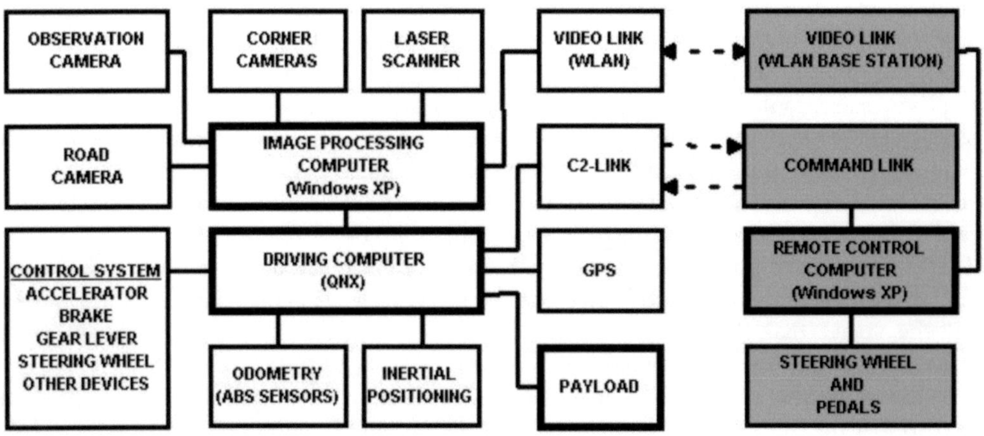

Fig. 2. Block diagram of the UGV demonstrator. Remote control station is represented by the gray boxes on the right while the white boxes represent the vehicle.

Locomotion system

The UGV demonstrator locomotion system consists of the vehicle platform together with its instrumentation. A decision to use a commercial off-road vehicle was made early in the project. This way sufficient off-road capability and reliability of the basic mechanics was ensured. Also, vehicles of this size can travel reasonably long distances and carry a wide range of payloads. Another decision made at that point was that the vehicle should remain street legal, which greatly simplifies the logistics in testing. This means that the computer control and the drive-by-wire capability of the platform had to be made using the control system of the vehicle. The actuators that drive the vehicle controls had to be detachable so that the vehicle can be safely driven on public roads by human driver.

It was decided that the drive-by-wire capability will be implemented entirely with electrical actuators due to their ease of use and relatively small size. Pneumatic and hydraulic actuators were also considered, especially for the brake pedal control which requires considerable force. However, the force required to fully apply the brake was measured and it was deemed possible to achieve the same force with a linear electric actuator. The decision that the vehicle must remain street legal dictated where the actuators would be installed, since no disturbance for the human driver was allowed. Achieving drive-by-capability with external actuators is not the optimal way since there are always unnecessary gaps in the installations and the control circuits are therefore hard to tune. An ideal platform for a UGV system would be a vehicle that has electric actuators integrated on it at the factory. This way it would be possible to connect straight to the vehicle bus and send drive commands to the on-board computer. Also, with a vehicle with a vehicle data bus it is possible to read all the information related to the vehicle from the bus. In our case there was no such bus in the vehicle.

Energy system
All the components in the figure 2 are electric so there is a clear need for an additional energy system. The additional energy system was part of the design from the start and the idea was that there would be enough capacity in the batteries to supply power for all the devices for several hours. Another design criterion was that that the energy system should be able to provide both 12 and 24 volts. Also, the batteries should be charged when the engine is running. A rectifier and additional charger that use electricity from an electric socket were also taken into consideration. With these components the batteries could be charged in a garage when the engine is not running and the UGV demonstrator could be powered through the rectifier, not charging the batteries.

Positioning system
Positioning the vehicle precisely enough is essential for autonomous operation. It was known that a GPS receiver by itself would not provide accurate enough information. Therefore it was necessary to add other positioning methods. Using an inertial navigation system (INS) based on gyroscopes and accelerometers together with dead reckoning calculations it is possible to get the position estimate accurate enough. So, the positioning system of UGV demonstrator, shown in figure 2, was designed to involve GPS-receiver, inertial positioning device and odometry sensors for travelled distance and steering angle. Fourth method for improving the position estimate is video navigation based on processing image flow and laser scanner data.

Sensor system
A dynamic model of the environment is acquired and updated through the sensor system. The sensor system of the UGV demonstrator consisted of three cameras and two single line laser scanners. Due to the limited resources, the plan was to use as few sensors as possible. Sensors are shown in the upper left corner of the figure 2. With these sensors it is possible to demonstrate the functionality of the system, but the reliability and performance will not be as great as they could be due to the inaccurate environment model. The environment model could be improved with the help of more sensors. For example, a Velodyne lidar would improve the model dramatically. The cameras were designed mainly for teleoperation purposes and the laser scanners for the autonomous mode for obstacle avoidance.

Navigation system
Navigation system is responsible for providing the route points for a UGV system. The navigation system of the UGV demonstrator was based on the idea that the vehicle was intended to use roads and open fields. There are accurate digital road maps of the Finnish roads available for a fee. Even the small forest roads are mapped. The operator can design the route for the vehicle with the help of digital map material. Only in case of an obstacle the navigation system would plan a route around it. In case there is need for off-road driving,

the operator would have to plan the route by hand or teleoperate the vehicle through the off-road part. No map is created in the vehicle during runtime. In our demonstrator, we deemed it simpler and cheaper to record a suitable digital map of roads than use a commercial one.

Motion control system

Motion control system is the low level system which executes the drive commands that are given from upper level system. In the teleoperation mode the motion control system executes the commands given by the remote control station. In the autonomous mode, the commands are given by the navigation system, which is trying to maintain the given velocity and position on the route. In the UGV demonstrator the motion control system includes the control loops of the actuators which drive the vehicle controls. Also the speed controller and path follower belong to the motion control system.

Communication system

Communication between the vehicle and the remote control station is necessary, even though fully autonomous vehicle would not necessarily need a communication link. The communication link is an essential part of the teleoperation capability and also very useful in the autonomous mode. A reliable, long range and high-bandwidth link was needed, but such radio modems were not yet available, at least not for a reasonable cost. This is of the biggest limiting factors in the UGV domain. A low-bandwidth but reliable tactical radio was used for the command and control (C2) data and a high-bandwidth video-link was used for transmitting the video feed.

Remote control station and HMI

The remote control station consists of a laptop computer running the user interface application for the vehicle, steering wheel and pedals and radio equipment. Another laptop can be added for payload control.

Human-machine interaction was a notable part of our research. Since fully autonomous task execution in hostile environments is not yet - if ever will be - practical, several modes of operation were employed. The idea was that there should be intermediate modes between direct teleoperation and the fully autonomous mode. The transfer or division of control between a human operator and computer - scalable level of autonomy - was one focus area of the project.

Safety issues

The starting point in the design of the UGV demonstrator was to construct such an emergency stop circuit (ESC) that the vehicle could be stopped at any time. All the thinkable malfunction scenarios were taken into account when designing the circuit. It is important that there is no software in the loop. The ESC was designed so that it activates if the software in the driving computer crashes, connection is lost or the system loses power. The ESC can also be activated manually, using emergency stop buttons or by software from the remote control station. When the ESC is activated, the brake pedal is applied fully and accelerator pedal is released. Stalling the engine was considered, but deemed not necessary.

4. Implementation

In this chapter the implementation of the UGV demonstrator is explained in detail. The chapter is divided into two main parts - the implementation of the hardware and the software systems of the vehicle.

4.1 Hardware Implementation

Drive-by-wire capability was realized with electric actuators that drive the control devices of the vehicle. Actuators were installed so that they can be detached from the vehicle controls when driving manually. This way the vehicle remains street legal and is safe to drive among normal traffic. With street legal vehicle there was no need for complicated and costly transportation arrangements.

All the actuators that drive the control devices were sized so that they fit the planned positions, but also that they have enough power to use the control devices in all situations. Measuring the needed force to use the control devices was done with standard scales. It was taken into consideration that the power steering and the brake booster are not in use when the engine is not running.

The vehicle is equipped with automatic transmission and the gear lever, situated in the middle console, was instrumented by a detachable linear actuator (see Fig. 3). The actuator has two limit switches and an incremental encoder integrated inside it. When the software is started the motor needs to be calibrated. Calibration is done by driving the motor to its limit where a limit switch stops the motor. That position becomes the reference point. Also, every time either of the limit switches is activated, the position is calibrated again in case some inaccuracies would have occurred. After the calibration, the position of the gear lever can be calculated from the encoder pulses. All gear lever positions are achievable with this arrangement.

The steering wheel is actuated using a brushed DC-motor attached to a stand built in place of the removed center console. The motor is attached to the steering wheel with a toothed belt and a belt-pulley arrangement shown in figure 3. The toothed belt can be detached from the steering wheel after which a human driver can turn the wheel normally. The steering angle is measured underneath the vehicle from the push-rod that mechanically turns the wheels. A linear micropulse tranducer with IP 67 (internal protection) was used. Linear potentiometer was tried but it did not endure the harsh conditions. Vibrations wore the conducting surface off from the middle point.

The brakes are actuated using the brake pedal to keep human operability at all times. The brake pedal is operated by a linear actuator. It is not attached to the pedal but it only pushes it allowing "safety driver" to step in any time and override the computer control. This greatly improves safety during testing. The throttle pedal is actuated similarly. Pushing the brake pedal to the floor required quite a lot of power and it was hard to find suitable actuator, because selecting spindle motor it is always a tradeoff between power and speed. The selected motor has just enough power to press the brake pedal to the floor when the brake booster is off. However, with the brake booster on, there is more than enough braking power. When driving on a public road, the actuators can be driven to the uppermost position and powered down so there is no danger of them interfering with the human driver's vehicle handling. The actuators are equipped with potentiometers that tell their absolute position. These signals are used as the feedback signal on the control loops. The motors are also equipped with limit switches that prevent the motors exceeding their range of movement.

Fig. 3. Steering wheel is instrumented with brushed DC-motor and belt-pulley arrangement. Gear lever is instrumented with a spindle motor. Instrumentation was tested and the controllers were tuned by using the steering wheel shown on the right.

Remote ignition of the engine is instrumented with relays. It is assumed that the key is in the ignition lock disabling the vehicle immobilizer and the steering lock. Feedback signal from the tachometer is used to determine whether the engine is running or not and the feedback signal from the annealing indicator reveals when the combustion chamber is warm enough to try starting (diesel engine). One relay switches on the power and another rotates the starter motor. Operator has to only press ignition button in UI, the software module takes care of everything else.

Also the windshield wipers and the windshield wash functions are instrumented. This is also done by connecting a computer controllable relays in parallel to vehicles own wires.

Safety

Instrumentation of the pedals was designed keeping also the safety issues in mind. When the ESC activates, the brake pedal is pushed to the floor and the throttle pedal is released. This is achieved by conducting electricity straight from the battery to the motors until limit switches break the circuit. In practice, the only definite way to stop the vehicle would require stalling the engine, but the ESC proved to be fully reliable so, for convenience's sake, the engine was left running. The ESC is connected to the driving computer with an external watchdog circuit. When the software is running, it feeds a rectangular wave signal to the watchdog circuit at a specific frequency. In case the software crashes, the watchdog circuit no longer receives the signal and activates the ESC. The ESC can also be activated within the software from the remote control station or manually using one of the six emergency stop buttons mounted on the vehicle. In case of a radio connection loss, the ESC is activated by the driving computers software. Also, in the event of electrical power loss, the ESC will be activated.

Sensors

Most of the necessary sensors for unmanned operation are mounted on the roof rack of the vehicle (see Fig. 1). The observation camera which turns 360° is mounted in the middle of the roof rack in a shielded dome which protects the camera from the weather. In the dome there is a heater element that keeps the camera functional even at freezing outside temperatures. Four corner cameras were installed for the observation of the immediate surroundings of the vehicle. Another laser scanner, used for obstacle detection, is installed in the middle of roof rack's front part and directed towards the ground approximately 9 m in front of the vehicle. The other scanner is installed on the bumper and it looks straight forward. Both scanners are also waterproof and equipped with heater elements. The roof rack features also two WLAN antennas, two radio antennas and a beacon. The road camera is mounted to the rear view mirror stand, looking straight ahead through the windshield. In addition, depending on the mission, some payload can be attached to the roof rack as well.

Positioning of the vehicle is done by fusing data from GPS receiver, inertial navigation system (INS) and dead reckoning calculations. Also, video navigation was used to detect wheel tracks and road boundaries. The GPS receiver is standard COTS receiver with SIRF star III -chip and it is installed to the front bumper of the vehicle. The GPS receiver used to be on the roof as well but was installed on the front bumper after it was discovered that it was interfered by the transmitting radio of the C2-link with full power. On the bumper, where the distance to the transmitting antenna was twice as much, the interference was manageable.

Inertial measurement unit is a military-grade equipment and it is bolted to the centre of the vehicle chassis. Travelled distance used in dead reckoning calculations is computed from the filtered signals of the ABS-sensors. The signal from the ABS-sensors is sine wave which amplitude and frequency depends on the speed of rotation. Therefore, some custom made electronics was needed to generate pulses that are readable by the I/O-board of the driving computer. Steering angle used needed in dead reckoning calculations is the same one used in the control loop of the steering wheel.

Electricity

All the electric components mounted in the vehicle are powered by an additional electricity system which is completely separate from the vehicles own electric system. Energy is provided by a 2 kW generator which is belt driven by the engine. The generator is installed under the hood replacing the air-con compressor. Two 12 V, 75 Ah lead-acid gel batteries connected in series are used as a buffer and storage. Three DC/DC converters, two for 24 volts and one for 12 volts, provide regulated power for the electrical systems. It was calculated that with all the electric components on, the batteries will last at least three hours. The batteries are also chargeable with a 24 volt charger installed to the poles of batteries. This charger is to be connected to 230 volts of alternating current. With a single switch it is possible to electrify the vehicle with external rectifier which also uses 230 Volts. Rectifier is needed because the 24 volt charger does not provide enough current to electrify all the electric components and charge the batteries at the same time.

Computers

The heart of the vehicle is a shock isolated rack behind the front passengers' seat where two computers and other electronics are mounted. Image processing computer is intended mainly for processing the camera and laser scanner data and driving computer for real-time drive command execution. The shock-isolation is needed because of vibration and trembling. In addition, the hard drives were installed in foamed plastic. The computers have been assembled from selected components and they feature ATX motherboards with 1.86 GHz Pentium-M processors. These mobile processors were selected because of their low power consumption and thus low heat dissipation.

Communication

Several options were considered when selecting the communication equipment. A reliable, long-range, but low-bandwidth command and control (C2) link and a high-bandwidth video link were needed.

A tactical, vehicle radio was selected as C2 link for its long range. For the video link, WLAN was chosen as the most practical possibility with the hope that superior solutions - such as mobile WiMAX or a tactical broadband solution - will replace it later.

The used WLAN for the image data is based on IEEE 802.11b, using the 2.4 GHz ISM (Industrial, Scientific and Medical) frequency band, which is quite a high frequency for non-line-of-sight operation in typical Finnish forest environment. Water molecules (e.g. in foliage, snow) absorb radio waves at this particular band, causing further attenuation to the signal. The band has fairly low transmitting power restrictions (20dBm, 100mW) in Finland. With special permission we were able to purchase and use high-power equipment (30dBm, 1W). The remote control station uses horizontally omnidirectional 15 dBi antennas. The vehicle has physically smaller, more convenient 8 dBi antennas. Both the vehicle and the remote control station use double antenna setup to minimize shadow zones.

4.2 Vehicle Software Architecture

Three or four computers, depending on the setup, are used in the system, two of which are in the vehicle. The driving computer takes care of vehicle real-time control, which is why it runs QNX Neutrino RTOS 6.3. The image processing computer runs Windows XP, taking care of grabbing the road camera image, reading the laser scanner data and processing of both. It is possible to use two computers in the remote control station, one for operating the vehicle and another for mission specific payload.

The software architecture of UGV demonstrator is shown in figure 4. Image processing computer is illustrated with a circle in the lower left corner of the figure and the remote control station with the rectangle in the same corner. The rest of the figure represents the software architecture of the driving computer.

There is also a possibility to link the remote control station to Command, Control, Communications, Computers and Intelligence (C4I) system, from where the missions are actually given to the operator who formulates the mission for use of one UGV. The system is designed so that one operator could control a fleet of vehicles at the same time.

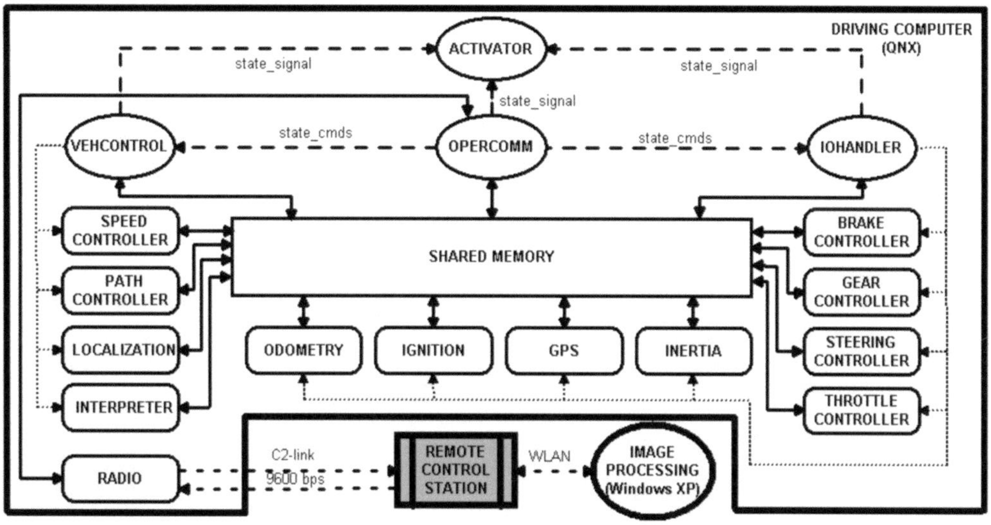

Fig. 4. Software architecture of the driving computer.

Driving computer

Software of the driving computer is based on processes, threads, signals and shared memory. A diagram of the software architecture is shown in figure 4. Parallel processes are represented by circles, and threads running inside them are represented by rounded rectangles. The core of the software, shared memory, is described with vertical rectangle. Every process and thread has access to its reading and writing routines.

First the ACTIVATOR process creates the shared memory and then starts all the other processes and monitors their state. In case the activator does not get a state signal from any of the processes within a specified time frame, it shuts down the software in a safe and controlled way. The OPERCOMM process handles the radio communication between the vehicle and the remote control station through the radio thread. It starts the radio connection thread, interprets the commands received from the remote control station and replies to each with an acknowledgment (ACK) message. The values received are written to the SHARED MEMORY, from where the controller threads read them periodically. If there are mode change commands in the message, opercomm will send appropriate signals to the other processes. In case the radio connection is lost, opercomm will also set on the emergency brake mode. The VEHCONTROL process is not in use while teleoperation mode is selected. It is designed with the threads it creates for autonomous use of the vehicle. The IOHANDLER creates and monitors the threads which control the actuators used to operate the vehicle's control system or read different sensors.

State machine

The whole software of the driving computer is based on state machine described in figure 5. When the software is started every thread and process are in a state called INITIALIZING. If something does not initialize right or software or hardware failure is detected, the software transits to SHUTDOWN state where the system is run down in a safe and controlled way. If the activator does not receive a state signal from any of the processes, regardless of the state, the software transits to the shutdown state. From initializing state, if everything is okay, the software goes to state MANUAL. In this state all the measurements are in use but actuator control is not available. This state is used for gathering data with a human driver. It is also a safe state to be in when the vehicle is stopped for a longer period of time, since the actuators are not in use. From MANUAL the software can transit either to TELEOPERATION, AUTOMATION or EMERGENCY_STOP. The POWER_SAVE state is not implemented but is an option designed for missions where the vehicle would need to stay in one place for days. This state is a sort of a stand by state in which all the unnecessary electric components are switched off. A timer signal would wake the system up. To implement this state some hardware changes would also be needed.

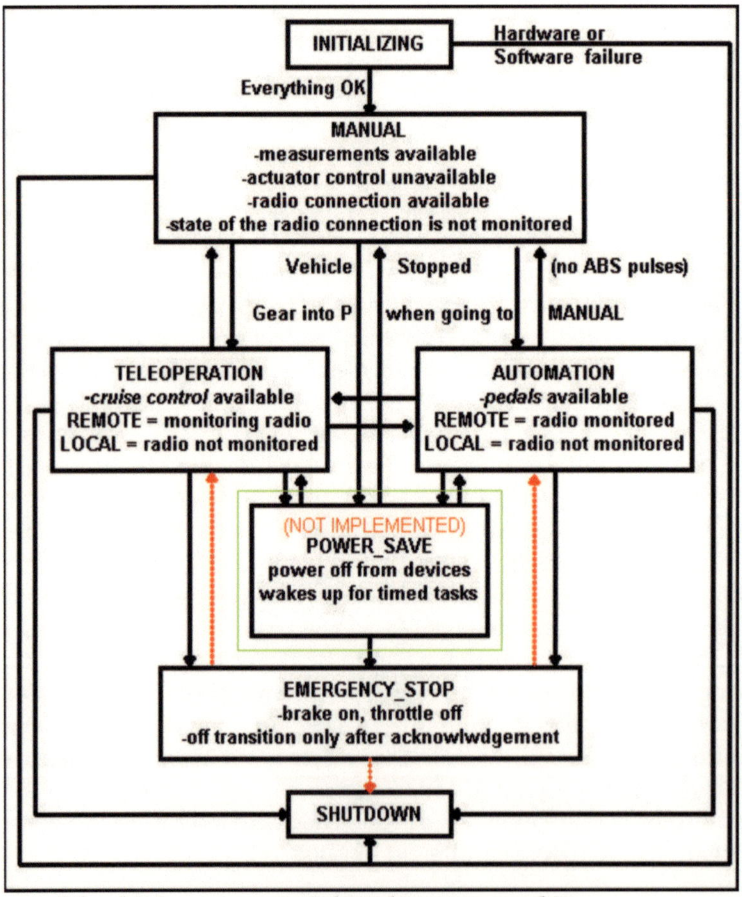

Fig. 5. Software of the driving computer is based on state machine.

The automation and teleoperation states, as their names suggest, are related to the degree of automation of the vehicle. In teleoperation state, the actuators are controlled directly from the remote control station. There is also an option where a cruise control can be set to assist the operator. The REMOTE and LOCAL flags tell whether the radio signal is monitored or not. In a local mode, the driving computer can be used to send control signals to vehicle. This sometimes handy state is only intended for testing purposes. In this state the radio signal is not monitored. When the remote flag is set, the radio signal is monitored and all the control signals come from the remote control station.

In the automation state, the local and remote flags are options as well. However, in this state the vehicle follows routes given by the operator, meaning that the vehicle follows some predefined waypoints. Only the supervision of the vehicle is left to the operator. In the automatic waypoint following mode it is also possible to leave the speed control for the operator: the operator controls the pedals of the vehicle but the steering wheel is under computer control. In the emergency stop state, the software control is overridden and the vehicle is stopped by activating the ESC with a computer controlled relay. To get out of this state a reset signal must be sent to the driving computer.

Image processing computer

The image processing computer acts as a server which the clients in the network can connect to and receive data from. The image processing computer grabs the image from the road camera, compresses it using a wavelet algorithm and transmits it to the client using UDP packets. The compression ratio can be selected at the remote control station. The observation camera is web browser based and it can be turned with buttons on the steering wheel at the remote control station. Image processing computer also processes the data from the laser scanner and is connected to the driving computer with a socket in order to exchange important information such as detected obstacles.

Remote control station and communication protocol

At the remote control station the vehicle can be controlled from a single laptop computer. Another laptop can be easily utilized for separate payload control. Vehicle control commands are transmitted through a military radio with a proprietary protocol that minimizes the necessary bandwidth and provides safety functions such as an emergency stop when the connection is lost. Communications between the vehicle computers and remote control station computers utilize a TCP and UDP packets over Internet Protocol. The higher level protocol to control the image processing functions is also proprietary. The observation camera uses an HTTP interface and vehicle payloads also have their own protocols.

The separate C4I system that was used to assign missions to the UGV demonstrator system and supervise their completion was connected through a local area network. The missions were defined in XML format and as such, could be transferred using a variety of methods such as low-bandwidth radios or a memory stick. In the demonstrations, a shared SQL database was used to record all mission-relevant data and allow the display of real-time data on the mission control system.

5. Milestone demonstrations and field tests

The UGV project plan included demonstrations meant to present the results of the work done and as an opportunity for the Finnish Defence Forces (FDF) to verify them and assess the potential usefulness of an unmanned vehicle. The demonstrations also served as project milestones and helped to stay focused on the current challenges.

Although to rough topics and dates were defined in the original project plan, the script of each demonstration was typically decided some months before. This decision was made on the basis of what was achievable, what was potentially most useful and interesting to FDF and also what was "demonstrable", meaning things that can be seen and not only explained. Even if some features were developed and tested along the span of the project, not all of them were demonstrated because they were not working reliably enough. So the decision was made to concentrate on features that - if not unseen in the current unmanned vehicles - would make successful demonstrations more likely. The demonstrations themselves were more like displays of achievements, but it was all the testing and preparing for them that provided the project team with valuable experience.

The three demonstrations held during the project all took place on an island which is being used exclusively by the FDF. In the middle of the island there is a fairly large open area that has some obstacles such as trees, rocks and pits. Surrounding the open area there are forested areas with some unpaved roads cutting through them. In the demonstrations the open area and the roads were used to simulate a typical Finnish rural landscape.

Delay in the communication chain was measured and analyzed since the delay experienced by the operator is a critical aspect in teleoperation. By measuring the delay we were able to analyze the system performance and evaluate future communication solutions. Also, in an effort to gain a better understanding of the requirements for an unmanned vehicle user interface, separate teleoperation tests were arranged where "real" users were used to review the user interface. The test subjects were officers of the FDF chosen in an effort to find users that would match the potential real users of a UGV system.

5.1 Demo 1 - Teleoperation

The aim of the first demonstration was to demonstrate the completion of the essential instrumentation of the vehicle and the teleoperation capability and it was held in April 2006. Teleoperation was seen as the first stepping stone on the way to autonomous driving and also a required function in the final system, sort of a backup if the autonomy cannot cope with a situation. The demonstration not only presented the teleoperation but also the whole system setup including the instrumentation in the vehicle. The demonstrator system essentially included the vehicle and the remote control station where it was controlled from. The remote control station was set up in a vehicle field maintenance tent, provided by the army base. The remote control station with a laptop computer, wheel and pedals was located so that there was no direct view to the vehicle but the driver had to rely on the video feedback and other data provided by the demonstrator system.

The temperature on the day of the first demonstration was around zero degrees Celsius and the ground was covered with snow, slush and water. The plants and trees had no leaves except for the evergreen trees. The conditions were typical early spring in Finland.

The demonstration itself was the vehicle being driven in the field and along the forest road along a defined course. The total length of the course was in the order of two kilometres, the furthest point being 500 meters from the remote control station. During the demonstration, driving speed was mostly around 20 km/h. The driver was one of the project members who had been responsible for developing the user interface.

After the demonstration, guests were given the opportunity to teleoperate the vehicle. This was to show that even though the system provided additional, more complicated functions, the basic driving is not very different from traditional driving.

The demonstration was a success as unmanned driving capability of the vehicle was achieved. The whole system was set up and made functional in a short time period of less than 9 months. However, there were notable restrictions, most of which were in some way related to the available resources. One of the main limitations was the communication link between the vehicle and the remote control station. For teleoperation, a fairly good quality video image is needed and in our setup, it was provided through 802.11 or WLAN connection. The connection is good as long as there is line of sight between the antennas but deteriorates with obstacles. Obstacles such as hills or buildings cut the connection instantly, but also vegetation and trees, even without foliage, becomes a factor within few hundred meters.

In our setup, the vehicle speed has no technical limitation. This means that the judgement is completely on the driver. The vehicle has been driven at a speed of 30-40 km/h in an area with no nearby obstacles. However, the real environment is very seldom free of obstacles and the speed has to be adjusted accordingly. The limiting factor in teleoperation systems is the delay in the control loop, or time that it takes from an driver action until the consequence of the action can be seen in the feedback. In our setup, this delay was in the order of 0.8-0.9 seconds (see section 5.4). This would allow somewhat higher speeds than what was used, but since the main focus was not on the speed, we opted for the increased safety margin instead.

5.2 Demo 2 - Path following and mission capabilities

The main goal of the second demonstration was to demonstrate the vehicle's ability to position itself, follow predefined paths and avoid obstacles. Demo 2 was held in April 2007. For autonomous navigation, the software in the vehicle had been improved to fuse positioning data from different sources: the GPS, the inertial navigation system (INS), odometry and laser scanner. Accurate positioning is the most essential requirement for autonomous driving.

Mission management was demonstrated by adding a separate C4I system with its own interface. The mission with its objective locations and tasks was planned using this interface. The mission was then transferred to the driver of the unmanned vehicle, who would plan the more precise execution of the mission. The demonstration mission included surveillance tasks but since the integration of the payload sensors was not yet complete, surveillance was done using the cameras already installed on the vehicle.

Several path segments in the demonstration area were defined and recorded by driving the vehicle along the paths and recording the integrated position information. These paths were stored in the vehicle and also copied to the remote control station to be displayed on the map interface. In the demonstration, the driver selected path segments to build up the complete route. The route also included points where the vehicle would stop. The vehicle drove through most of the path autonomously. The route was defined to include a more challenging part with narrow passages and nearby obstacles. To demonstrate the transitions between autonomous driving and teleoperation, the driver took over the control of the vehicle, drove it through the challenging part and then handed control back to the vehicle computer. The transitions were done while the vehicle was moving to demonstrate that as long as both parties are aware of the current situation, the handover does not require a full stop and a new start.

The positioning turned out to be the main limiting factor for autonomous driving at this phase. The GPS is accurate only within few metres and has sometimes difficulties on the forest roads, where trees block a significant part of the sky. The INS device used in this project was a military grade device, but was missing its original odometer. The unavailability of odometer signal resulted in problems with the device creeping. The creeping could be controlled by making intermittent stop to allow the INS to make a "zero update" (set the speed to zero and remove much of the position error). However, if this update was still in progress when the vehicle started moving again, the INS position data would often become useless. At that point, a new start up was required. Even though in the vehicle computer we had the odometry information available, we were unable to feed it into the INS.

In the demonstration the vehicle was able to follow the defined course. On the forest road the positioning was greatly improved by using the tree trunks as landmarks. Their positions had been recorded and on consequent passes, the vehicle could achieve greater path following precision.

In this second demonstration the user interface had also been upgraded to include a map interface (see Fig. 6). The map was actually an aerial image of the demonstration area and on top of it were drawn the available route segments the vehicle was able to follow. With the same view, the driver could plan the path including setting stops and different reference speeds for different parts of the path. The position of the vehicle was seen on the map and the video frame showed real time video. The driver could fine tune the autonomous path following by setting an offset the vehicle would take into account. The driver could also take over the control or stop the vehicle in case the vehicle was not performing as it was supposed to, but no such case occurred in the demonstration. Obstacle avoidance was demonstrated by placing objects on the path of the vehicle. The vehicle detected and avoided these objects by planning a path around them. The second demonstration was again successful by achieving the main goal of driving autonomously through a pre-defined course. The main source of problems and limitations with autonomous driving were the positioning issues.

5.3 Demo 3 - Unmanned mission scenario

The aim of the third demonstration was to demonstrate the capability of the whole UGV system to perform a complete mission. The final demonstration was held in October 2007. The area used was the same as in previous demos, but the exact location was changed to allow a better view of the events to the audience. Essentially the new content of the final demonstration was to put together all the new and all previously seen features of the unmanned vehicle system. As in the previous demo, a reconnaissance/surveillance mission was assigned using a separate command and control system. The first objective was to search a designated area for enemy activity using the cameras on board the vehicle. The second objective was to measure the level of radiation and test the presence of harmful chemical and biological aerosols using a dedicated NBC detector payload. The third objective was to park the vehicle at a suitable location and use it as a remote observation platform using the ground surveillance radar fitted on the roof of the vehicle. To add interest for the audience, a number of different vehicles and soldiers on foot were acting as the enemy. Also, arrangements were made so that the NBC detector could detect chemicals and radiation and raise an alarm. This demonstration also introduced a payload operator, who was in charge of operating the observation camera, the NBC detector and the ground surveillance radar.

After receiving the mission orders, the driver plotted the path for the vehicle and sent it on its way. The vehicle followed the designated path using the autonomous path following function and cruise speed controller. The payload operator was using the camera to check for "enemy" presence. For the remote observation task, the vehicle was driven and parked at one edge of an open field. The "enemy" soldiers and vehicles would then cross the field and the payload operator would pick up them on the radar and camera.

In this demonstration we experienced the most notable demonstration setback as just before starting, the network card of the NBC device broke and with no replacement part available, communication with the device became impossible. The demonstration still followed the original script, but without the NBC alarms. Apart from that, the demonstration was a success as the mission was accomplished as planned.

Fig. 6. User interface of the UGV demonstrator. On the left the map view of the demo area and on the right the video frame of the driving camera. Information about the vehicle state can be seen on the bottom right corner.

5.4 Analysis of the communication delay

Accumulation of the delay in the communication chain was measured for benchmarking purposes to analyze the system performance and to be able to evaluate future communication solutions. The delay experienced by the operator is a critical aspect in teleoperation. The total delay experienced by the operator is the sum of delays from two sources: the forward-loop delay, which is the time that passes from the moment the operator gives a command until the moment the command is executed and feedback delay, which is the time that passes from the moment something happens at the remote site, until the moment the operator gets information about the event.

Different methods have been considered to overcome the problems associated with teleoperating with long delays, including raising the level of autonomy of the system being controlled and using a model of the real system at the operator side of the system. In some cases the response of the real system is predicted and the predicted state is visualized for the operator.

In the case of this project, the forward-loop delay can be further divided into processing delay within the remote control station, transmission delay and the processing and actuation delay in the vehicle. In the feedback channels (data and video use slightly different methods) there are also delays in processing, transmission and display of information.

To measure the performance of the teleoperation system and to identify the sources of largest delays, a measurement setup was arranged. The remote control station and vehicle brake lights were recorded with a single video camera so that the delay between operator brake pedal press and brake light illumination could be calculated from the recording. The camera records 25 fps so the delay calculated from the number of frames is accurate to ± 40 ms. In an effort to improve this, we measured several runs and averaged the results.

To measure the delay in the mechatronic system, a light emitting diode was connected to another I/O output and set to illuminate simultaneously with an electrical signal that causes the brake pedal to be pushed. Again, the lights were videotaped and delay calculated from the frames. The delay between the diode lighting up and the vehicle brake light doing the same is essentially the time it takes to accelerate and move the actuating motor enough to push the brake pedal. There are small delays associated with the electronics (e.g. H-bridge driving the motor, led lights up faster than a bulb), but these can be omitted in the time scale we are working in.

The delay in the video image was measured similarly. A person standing in the view of the vehicle cameras waved an arm and the video camera recorded the person and the remote control station display showing the person on screen. The delay was again calculated from the frames in the recording.

Total forward-loop delay from pushing the teleoperator's brake pedal until vehicle brake lights lit up averaged to 560 ms (see Fig. 7.). The delay in the mechatronic system was measured to be 224 ms on average. All processing delays proved insignificant. The longest delays seemed to be associated with the radio. It took an average of 665 ms, which seemed surprisingly long, but was verified to be correct. The video delay was measured with different image compression settings, but the differences in delays were not statistically significant. The video delay averaged to 340 ms for road camera and 550 ms for observation camera.

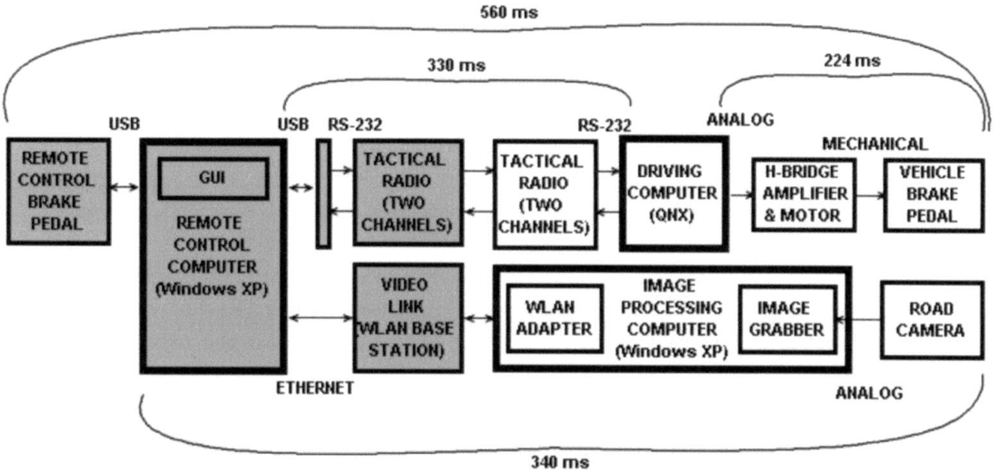

Fig. 7. Data flow and delays in teleoperation. The upper part depicts the forward-loop delay and the lower part the video delay of the road camera.

5.5 Teleoperation tests

In an effort to gain a better understanding of the requirements for an unmanned vehicle user interface, to somehow measure the quality of the current implementation and enable its further development, a separate project was launched. The aim was to find out what requirements there should be for the interface and how they could be measured. Three most important usability attributes out of Nielsen's total of five (Nielsen, 1993) were identified. They were low error rate, efficiency and memorability.

People involved in the development process of a system often adapt to using it and become blind to its problems and shortcomings. This is why "real" users were used to review our user interface. The test subjects were officers of the FDF chosen in an effort to find users that would match the potential real users of a UGV system. The test setup is shown in figure 8.

The user interface used in the experiment was essentially the same as in the final third demonstration. Much of the setting was also the same: the vehicle was driven from a tent mostly at the open field. The operator could see the vehicle at its starting position, but during the experiments it was driven so far and also to places where it could not be directly seen. The test subjects were driving the vehicle first assisted and after becoming comfortable with the controls, unassisted. They were also utilizing some of the automatic functions such as the cruise controller. They were asked to think aloud their thoughts during the experiment. The user interface, user actions and speech were recorded using a video camera and logging user actions by software. After the experiment the subjects were interviewed. In addition to user testing, formal evaluation without users (Lewis & Rieman, 1993) - also known as expert methods - such as cognitive walkthrough and usability heuristics were used.

Problem spots and ideas for improvement

The video images in the user interface are essential. The observation camera was found to provide better view of the terrain ahead. This is somewhat surprising since it is located higher and contours of the terrain are harder to see. The colors are the likely quality factor since the lower camera is a monochrome one. Some of the most significant problems and ideas for improvement were:

- It was difficult to understand which way the observation camera was pointing. A compass or preferably a 3D-arrow or vehicle should be overlaid to visualize the direction.
- The fixed camera's viewing angle was not wide enough to see the road while making sharp turns. Wider lens would improve this, but also introduce distortion, which is an additional burden for image processing in case the camera is used also for that.
- The vehicle traveling direction nor planned path could not be seen in the view. These should be overlaid to aid the driver to see where he should drive and where the vehicle is going. The steering wheel angle gives a rough idea, but to see the path in the image would provide additional help.
- Obstacles could be highlighted and/or distance coded. Judging the distance to an obstacle becomes harder with non-stereoscopic vision.
- The direction of the cameras and vehicle and projected path and speed of the vehicle would improve the map view. The map should (optionally) turn vehicle traveling direction up.
- Vehicle is often driven at same setting for long periods. A hand-operated throttle similar to tractors or airplanes would ease strain on foot.
- Emergency stop buttons were available both in software and as a small button on the wheel. A better option would be a separate big, fixed switch.
- Turning the observation camera was done by buttons on the wheel, but as the wheel was turned, it was difficult to see which way they worked. A separate controller would be better.
- The system has a number of modes to allow direct teleoperation, some automatic functions and autonomous driving. The current mode was displayed as big text and color, but this was unintuitive.
- The number of modes should be low to make a simple interface, but more modes would allow better control and freedom for the driver. Selecting the mode could be automatic, but the user should be kept aware of the mode currently in use. Even when the user specifically selects the mode, different modes are known to cause mode confusion, which is a known issue in aviation psychology (Bredereke & Lankenau, 2002) and also a significant problem area in the control of automatic systems.

The teleoperation tests showed that test subjects were able to learn to drive the vehicle in a relatively short time. They were generally comfortable operating it and said they were positively surprised of the general easiness of driving. On the other hand, as the user tests and expert methods revealed, the system and the user interface were still quite limited and there was much room for improvement. Even though our UGV project was not aiming to build a product, the usefulness of involving outside users in testing its user interface was clear. In a bigger scope, the robustness and reliability of autonomous systems in unstructured environment still remains a challenge.

Fig. 8. Setup for the teleoperation tests. Observing researcher sitting on the left and test operator using the remote control devices on the right. The UGV starting the mission can be seen in the background.

6. Summary and conclusions

The presented UGV technology demonstrator was successfully instrumented using mainly affordable COTS components and thus represents a minimum realization able to realistically demonstrate capability potential with the given missions. The test tasks were related to tactical reconnaissance and surveillance and the UGV was able to cope with the predefined tasks largely autonomously, requiring operator support only with demanding decision making and in unexpected problem situations. Capability for task-oriented operation was enabled by the systems level solution that combines autonomous motion and task execution capabilities of the UGV, as well as the scalable level of autonomy in the HMI supporting the remote operator. Also, systems integration and information flow between the C4I system and the UGV remote control station was essential part of the whole system.

In the first phase of the UGV project, unmanned driving capability of the vehicle was achieved. It was noticed that a fairly good quality video image is needed for the teleoperation. This requires a lot from the communication link, not only range but also delay characteristics are important. In the UGV demonstrator case, the maximum distance was limited to approximately one kilometre by the WLAN connection at our test site. In our setup, the vehicle speed had no technical limitation but the judgment depending on the sensed delay was left completely for the teleoperator. Real environment is very seldom free of obstacles and the speed has to be adjusted accordingly. The limiting factor for speed in teleoperation systems is the delay in the control loop, or time that it takes from a driver action until the consequence of the action can be seen in the feedback. Another observation made was that more feedback - an audio signal for example - from the vehicle would have been useful.

The second phase demonstrated more autonomous features of the vehicle. Autonomous driving capability was achieved and the HMI was developed especially for the mission execution. Positioning of the vehicle was accurate enough but it was discovered that the environmental model was clearly inadequate for reliable autonomous operation. Obstacle avoidance, for example, was not to be trusted since some obstacles remained unobserved. Therefore, the operator had to supervise the vehicle at all times, which is not convenient in the long run. More sensors should be added to the sensor system in order to enhance the environmental model of the UGV and situation awareness of the operator.

In the final demonstration the mission capability was demonstrated. A fictitious reconnaissance mission was assigned from the C4I system to the operator, who formulated the mission for the vehicle. Payload was teleoperated, other tasks were performed autonomatically under the supervision of and occasional correction by the operator.

Autonomous task execution with UGVs is still relatively new area of research and it is still in its infancy. Fully autonomous task execution, especially in hostile environment, is not yet - if ever will be - practical. There is a growing research interest (both civilian and military) in the higher level control of mobile robotics that deals with task execution and related HMIs to assist and supervise this capability. Therefore, scalable level of autonomy is needed to enable smooth task execution and seamless operator involvement.

At the same time, the development of navigation and positioning systems has gone a long way and it has been shown that those problems can be solved. As an example, there are already commercial devices and features which can be added to vehicles to improve the performance and situation awareness of the driver.

Although UGVs coping with dynamic environment have been under research and development only for the past few decades, the results of the development of subsystems are already visible. Nowadays, it is possible to set up a functional UGV system, relatively cost-effectively, by integrating commercial subsystems. In other words, one does not have to design and produce every single subsystem from scratch anymore. However, a system integrated from commercial components might be functional, but it is not in any way optimized or fit for military operations. To achieve reliability needed in field operations, a great deal of platform specific development and optimization is needed.

The most difficult problems within current UGV technology are qualitative environment sensing, context awareness and communications. Also, it is essential to observe humans from the vicinity of the UGV for safe operation. It can be estimated that the role of the driver in both commercial and military vehicles will be more of a supervisory one in the future. There will be fully autonomous vehicles for specific tasks and applications, but for most vehicles, the driver - either on or off the vehicle - will be in charge and take over the control when necessary.

7. Acknowledgements

Our research was funded by the Finnish Defence Forces' technology program and additionally by the Scientific Advisory Board for Defence. The authors also want to thank other project partners, especially Patria Group and VTT (Dr. Hannu Lehtinen, Dr. Ilkka Kauppi, Petri Kaarmila and Jarmo Prokkola).

8. References

Armchair General online. (2004). Soviet Army (RKKA) in World War II. [Online] Available at: http://www.armchairgeneral.com/rkkaww2/weapons/tanks1.htm [Accessed 29 April 2009]

Bredereke, J. & Lankenau, A. (2002). A Rigorous View of Mode Confusion, *Proceedings of SafeComp 2002*, pp.19-31, ISBN 3-540-44157-3, Catania, Italy, September 2002, Springer Verlag

Cage, D.W. (1995). UGV HISTORY 101: A Brief History of Unmanned Ground Vehicle (UGV) Development Efforts. *Unmanned Systems Magazine*, Special Issue on Unmanned Ground Vehicles. Vol. 13, No. 3, Summer 1995 Available at: http://www.spawar.navy.mil/robots/pubs/ugvhist95-nopix.pdf [Accessed 29 April 2009]

DARPA (not dated). DARPA Urban Challenge. [Online] Available at: http://www.darpa.mil/grandchallenge/index.asp [Accessed 27 April 2009]

ELROB (not dated). European Land-Robot Trial. [Online] Available at: http://www.elrob.org/ [accessed 28 April 2009]

GDRS (not dated). Autonomous Robotics Programs,Mobile Detection Assessment and Response System (MDARS). [Online] Available at: http://www.gdrs.com/robotics/programs/program.asp?UniqueID=27 [Accessed 28 April 2009]

IAI (not dated). Guardium - Autonomous Security Vehicle. [Online] Available at: http://www.iai.co.il/17483-31663-en/default.aspx [Accessed 5 May 2009]

John Deere (2008). R-GATOR. A stand-alone autonomous vehicle and a rugged, versatile platform for tactical payloads. [Online] (updated September 2008) Available at: http://www.deere.com/en_US/contractsales/fedmilitarysales/media/pdf/rgator_08_3480.pdf [accessed 4 May 2009]

Kairos Autonomi (not dated). Pronto4 Strap-on Autonomy System for Existing Vehicles or Vessels. [Online] Available at: http://www.kairosautonomi.com/pronto4_system.html [Accessed 4 May 2009]

Kaleun, T. (not dated). Light demolition carrier "Goliath" (Sd.Kfz.302/303a/303b) [Online] Available at: http://www.geocities.com/CapeCanaveral/Lab/1167/egoliath.html [Accessed 29 April 2009]

Lewis, C. & Rieman, J. (1993) Task-Centered User Interface Design: A practical introduction. Shareware, published by the authors. Available at http://hcibib.org/tcuid/ [Accessed 26 April 2009]

Lightoller, C.H. (1935). Titanic and other ships. [Online] Project Gutenberg of Australia (updated April 2005) Available at: http://gutenberg.net.au/ebooks03/0301011h.html [Accessed 28 April 2009]

Lohmann, R.H.C. (2007) About Group Rapid Transit and Dual-Mode Applications [Online] Available at: http://www.2getthere.eu/media/apm07_paper_-_about_group_rapid_transit_and_dual-mode_applications.pdf [Accessed 29 April 2009]
NASA (2009). NASA's Mars Exploration Program. [Online] (updated Mar 2009) Available at: http://marsprogram.jpl.nasa.gov/missions/ [Accessed 29 April 2009]
Neumann, G.P. (1921). *The German air force in the Great War*. Translated from German by J.E. Gurdon. Hodder and Stoughton limited, London. Available at: http://www.archive.org details/germanairforcein00gurduoft [Accessed 28 April 2009]
Nielsen, J. (1993). *Usability Engineering*. Academic Press. San Diego, CA. ISBN-13 978-0-12-518406-9
NREC (not dated). Unmanned Vehicle Design. [Online] Available at: http://www.rec.ri.cmu.edu/projects/unmanned/index.htm [Accessed 4 May 2009]
PATH (not dated). AVCSS Vehicle Demonstrations, Vehicle Control.[Online] Available at: http://www.path.berkeley.edu/PATH/Research/demos [Accessed 27 April 2009]
Sandia (2008). Accident Response Mobile Manipulator System (ARMMS) [Online] (updated Jan 2008) Available at: http://www.sandia.gov/isrc/ARMMS.html [Accessed 26 April 2009]
Sandvik (2008). Automine [Online] (updated Mar 2008) Available at: http://www.miningandconstruction.sandvik.com/ (products - mine automation systems - Automine) [Accessed 4 May 2009]
Schmidhuber, J. (2005). Prof. Schmidhuber's highlights of robot car history. (updated 2007) Available at: http://www.idsia.ch/~juergen/robotcars.html [Accessed 4 May 2009]
Sheridan, T.B. (1992), *Telerobotics, Automation, and Human Supervisory Control*, The MIT Press. ISBN 0-262-19316-7, Cambridge, Massachusetts
SPAWAR (2007a). Evolving Paradigms of Human-Robot Interaction [Poster]
SPAWAR (2007b). Unmanned Systems Branch Honors WWII UAV Pioneer, In: *Robotics update*. Winter 2007 / Vol. 7, No. 2. [Online]. Available at: http://www.spawar.navy.mil/robots/newsletter/RoboticsUpdate_7_2.pdf [Accessed 26 April 2009]
Tekes (2006). Kalmar Industries Oy Ab: The automatic port of Brisbane. [Online] (updated Apr 2007) Available at: http://www.tekes.fi/eng/success_stories/Menestystarina_tiedot.asp?id=4916 [Accessed 4 May 2009]
UniBwM (not dated). Vorarbeiten UniBwM: Die Versuchsfahrzeuge VaMoRs und VaMP. [Online] Available at: http://www.unibw.de/lrt13/tas/mitarb/prof/vamp.pdf?searchterm=VaMP [Accessed 16 November 2008]
U.S.Army (2008). Future Combat Systems. [Online] (updated May 2008) Available at: https://www.fcs.army.mil/systems/index.html [Accessed 4 May 2009]
Wade, M. (2008). Lunokhod. [Online] Available at: http://www.astronautix.com/craft/lunokhod.htm [Accessed 29 April 2009]
Washington Technology (2008). General Dynamics to build robot vehicles. [Online] (updated Feb 2008) Available at: http://washingtontechnology.com/articles/2008/02/25/ general-dynamics-to-build-robot-vehicles.aspx [Accessed 29 April 2009]
Velodyne (not dated). Velodyne High Definition Lidar. [Online] Available at: http://www.velodyne.com/lidar [Accessed 27 April 2009]

16

Unidirectional feeding of submillimeter microparts along a sawtooth surface with horizontal and symmetric vibrations

Atsushi Mitani[1] and Shinichi Hirai[2]
[1]*Department of Design, Sapporo City University*
[2]*Department of Robotics, Ritsumeikan University*
Japan

1. Introduction

Devices to feed along microparts, such as ceramic chip capacitors and resistors, have become more common, due to their use in sorting, inspecting, and shipping mass produced microparts. In microparts feeding, to feed along microparts in one direction, the driving force applied to each micropart must vary according to the direction of movement of the micropart. Especially, the movement of microparts smaller than submillimeter can be affected not only inertia but also adhesion which is caused by electrostatic, van der Waal's, intermolecular, and surface tension forces (Ando, 1997). Therefore, we need to derive dynamics including adhesion to evaluate the movement of microparts.

We have previously shown that a sawtoothed surface with simple planar and symmetric vibrations can be used to feed along microparts (Figure 1) (Mitani, 2006). In this case, contact occurs in one of two ways: point contact, the point of the tooth contacts the fed part, and slope contact, the sloping side of the tooth contacts the micropart. Because of the difference in contact area of micropart with the sloping side of a tooth and with the other side, microparts adhere more strongly in one direction than in the other. Also, the driving forces transferred from vibrations of feeder surface vary according to contact. These result in the microparts moving in one direction with simple planar symmetric vibrations.

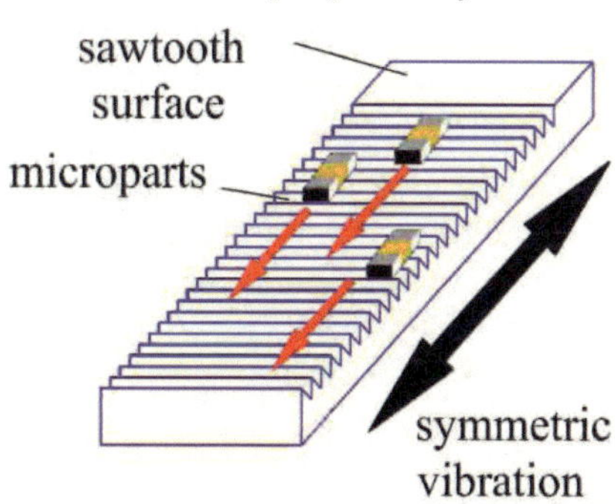

Fig. 1. Diagram of microparts feeding using a sawtoothed surface

We assessed the effect of sawtoothed silicon wafers for feeding of 0603 capacitors (size, 0.6 x 0.3 x 0.3 mm: weight, 0.3 mg). Using these experimental results, we verified relationship among feed velocity, driving frequency, and sawtooth pitch. Analysing contact between feeder surface and a micropart based on measurements using a microscope, we developed feeding dynamics including adhesion. Comparing experiments with feeding simulation using the dynamics derived, we found large errors between both results. To examine these errors, we observed the movement of a micropart when the micropart moved in one direction using a high speed video camera. We then found that the micropart rotated around vertical axis against the feeder surface and swung around the axis parallel to the tooth groove, thus reductions of feed velocity occurred. Consequently, the feeding dynamics considering these movements were needed for more accurate simulations.
The objective of this work was to examine the dynamics of microparts tens or hundreds of micrometers in size. We found that the movement of these parts depends on both inertia and adhesion.

2. Related Works

Partsfeeder is a key device in factory automation. The most popular feeders are vibratory bowl feeders (Maul, 1997), which use revolving vibrators to move parts along a helical track on the edge of a bowl. Linear feeders as well as an inclined mechanism and oblique vibration for unidirectional feeding (Wolfsteiner, 1999), have also been developed. In all of these systems, the aspect ratio of the horizontal/vertical vibrations must be adjusted to prevent parts from jumping. In our system, however, this adjustment is not necessary because only horizontal vibration is used.
A parts feeding that employs non-sinusoidal vibrations (Reznik, 2001) has been developed. The part moves to its target position and orientation or is tracked during its trajectory by using the difference between the static and sliding friction. Our system realizes unidirectional feeding by symmetric vibration of a sawtoothed surface, which yields different contact forces in the positive and negative directions.
Designing have been tested by simulation (Berkowitz, 1997 & Christiansen, 1996). The focus was mainly on the drive systems such as the structure and actuator, the movement of fed parts was generally neglected. In contrast, the movement of the microparts are considered in the present study.
Attempts have been made to improve the drive efficiency by feedback control systems (Doi, 2001) and nonlinear resonance systems (Konishi, 1997). Our system depends only upon contact between the feeder surface and the micropart. So the driving system is simple and uses an open loop system for feeding.
Micro-electro-mechanical systems (MEMS) technology has been used to mount on a planar board arrays of micro-sized air nozzles which, by turning on or off their air flow, have been used to control the direction of moving microparts (Fukuta, 2004 & Arai, 2002).
It is possible to perform manipulation with ciliary systems (Ebefors, 2000) and vector fields (Oyobe, 2001) without sensors. In this case, there are many actuator arrays on a vibratory plate. Actuator arrays enable control of contact between the vibratory plate and micropart in order to accomplish the target manipulation. However, these studies did not mention the dynamics of the micropart, especially the effects of adhesion forces on its motion. Other various feeding systems using electric-field (Fuhr, 1999), magnetic (Komori, 2005), bimorph

piezoelectric actuators (Ting, 2005), and inchworm systems (Codourey, 1995) have been developed. These studies, however, have also not investigated the contact between the feeder surface and the micropart.

3. Principe of unidirectional feeding

Let us first look at a typical micropart, a 0603 ceramic chip capacitor used in electronic devices (Figure 2). Then let us analyse feeding by developing a model for contact between a micropart and a sawtooth.

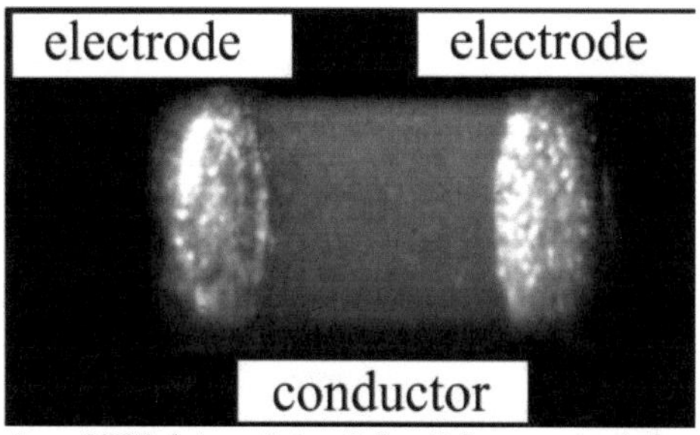

Fig. 2. Ceramic chip capacitor 0603 (size, 0.6 x 0.3 x 0.3 mm: weight, 0.3 mg)

A capacitor consists of a conductor and electrodes with convexities on each end surface. We obtained representative contours along a capacitor using a Form Talysurf S5C sensing-pin surface measurement tool (Taylor Hobson Corp.) (Figure 3). Electrodes contact the feeder because they protrude 10 μm higher than the conductor.
Assuming that convexities are perfectly spherical (Figure 4 (a)), let r be the radius of a convexity (Figure 4 (b)). The feeder surface is sawtoothed (Figure 5), let θ be sawtooth elevation angle, p sawtooth pitch, and d the groove depth. The sawtooth contacts the electrode in one of two ways (Figure 6) - at the tooth point or at the tooth slope. To drive the microparts unidirectionally, driving must depend on the contact and direction of movement.

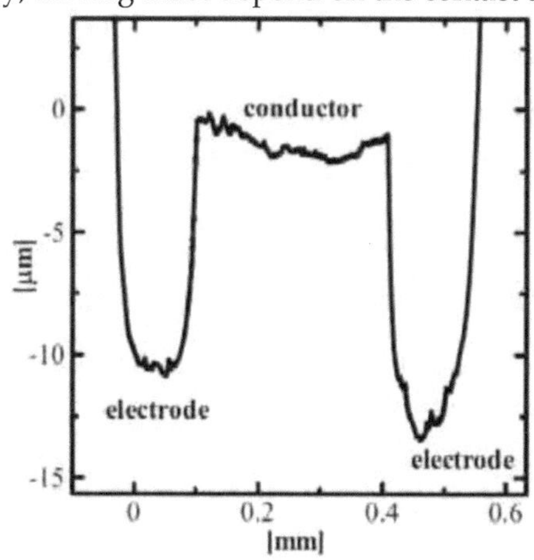

Fig. 3. A section of 0603 capacitor

(a) surface model (b) convexity

Fig. 4. Model of surface convexity on an electrode

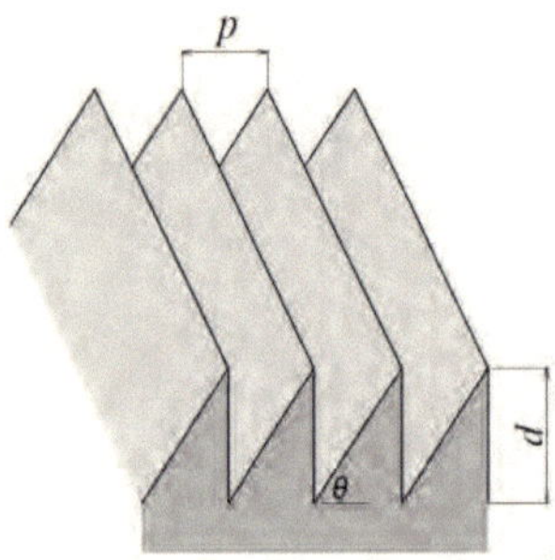

Fig. 5. Model of sawtooth surface

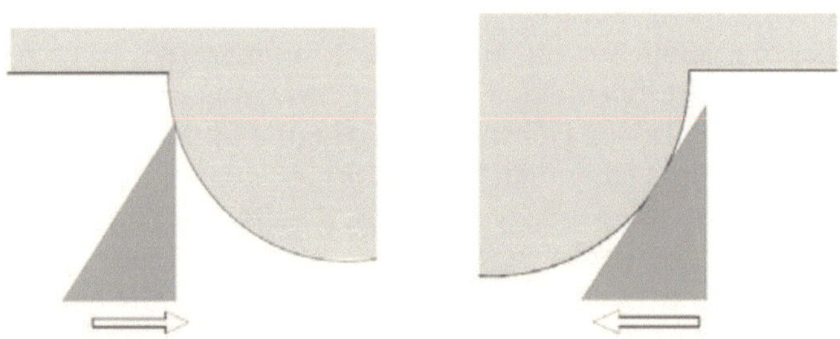

(a) at tooth point (b) at tooth slope

Fig. 6. Two contacts between micropart and sawtooth

4. Feeding experiments of 0603 capacitor

4.1 Experimental equipments

In micropart feeder (Figure 7), a silicon wafer is placed at the top of the feeder table, which is driven back and forth in a track by a pair of piezoelectric bimorph elements, powered by a function generator and an amplifier that delivers peak-to-peak output voltage of up to 300 V.

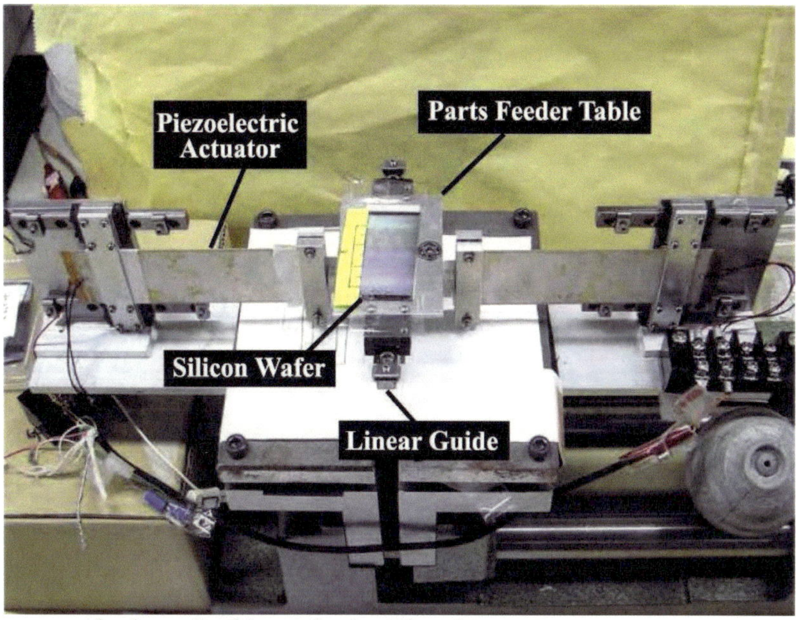

Fig. 7. Microparts feeder using bimorph piezoelectric actuators

4.2 Sawtooth surfaces

We used a dicing saw (Disco Corp.), a high-precision cutter-groover using a bevelled blade to cut sawteeth in silicon wafers. Figure 8 shows a microphotograph of a cut silicon wafer with sawteeth of $p = 0.1$ mm, $\theta = 20$ deg, and $d = p \tan \theta = 0.0364$ mm. We prepared sawtoothed silicon wafers with pitch $p = 0.01, 0.02, \cdots,\ 0.1$ mm and elevation angle $\theta = 20$ deg.

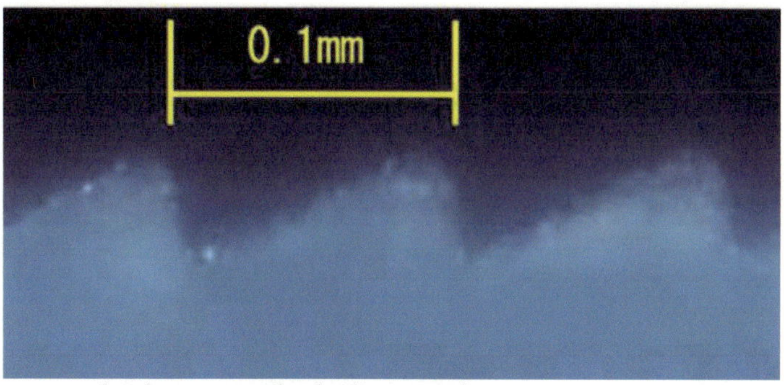

Fig. 8. Microphotograph of a sawtoothed silicon wafer

4.3 Experiments

Using the microparts feeder and these sawtoothed surfaces, we conducted feeding experiments with 0603 capacitor. Micropart movement was recorded using a digital video camera at 30 fps. Velocity was measured by counting how many frames it took for a micropart to move 30 mm along the sawtooth surface. Microparts moved at a drive frequency f = 98 to 102 Hz and feeder table amplitude was about 0.20 mm. Each value is the average of three trials, each trial using five capacitors (Figure 9).

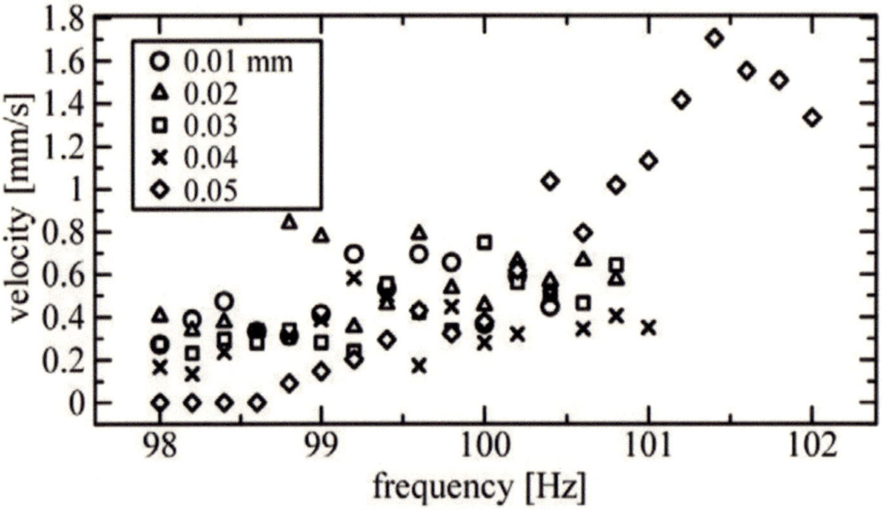

(a) p = 0.01 to 0.05 mm

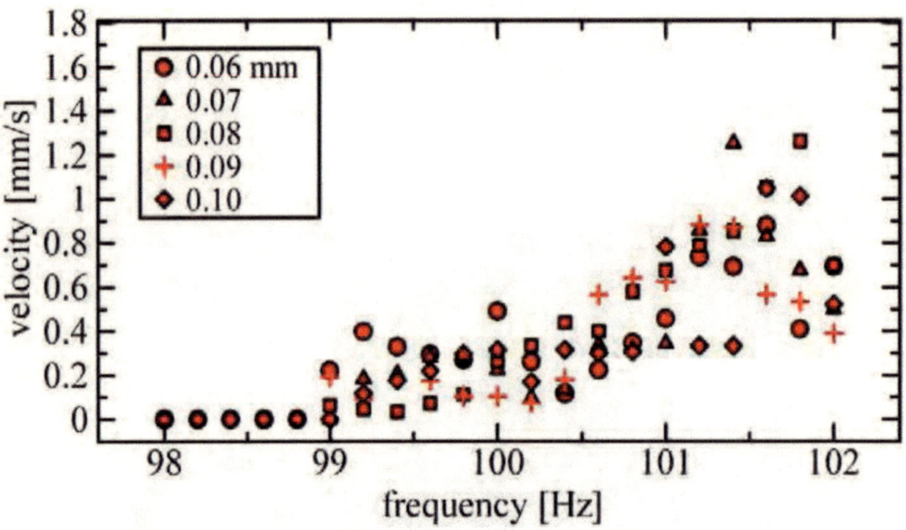

(b) p=0.06 to 0.10 mm

Fig. 9. Experimental results of 0603 capacitor

Table 1 shows the drive frequency that realized maximum velocity for each pitch, and its maximum velocity. When the pitch was 0.04 mm or less, velocity was 0.6 mm/s at drive frequency f = 98 to 101 Hz, but movement was jittery. At higher drive frequency, the microparts jumped. Fastest feeding was 1.7 mm/s, realized at f = 101.4 Hz with p=0.05 mm. When the pitch was 0.06 mm or greater, maximum feed velocity on a surface was realized

when drive frequency was 101.4 Hz. The maximum velocity decreased with increasing pitch, indicating the appropriate pitch for 0603 capacitors is p = 0.05 mm.
Figure 9 shows velocity dispersion at the maximum feed velocity on each sawtooth surface. Feed velocity dispersed within 6.7 to 23.5 %, averaging 15.8 %. The smallest dispersion occurred at a sawtooth pitch of 0.05 mm. Consequently, the sawtooth surface with pitch p = 0.05 mm was most appropriate for feeding 0603 capacitor.

pitch, mm	velocity, mm/s	frequency, Hz
0.01	0.695	99.2
0.02	0.839	98.8
0.03	0.749	100.0
0.04	0.582	99.2
0.05	1.705	101.4
0.06	0.880	101.6
0.07	1.253	101.4
0.08	1.262	101.8
0.09	0.883	101.2
0.10	1.049	101.6

Table 1. Maximum feed velocity of 0603 capacitor and drive frequency

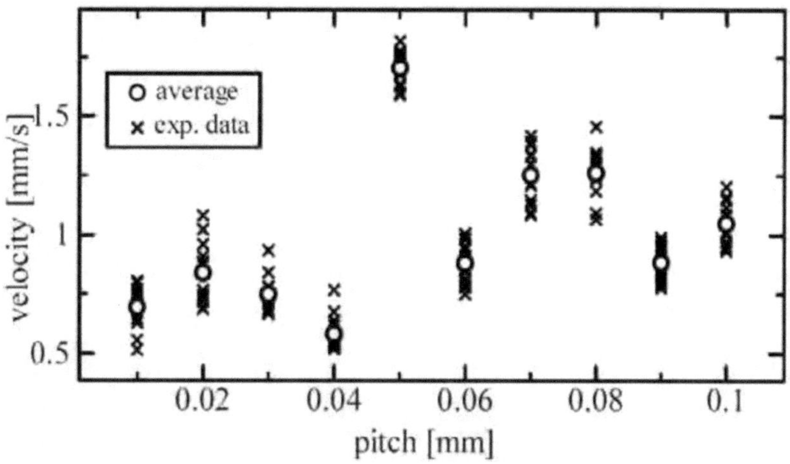

Fig. 10. Relationship between feeding velocity and sawtooth pitch

5. Analysis of 0603 capacitor

5.1 Measurement tools

As in the previous work (Mitani, 2006), the sawtooth surface profile should be selected according to the convexity size on the surface of the capacitor electrodes. To observe them, we used AZ-100 multi-purpose zoom microscope (Nikon Instruments Co.) (Figure 11), which can take pictures at up to 16 times magnification. The microscope also has an automatic stage to control focus height at a resolution of 0.54 μm. Each image is forwarded to a personal computer and saved as a bitmap file. We used DynamicEye Real focus image synthesizing software (Mitani Corp.) to analyse these convexities. The software can synthesize a three dimensional (3D) model from these pictures according to focus height. Sections of the 3D model are analysed to obtain a convexity size and position.

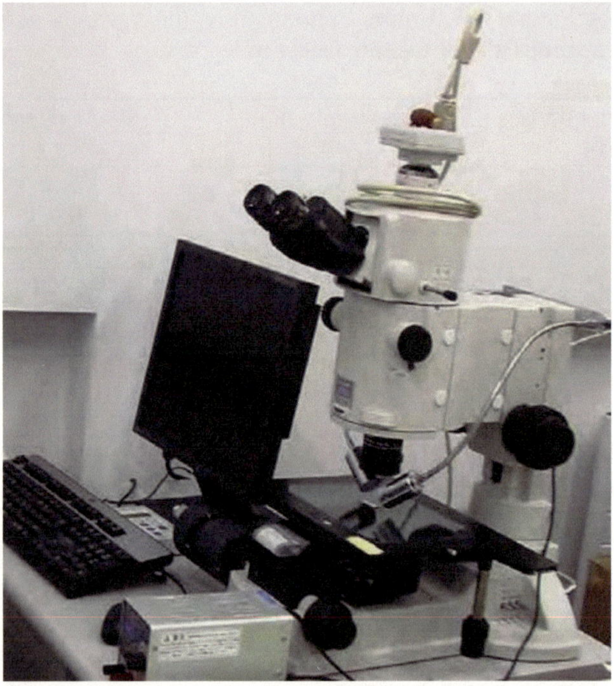

Fig. 11. AZ-100 multi-purpose zoom microscope (Nikon Instruments Co.)

5.2 Convexity size and position

We assumed that each convexity on the electrodes of capacitor was defined as a half sphere. The radii of each convexity and its position were analysed from the 3D model. Analysing a synthesized model (Figure 12), we obtain a contour line of the synthesized model, defining the micropart coordinate G-xy (Figure 13). In this figure, the arrowed convexities could be disregarded because the convexities labelled as **A** occurred besides the capacitor, and the convexities labelled as **B** did not occur on any electrode of the capacitor. We thus defined four convexities on the surface of the 0603 capacitor.

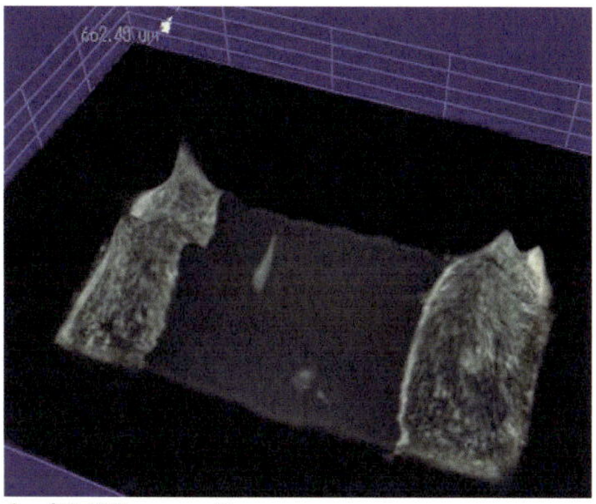

Fig. 12. Synthesized model of 0603 capacitor

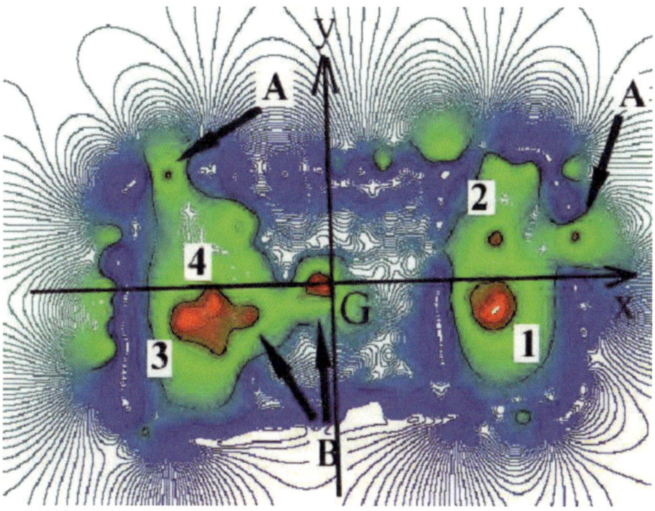

Fig. 13. Contour model

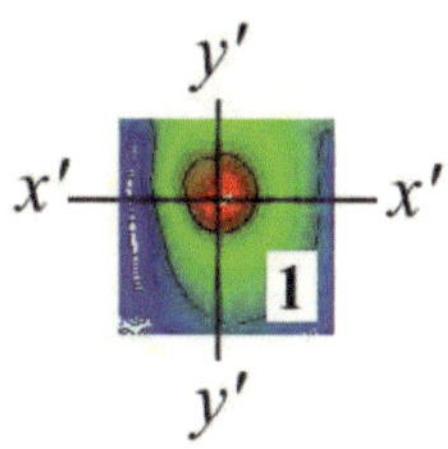

Fig. 14. Analysis line of convexity #1

Let us analyse convexity size from the 3D model. We first analysed the convexity #1 along a line x′x′ parallel to the x axis, and a line y′y′ parallel to the y axis, both lines pass the top of the convexity (Figure 14), and then we obtained two section models shown in Figure 15. Similarly, we analysed and obtained each section of convexities #2, #3, and #4, (Figures 16 to 18). Each convexity was approximated in a half sphere from the top to less than 18 μm. The radii of each convexity were assumed to be the mean value of radii along both directions.

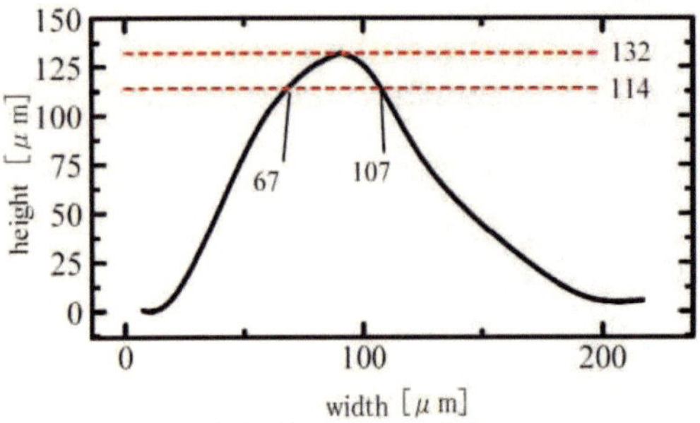

(a) along line x′x′

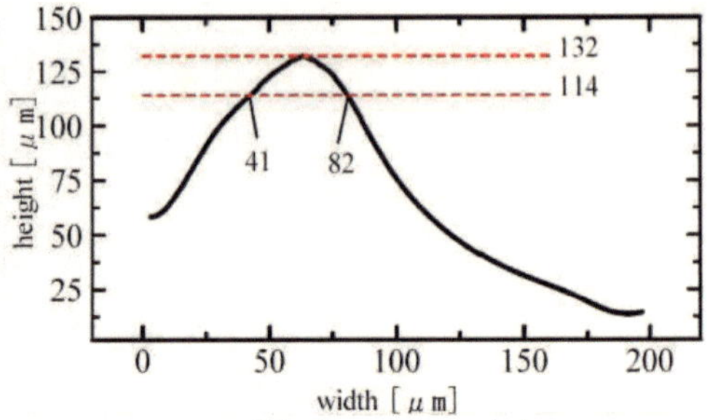

(b) along line y′y′

Fig. 15. Sections of convexity #1

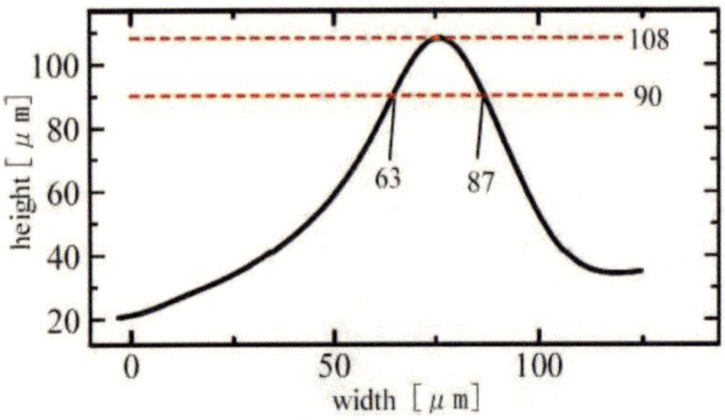

(a) along line x′x′

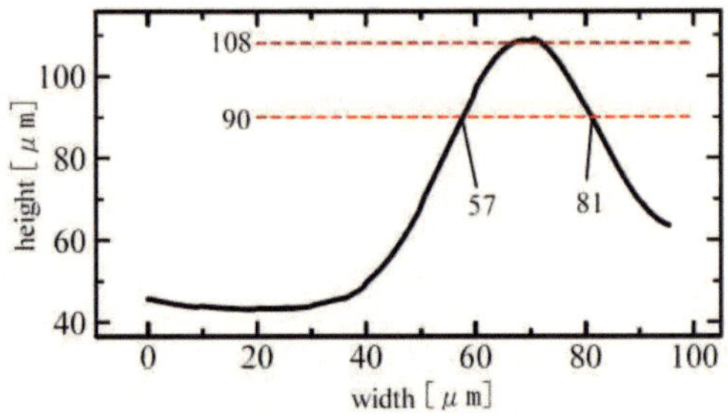

(b) along line y′y′

Fig. 16. Sections of convexity #2

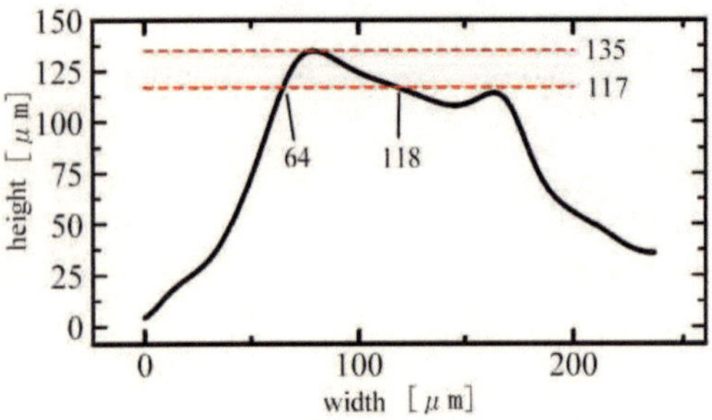

(a) along line x′x′

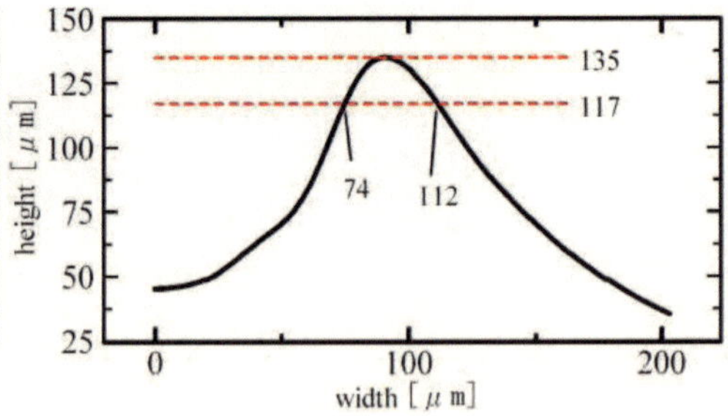

(b) along line y′y′

Fig. 17. Sections of convexity #3

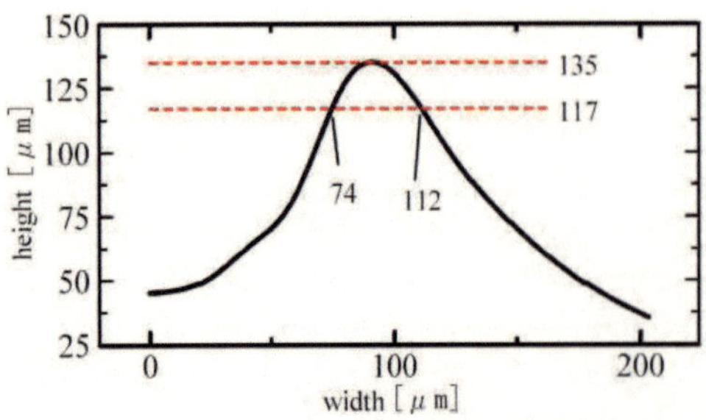

(a) along line x′x′

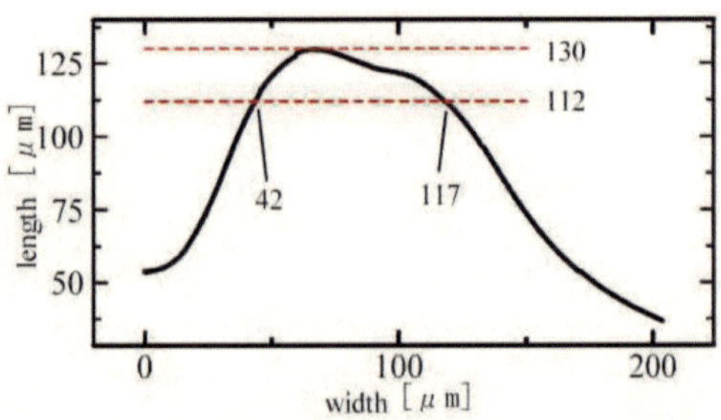

(b) along line y′y′

Fig. 18. Sections of convexity #4

From Figure 13, we measured position of each convexity with the top of each convexity on G-xy. Finally, we obtained convexity size and position appeared in Figure 13 (Table 2), and defined surface model of a 0603 capacitor (Figure 19).

no.	cordinate (x, y), μm	radus, μm
1	(207, -37)	20
2	(216, 51)	13
3	(-241, -36)	24
4	(-200, -6)	36

Table 2 Coordinate and radius of convexity

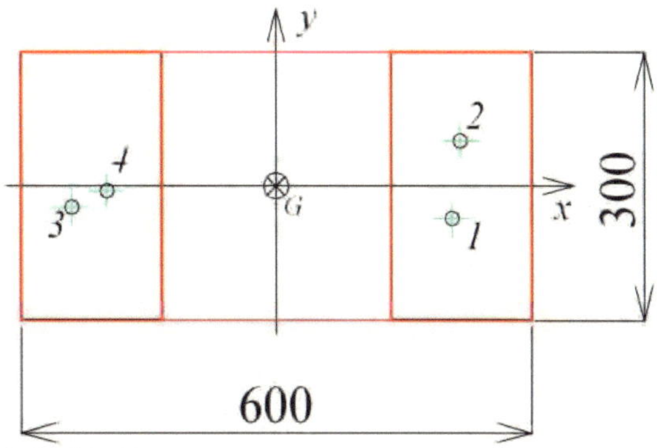

Fig. 19. Convexity model of 0603 capacitor

6. Feeding simulation and comparison

6.1 Feeding dynamics

We have already derived the dynamics of micropart when a convexity exists on the surface of micropart (Mitani, 2006). We extended these results to plural convexities. We defined the feeder coordinate O-x_0y_0 and micropart position and posture on its coordinate P = (x_c, y_c, φ).

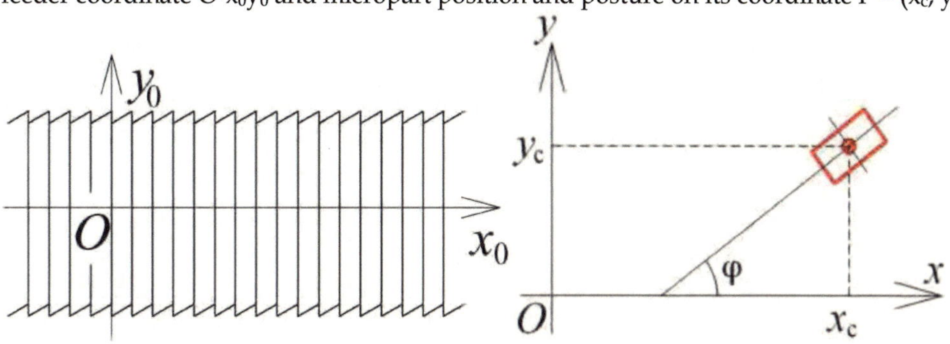

Fig. 20. Position of micropart on coordinate

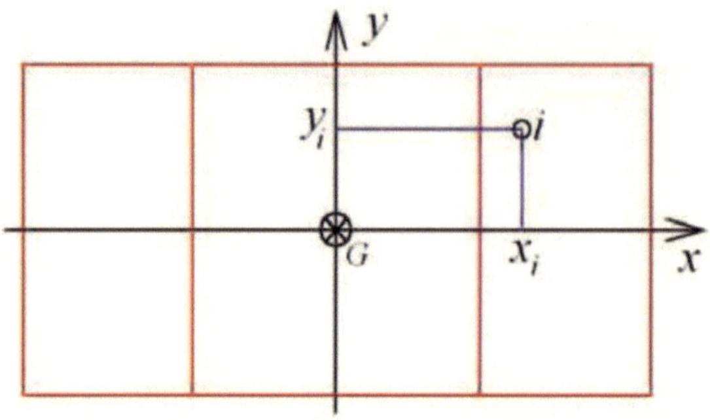

Fig. 21. Position of convexity i on the coordinate G-xy

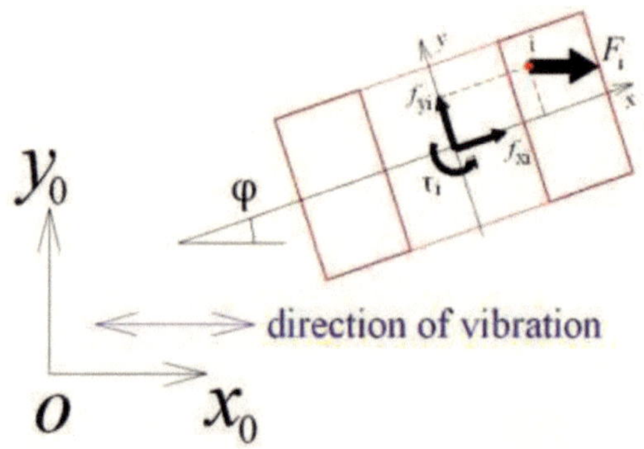

Fig. 22. Driving force of micropart transferred from convexity

We also defined potion of the i-th convexity as $c_i = (x_i, y_i)$ on the coordinate G-xy. Dynamics of micropart is represented as:

$$\begin{bmatrix} F_x \\ F_y \\ \tau \end{bmatrix} = \begin{bmatrix} m & 0 & 0 \\ 0 & m & 0 \\ 0 & 0 & I \end{bmatrix} \begin{bmatrix} \ddot{x}_c \\ \ddot{y}_c \\ \ddot{\varphi} \end{bmatrix} + \begin{bmatrix} c & 0 & 0 \\ 0 & c & 0 \\ 0 & 0 & d \end{bmatrix} \begin{bmatrix} \dot{x}_c \\ \dot{y}_c \\ \dot{\varphi} \end{bmatrix} \tag{1}$$

where m indicates mass of micropart, I intertia, c attenuation coefficients of motion, and d attenuation coefficients of rotation. Driving force and torque, $f \equiv (F_x, F_y, \tau)^T$, is calculated by the sum of driving force transferred from each convexity. Force generated by vibration of feeder surface occurs along direction of vibration. Considering the driving force $f_i \equiv (f_{xi}, f_{yi}, \tau_i)$ generated by contact force F_i, vibration force at i-th convexity shown in Figure 22, we found:

$$f_i = \begin{bmatrix} F_i \cos\varphi \\ -F_i \sin\varphi \\ -F_i \sqrt{x_i^2 + y_i^2} \sin(\varphi + \tan^{-1}(y_i/x_i)) \end{bmatrix} \tag{2}$$

Assuming that 1st, 2nd, …, and n-th convexities appear on the surface of a micropart, driving force f is represented as follows:

$$f = \sum_{i=1}^{n} f_i \tag{3}$$

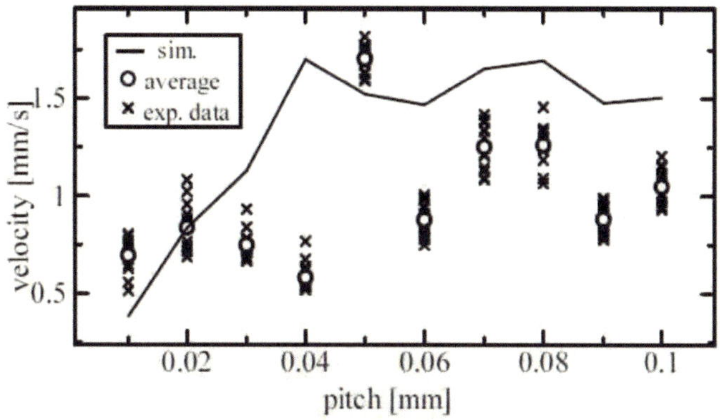

Fig. 23. Comparison of experimental and simulation results

6.2 Feeding simulation

In equation (2), each contact force F_i is decided according to its contact between a sawtooth and a convexity (Mitani, 2006). We conducted feeding simulation of the 0603 capacitor model shown in Figure 21, using the same parameters as feeding experiments, and then compared with experimental results (Figure 23). In the simulation, feed velocity peaked at p = 0.04 mm, whereas it peaked at p = 0.05 mm in the experiments. At the pitch of 0.01 to 0.04 mm, velocities were proportional to the sawtooth pitch. At the pitch of 0.07 to 0.1 mm, the experimental results were about 0.5 mm/s lower than simulation though the tendency was the same. Consequently, there were large differences between the simulation and experimental results. In the next section, we examine these differences by analyzing the micropart movement and feeder surface.

7. Examination of simulation error

7.1 Observation of micropart movement

We used Fastcam-1024PCI highspeed video camera (Photron) to capture micropart movement at 1000 fps. A 0603 capacitor was initially placed lengthwise on the feeder and the video camera was set to the side of the capacitor (Figure 24).

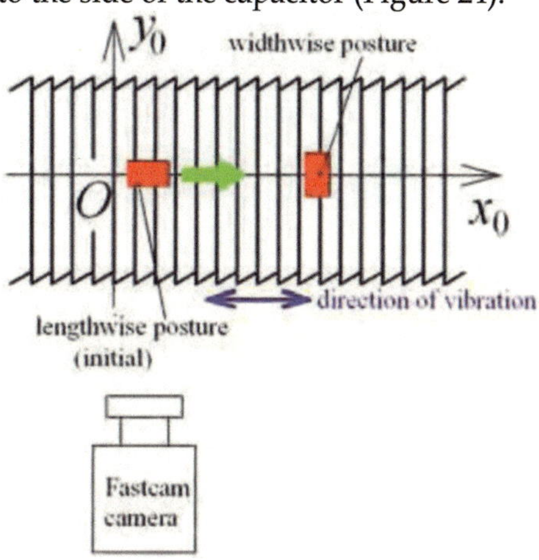

Fig. 24. Capture setup of micropart movement using a Fastcam video camera

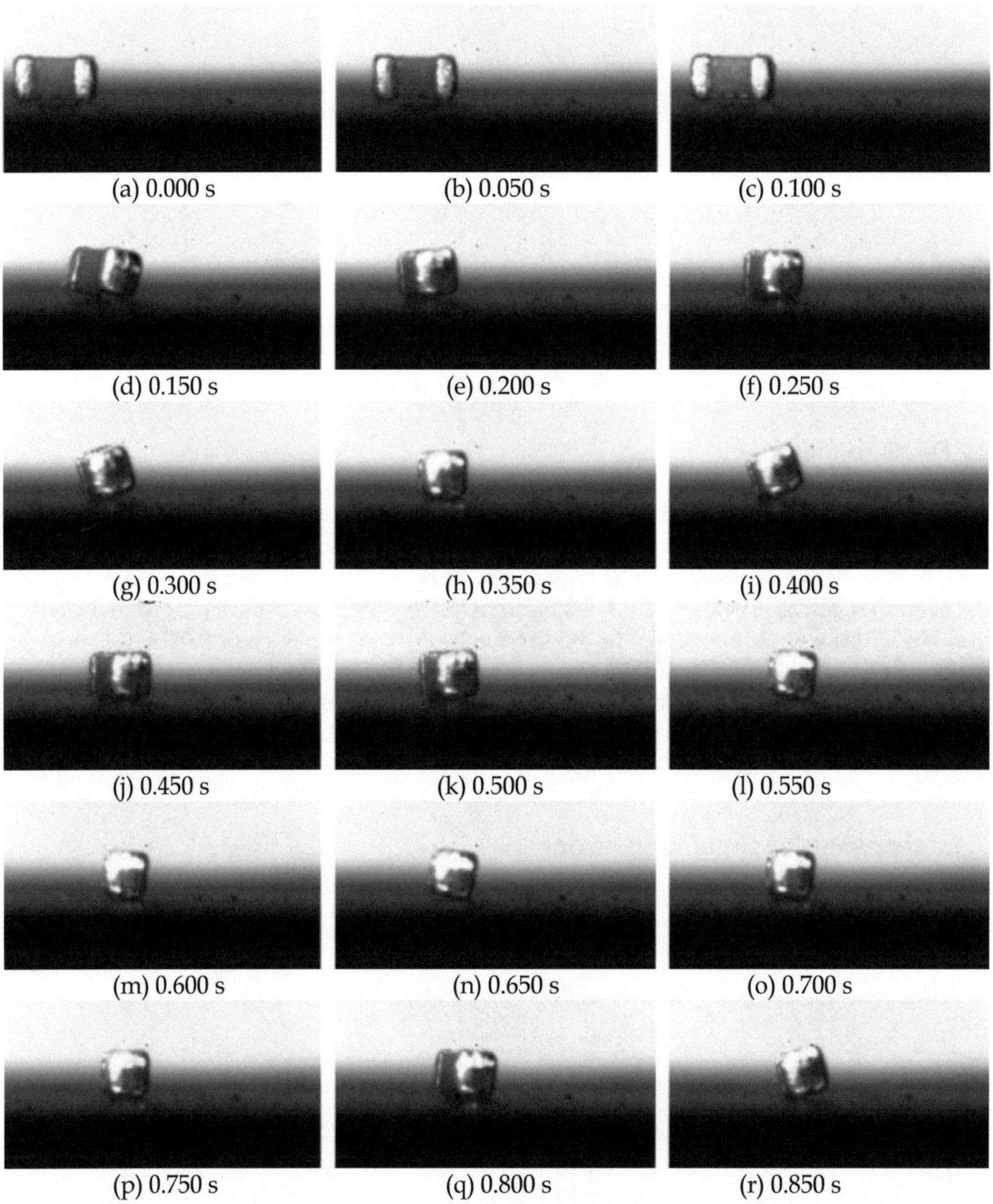

Fig. 25. Micropart movement

We obtained successive pictures of the capacitor movement from t = 0.000 to 0.850 s with an interval of 0.050 s (Figure 25). Beginning the feeder vibration at t = 0.000 s, the capacitor started to move along the feeder in the right direction upon feeder vibration. During the micropart moved in the right direction, the capacitor rotated around its vertical axis against the feeder surface (t = 0.150 s) and became oriented to widthwise at t = 0.300 s. When moving along the feeder in this widthwise posture, the capacitor began to rotate around the y_0 axis. Rotation angles were 17° at t = 0.300 and 0.400 s, and -3° at t = 0.550 s.

7.2 Analysis of micropart rotation

Let us formulate this rotation at this widthwise posture. We added the z axis to the coordinate G-xy defined in Figure 19: the z axis is perpendicular to the xy plane (Figure 26). Considering the capacitor rotation around the point of contact when a tooth contacts a convexity C_i with contact force F_i, force F_τ, generated by torque τ_i, is represented as:

$$F_\tau = \frac{\tau_i}{r_i}, \qquad (r_i \equiv |C_1 G|) \tag{4}$$

If β is angle between C_1 G and y axis, force F′ along the y axis can be formulated as:

$$F' = F_i - F_\tau \cos\beta \tag{5}$$

This suggests that drive force reduces by rotation of the micropart. Consequently, we need to derive dynamics considering rotation to simulate the movement of microparts more accurately.

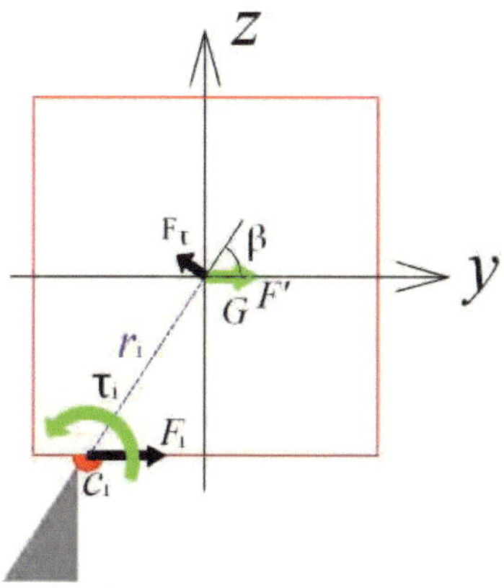

Fig. 26. Micropart rotation at widthwise posture

7.3 Analysis of feeder surface

Using the AZ-100 microscope (Figure 11), we obtained a synthesized model (Figure 27) and its contour model (Figure 28) of a sawtoothed surface. From these figures, feeder surface had many cracks and errors, not perfectly sawtoothed, which caused instable contact between the surface and a micropart, and affected the movement of micropart. Therefore, we need to formulate a feeder surface profile model based on measurements, and consider contact and adhesion using this model.

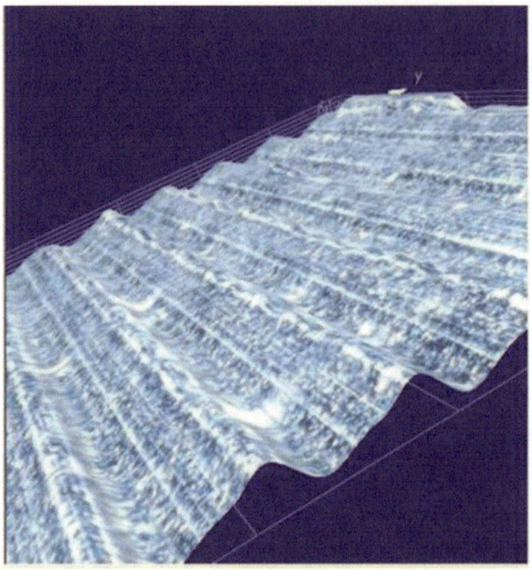

Fig. 27. Synthesized model of sawtoothed surface (p = 0.1 mm and θ=20 deg)

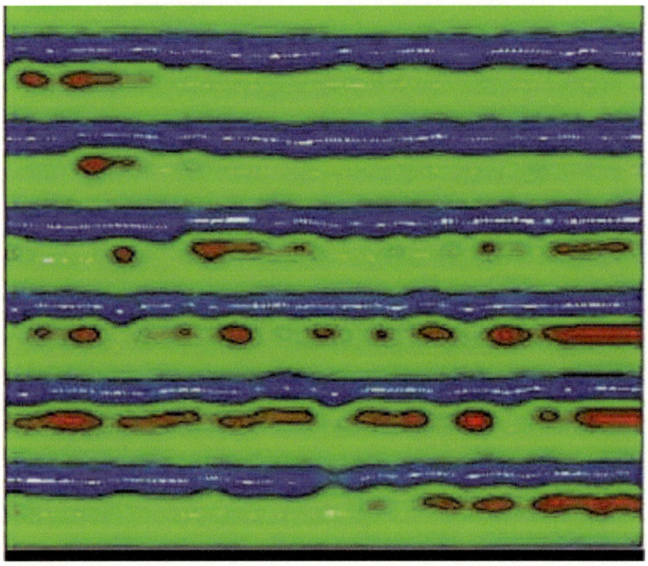

Fig. 28. Contour model

8. Conclusion

We examined a surface model of 0603 capacitor based on measurements. A microscope was used to analyse convexity sizes in the electrode surface. Each convexity was approximated as a half sphere model. These models were then applied for feeding simulation proposed in the previous work. Comparing with feeding experiments, we found large differences between the simulation and experimental results. We examined these differences by analyzing the movement of parts using a high speed video camera and found an error of oversight in our simulation. Capacitors rotated around the vertical axis against the sawtooth surface from a lengthwise to widthwise posture and continued to move along the feeder in

the desired direction while swinging around the axis along the sawtooth. This movement reduced the actual feeding velocity of a capacitor in contrast to the simulation. We also inspected a feeder surface profile using a microscope, and found many cracks and errors at the top of sawteeth, whereas feeder surface was perfectly sawtoothed in simulation. We concluded to need analysis of micropart rotation and a strict contact model between feeder surface and micropart based on measurements to simulate the feeding more accurately.
In future studies, we will try to:

· Identify dynamics of micropart including rotation,

· Formulate surface profile model of sawtoothed surface based on measurements, and analyse contact and adhesion using the model derived.

· Develop new feeder surfaces for smaller microparts, and,

· Verify the effect of ambient humidity on feeding.

This research was supported in part by a Grant-in-Aid for Young Scientists (B) (20760150) from the Ministry of Education, Culture, Sports, Science and Technology, Japan, and by a grant from the Electro-Mechanic Technology Advancing Foundation (**EMTAF**), Japan.

9. References

Mitani, A., Sugano, N. & Hirai, S.(2006). Micro-parts Feeding by a Saw-tooth Surface, *IEEE/ASME Transactions on Mechatronics*, Vol. 11, No. 6, 671-681.

Ando, Y. & Ino, J. (1997). The effect of asperity array geometry on friction and pull-off force, *Transactions of the ASME Journal of Tribology*, Vol. 119, 781-787.

Maul, G. P. & Thomas, M. B. (1997). A systems model and simulation of the vibratory bowl feeder, *Journal of Manufacturing System*, Vol. 16, No. 5, 309-314.

Wolfsteiner, P. & Pfeiffer, F. (1999). The parts transportation in a vibratory feeder, *Procs. IUTAM Symposium on Unilateral Multibody Contacts*, 309-318.

Reznik, D. & Canny, J. (2001). C'mon part, do the local motion!, *Procs. 2001 International Conference on Robotics and Automation*, Vol. 3, 2235-2242.

Berkowitz, D.R. & Canny, J. (1997), A comparison of real and simulated designs for vibratory parts feeding, *Procs. 1997 IEEE International Conference on Robotics and Automation*, Vol. 3, 2377-2382.

Christiansen, A. & Edwards, A. & Coello, C. (1996). Automated design of parts feeders using a genetic algorithm, *Procs. 1996 IEEE International Conference on Robotics and Automation*, Vol. 1, 846-851

Doi, T, (2001), Feedback control for electromagnetic vibration feeder (Applications of two-degrees-of-freedom proportional plus integral plus derivative controller with nonlinear element), *JSME International Journal*, Series C, Vol. 44, No. 1, 44-52.

Konishi, S. (1997). Analysis of non-linear resonance phenomenon for vibratory feeder, *Procs. APVC '97*, 854-859.

Fukuta, Y. (2004). Conveyor for pneumatic two-dimensional manipulation realized by arrayed MEMS and its control, *Journal of Robotics and Mechatronics*, Vol. 16, No. 2, 163-170.

Arai, M (2002). An air-flow actuator array realized by bulk micromachining technique, *Procs. IEEJ the 19th Sensor Symposium*, 447-450.

Ebefors, T. (2000), A robust micro conveyer realized by arrayed polyimide joint actuators, *Journal of Micromechanics and Microengineering*, Vol. 10, 337-349.

Böhringer, K.-F. (2003). Surface modification and modulation in microstructures: controlling protein adsorption, monolayer desorption and micro-self-assembly, *Journal of Micromechanics and microengineering*, Vol. 13, S1-S10.

Oyobe, H. & Hori, Y. (2001). Object conveyance system "Magic Carpet" consisting of 64 linear actuators-object position feedback control with object position estimation, *Procs. 2001 IEEE/ASME International Conference on Advanced Intelligent Mechatronics*, Vol. 2, 1307-1312.

Fuhr, G. (1999), Linear motion of dielectric particles and living cells in microfabricated structures induced by traveling electric fields, *Procs. 1999 IEEE Micro Electro Mechanical Systems*, 259-264.

Komori, M. & Tachihara, T. (2005). A magnetically driven linear microactuator with new driving method, *IEEE/ASME Transactions on Mechatronics*, Vol. 10, No. 3, 335-338.

Ting, Y. (2005), A new type of parts feeder driven by bimorph piezo actuator, *Ultrasonics*, Vol. 43, 566-573.

Codourey, A. (1995). A robot system for automated handling in micro-world, *Procs. 1995 IEEE/RSJ International Conference on Intelligent Robots and Systems*, Vol. 3, 185-190.

17

Palletizing Simulator Using Optimized Pattern and Trajectory Generation Algorithm

SungJin Lim, SeungNam Yu, ChangSoo Han and MaingKyu Kang
Hanyang University
South Korea

1. Introduction

Collision avoidance and robot path planning problems have emerged as a potential domain of robotics research of late because of their indispensable requirements in the field of manufacturing vis-à-vis material handling, such as picking and placing an object and loading/unloading a component to/from a machine or storage bins. This chapter focuses on palletization, a form of unitization in which a uniform load is stacked on a wooden pallet using a predetermined case pattern sequence and a given number of layers. In many kinds of proposed C-space construction approaches, several algorithms deal with the boundary of the C-obstacle analytically. Lozano Perez proposed the fundamentals of the C-space approach. When both the robot and obstacles have the shape of convex polygons, the C-obstacle boundary for an n-DOF manipulator is approximated by sets of n-1 dimensional slice planes, which are made from a one-dimensional slice plane. C. Zhao and his colleague proposed an algorithm to describe the C-obstacle as a set of parametric equations formulated from the mapping of the boundaries of the obstacles in a workspace. They use inverse pseudo kinematics to convert the obstacles in a workspace into a C-space. Debanik Roy studied path planning algorithms and their heuristics using the concept of visibility graph, and he presented an overview of the case study of robot path planning in an industrial environment in real time. Xiaojun and his colleague proposed a two-phase approach for C-obstacle construction and the collision detection of manipulators. This method is applicable to manipulators with various types of kinematic structures and geometric shapes. M. Pettersson and his coworkers proposed trajectory optimization method considering fatigue and thermal load of real robot. They also referred that the proposed method could be directly adapted to palletizing system. The issues of these papers, however, are for the operation of real industrial robot, and it is a part of entire palletizing system. Many other latest researches solely oriented towards path planning or modified apparatuses to improve the specified handling task have also been conducted. Studies on the total robot palletizing system, however, which integrates loading pattern optimization, robot OLP simulation, and path optimization, have yet to be systematically conducted.

This study was dedicated to the development of OLP Simulation S/W for a robot palletizing system (Non-vision system), which means that this study prioritized the reflection of the

characteristics of the palletizing task to realize the path generation algorithm. Undoubtedly, the shape of stacked boxes is changed after the unit-stacking step and palletizing robot encounters shape-changed obstacles in every step. This study pays attention to this characteristic of the palletizing task and proposes a simple and efficient methodology to generate a proper path generation algorithm for palletizing robots.
For the practical use of the proposed palletizing OLP Simulation S/W, a user-defined task design is possible in this simulator, and the user can store his own data. This application deals with rectangular boxes only as they make up the vast majority of objects involved in palletizing applications.
This chapter is organized as follows. In section 2, the layout of the task space and its coordinates, which are used consistently throughout this study, are defined. Subsequently, the proposed Fast algorithm and its application are introduced. Section 3 deals with the newly designed 3D robot simulator and its combination, with the application of the Fast algorithm. Section 4 presents the simulation of general C-space and A* algorithm for specified path generation using proposed simulator described in the previous section. Section 5 deals with the simulation of modified C-space to fit it into the real robot system. Based on the approaches of Section 5, Sections 6 is devoted to the newly proposed path generation method (overlap method) as well as to the simulation and its results. Finally section 7 concludes the chapter.

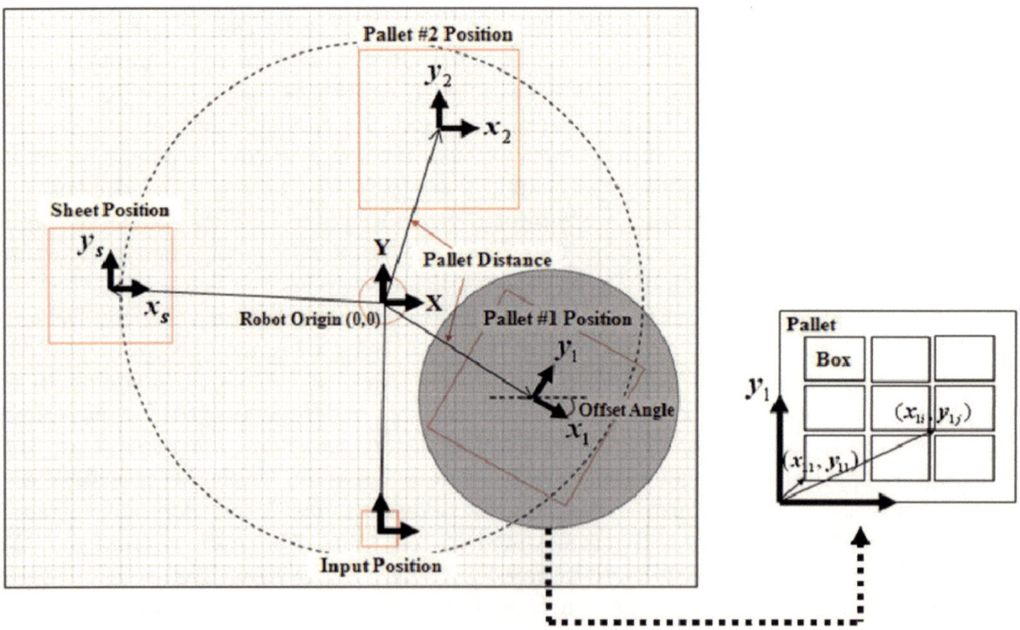

Fig. 1. Definition of task space

2. Fast Algorithm

2.1 Definition of the System Layout

As stated above, this application is a combination of an optimized pallet pattern generation algorithm, an industrial robot simulator, and a modified trajectory optimization algorithm.

To integrate these modules and to define the positions of the boxes, the robot, and its peripherals, the system layout and its coordinates have to be defined.
Fig. 1. describes the system layout and coordinates of the proposed palletizing OLP S/W. The robot is located in an origin of the total system, and other components (pallet, sheet, input facility, etc.) are expressed using this coordinate. Stacked boxes belong to the coordinate of the corresponding pallet. Consequently, the position and rotated angle of the boxes are expressed by the robot origin and its coordinate using relative coordination. By using this coordinate system and layout, the following chapter presents a pair of pattern

2.2 Steudel's Heuristic Algorithm

The objective of the pallet loading problem is to maximize the number of products that are loaded onto a pallet used for the transportation and storage of products. The distribution and storage costs of the product can be reduced by increasing its pallet utilization. The Fast algorithm presented in this study is an improved version of the 4-block pattern heuristic algorithm proposed by Steudel. The typical pattern of Steudel's heuristic algorithm is presented in Fig. 2. This heuristic finds the four-block pattern, in which each block is in a homogeneous pattern with the same box orientation. This heuristic consists of two phases. First, an initial solution is made with the combination of L_i and W_i, which maximizes the utilization of all four pallet edges.

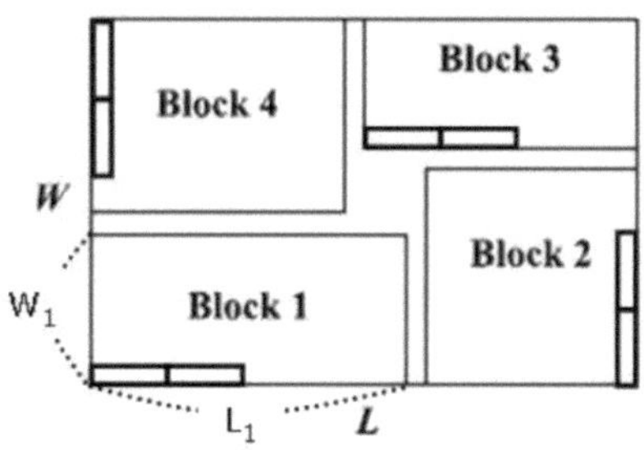

Fig. 2. Steudel's heuristic algorithm

Dynamic programming is applied to find the combination, and the initial solution has one of the four patterns shown in Fig. 3. The initial solution has a central hole that is sufficiently large to load more than one box in the case of P1 and P3, or an infeasible pattern, such as P2 and P4, due to the overlapped area.

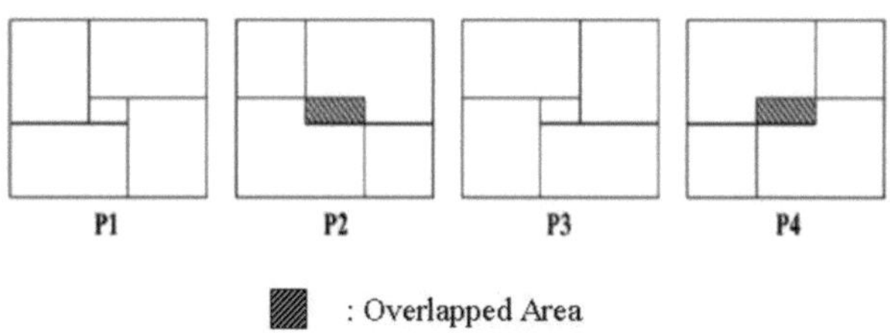

Fig. 3. Four patterns of the initial solution

This case involves the second phase. In the second phase, Treatment 1 fixes (L_3, W_3) and (L_4, W_4) and resizes L_1 and L_2, and Treatment 2 fixes (L_1, W_1) and (L_4, W_4) and resizes W_2 and W_3. Then, the first and second methods are compared and the better solution is chosen (Fig. 4.).

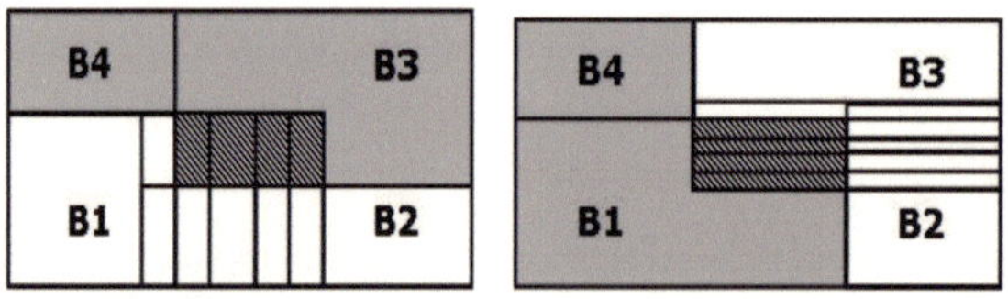

Fig. 4. Treatment of Steudel's algorithm

2.3 The Fast Algorithm

2.3.1 Definition

The Fast algorithm has similar processes with which to generate the initial four solution patterns. In addition, Treatment 3 is adapted to apply the heuristic recursively to the central hole in the following three methods so as to remove the overlapped area (Fig. 5.).

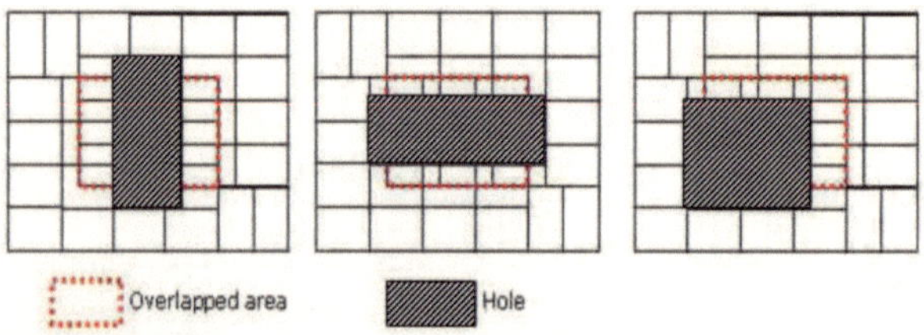

Fig. 5. Treatment of the Fast algorithm

(1) In the first method, the boxes are cut by the two horizontal edges of the overlapped area.
(2) In the second method, the boxes are cut by the two vertical edges.
(3) In the third method, the boxes are cut by the left vertical edge and the lower horizontal edge.

2.3.2 Schematic Diagram of the Fast Algorithm

As this algorithm does not consider all block sizes, it has a more rapid calculation time. The initial solutions of the first phase find the combination rather than using DP, and define the four parameters (Fig. 6.).

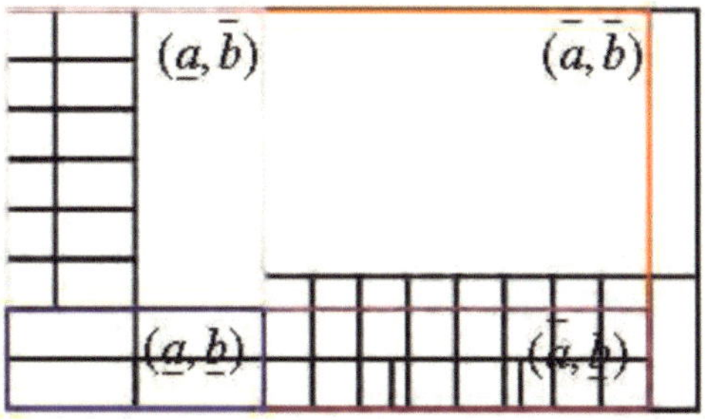

Fig. 6. Parameters of the Fast algorithm

- $\bar{a}$: When maximizing the length of the block and disposing the boxes lengthwise, the maximal possible number of boxes $= 5l$.
- $\underline{a}$: When maximizing the length of the block and disposing the boxes lengthwise, the minimal possible number of boxes $= 2l$.
- $\bar{b}$: When maximizing the width of the block and disposing the boxes lengthwise, the maximal possible number of boxes $= 8w$.
- $\underline{b}$: When maximizing the width of the block and disposing the boxes lengthwise, the minimal possible number of boxes $= 2w$.

In the first phase, (L_1, W_1), such as $(\bar{a},\bar{b})$, $(\bar{a},\underline{b})$, $(\underline{a},\bar{b})$, and $(\underline{a},\underline{b})$, are combined, and ($L_1, W_1$), the width and length of the other blocks, can be determined.

$$(L_2,W_2)=(L_4,W_4)=\left(\left[\frac{L-L_1}{w}\right]w,\left[\frac{W-W_1}{l}\right]l\right) \tag{1}$$

$$(L_3,W_3)=(L_1,W_1) \tag{2}$$

After obtaining the four initial solutions in the first phase, these solutions are redefined by applying the three treatments in the second phase.

Procedure ***FindBlockLayout*****(L,W,depth)**
  bestSolution **← 0**
  Find $\bar{a}, \underline{a}, \bar{b}$, **and** $\underline{b}$
  Make four initial Solutions.

    s_i (i=1,2,3, and 4), using them

For all s_i (i=1,2,3, and 4)
  s_i' ← Number of boxes after the first treatment
  s_i'' ←Number of boxes after the second treatment
  If max{ s_i', s_i'' }>*bestSolution,* then
  bestSolution←max{ s_i', s_i'' }
  End If
  If depth>>*MaxDepth* then
  Return *bestSolution*
  End If
  For all central holes
    s_i''' ←Number of boxes in the area
    excluding central hole
    Let(L_h, W_h)=size of central hole
    $s_i''' \leftarrow s_i'''$ +*FindBlockLayout*(L_h, W_h ,*depth*+1)
    If s_i''' >*bestSolution,* then
    bestSolution← s_i'''
    End If
  End For
End For
Return *bestSolution*
End Procedure

Algorithm *SolvePLP*(L, W, l, w)
  bestSolution←0
  For all(L_1, W_1, l, w) that satisfy the inequality (2) or (3) and $L_1 \in C_1, W_I \in C_w$
    Calculate all size of the five blocks
    Call *FindBlockLayout*($L_1, W_1, 0$) for all
      i=1,2,3,4 and 5
    If $\sum_{i=1}^{5} n(B_I)$ >*bestSolution* then
      bestSolution← $\sum_{i=1}^{5} n(B_I)$
    End If
  End For
End Algorithm

Fig. 7. The Fast algorithm

2.3.3 Computing Experience

The proposed algorithm was implemented in Visual C++ 6.0 and was compiled with the maximized-speed option. This algorithm test generated a 2D pattern of boxes and its calculation speed. As a hypothesis, the load balancing of a box and its stability were not considered.

(L,W,l,w)	Amount of boxes loaded
(1000,1000,205,159)	30
(1000,1000,200,150)	33
(22,16,5,3)	23
(30,22,7,4)	23
(14,10,3,2)	23
(53,51,9,7)	42
(34,23,5,4)	38
(87,47,7,6)	97
(1200,800,176,135)	38

(L: Length of Pallet, W: Width of Pallet, l: Length of Box, w: Width of Box)

Table 1. Test results of The Fast Algorithm (2D)

The above results were acquired by a computer with a K6-350-MHz CPU and 64MB RAM. All problems were calculated within 1 s and resulted in optimal solution. To use this algorithm practically, one dimension of height is applied additionally, and the 3D pallet loading simulator is realized, as shown in Fig. 8.

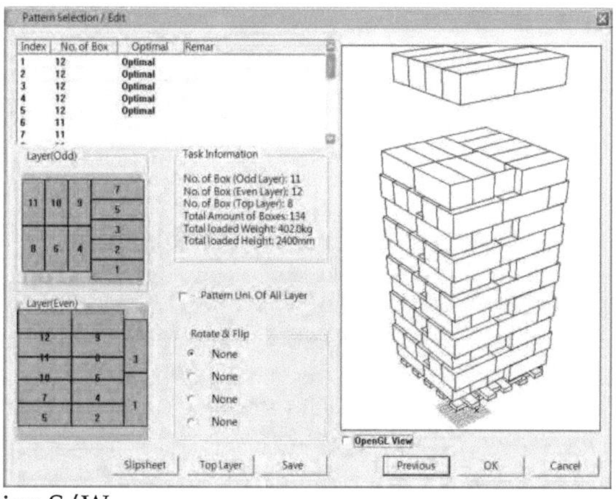

Fig. 8. pattern generation S/W

3. Development of the 3D Robot Simulator

Several methods have been introduced to make industrial robots perform the palletizing task. The first involved an online tutorial for the robot, which used a teach pendant to enable the robot to mimic and memorize the worker's motion. The second method is an offline method that generates task data using a computer, and that downloads it onto the robot controller. This chapter focused on offline task generation and simulation using a robot simulator. In this phase, the 3D robot simulator is presented based on the dimensional data of a real target machine, the HX300, which is a six-axis industrial robot of Hyundai Heavy Industrial Co. This robot model was realized by a commercial CAD modeler, and the GUI was developed using OpenGL® and MFC of Microsoft Visual C++®. To solve and analyze the forward and inverse kinematics equations, a general D-H parameter and the Lagrangian dynamic equation were used. With this simulator, it was possible to compute and display the joint torque, angle, and angular acceleration simultaneously. Fig. 9. shows the realized 3D robot simulator that was developed using Microsoft Visual Studio® and OpenGL®. It was possible to functionally calculate the velocity and acceleration of the gripper and to simulate the user-defined motion. The coordinates, which are generated by the pattern of loaded boxes on the pallet and the initial position of the box coming through an in-feeder, are passed to the simulator, and using these coordinates, it was possible to simulate the specified motion.

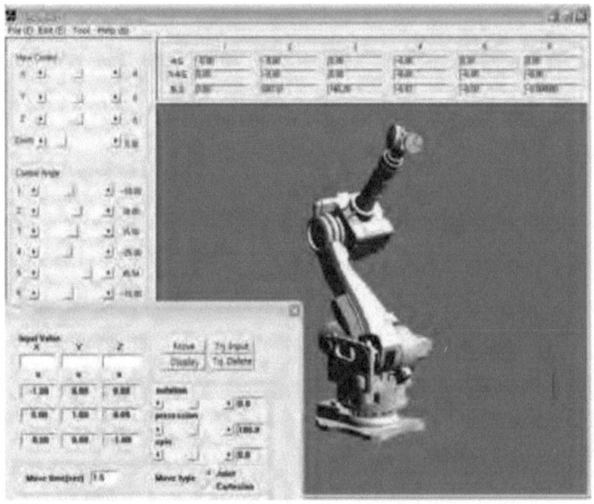

Fig. 9. Robot simulator for a palletizing task

4. C-Space and A* Algorithm for Trajectory Generation

4.1 C-Space Mapping of Obstacles

The palletizing task is generally composed of several palletizing components. These are auxiliary but are nevertheless obstacles for the palletizing robot. The important part of this study was to find the optimal path, considering the obstacles; hence, the concept of C-space (Configuration Space) to solve this problem was applied. The configuration defined the variables that exactly express the position and direction of an object, and the C-space represented all of the spaces where configurations may be acquired. Using this concept, a coordinate for each configuration was defined. In this coordinate, each point that was approached by the robot gripper was expressed by joint angles (configuration, posture) of the palletizing robot.

Fig.10. shows an example of the generation of the configuration space. First, on the basis of the joint of the base frame, the imaginary plane was rotated 360 degrees like Fig.10.(a).

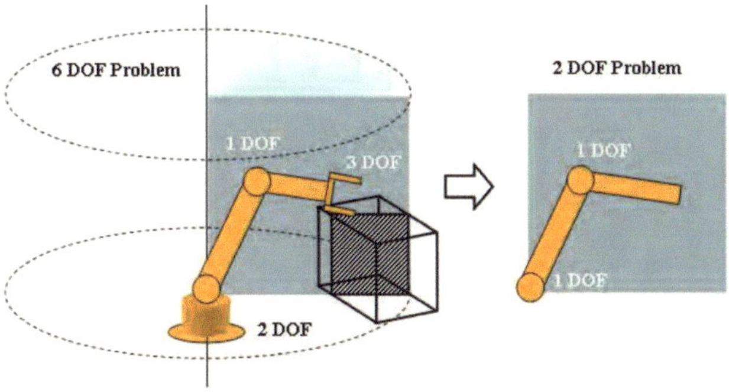

(a) Slice plane

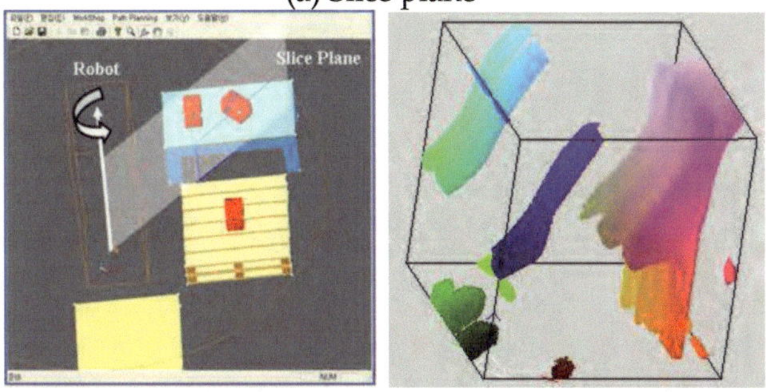

(b) Apply the slice plane to the workspace to generate the C-space.

Fig. 10. Obstacles expressed in C-space

Step	Task Layout	C-space	Enlarged Image	Elapsed Time (sec)
1	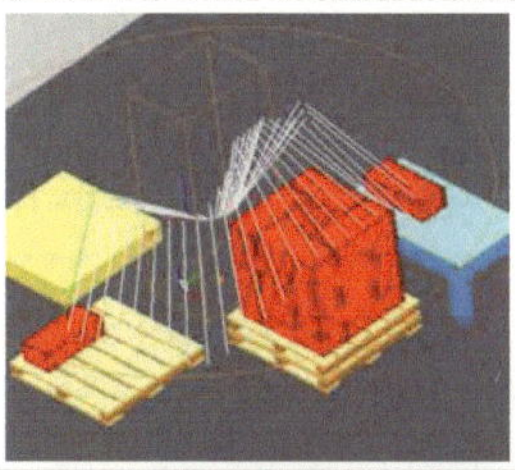		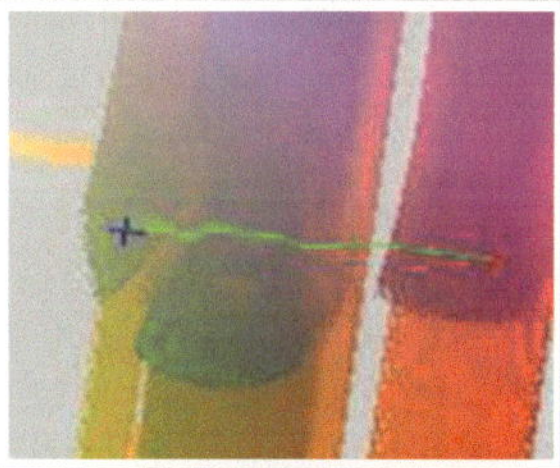	3.132
2	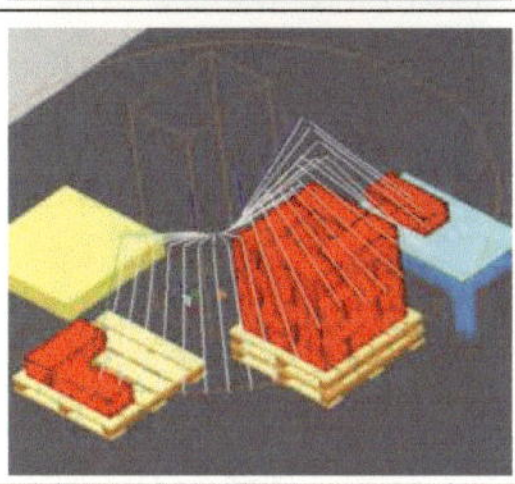	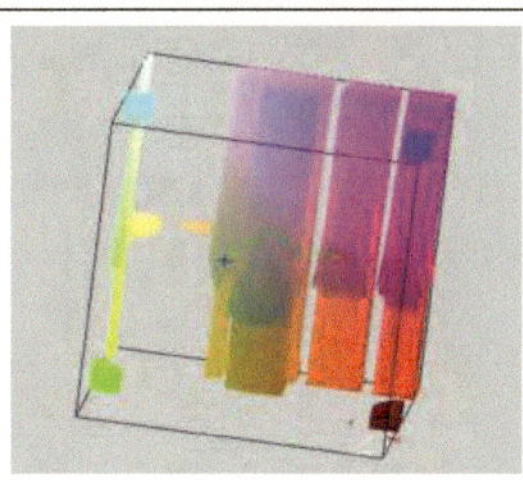	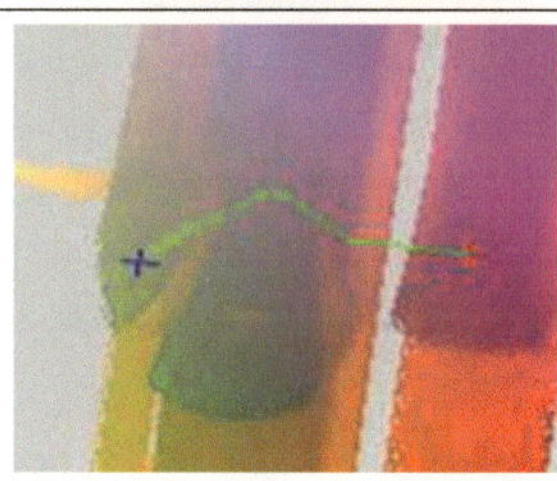	0.384

3	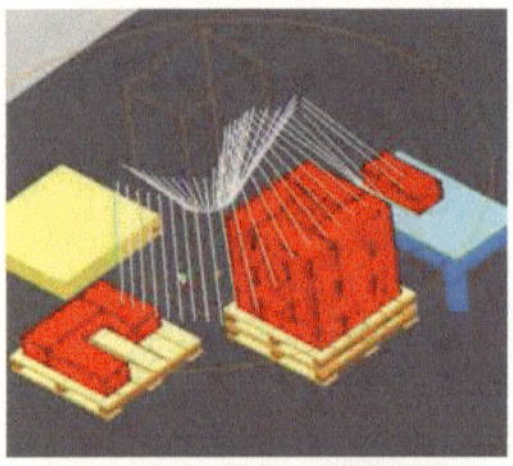	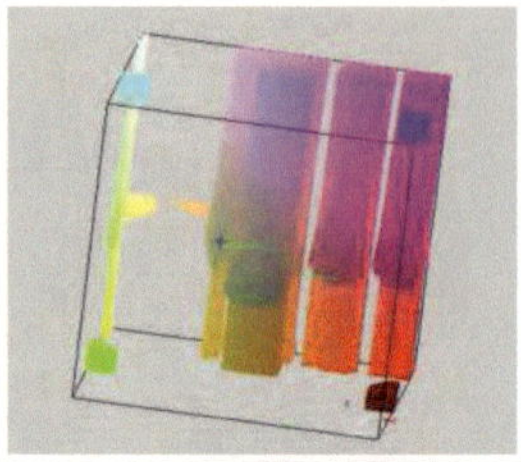	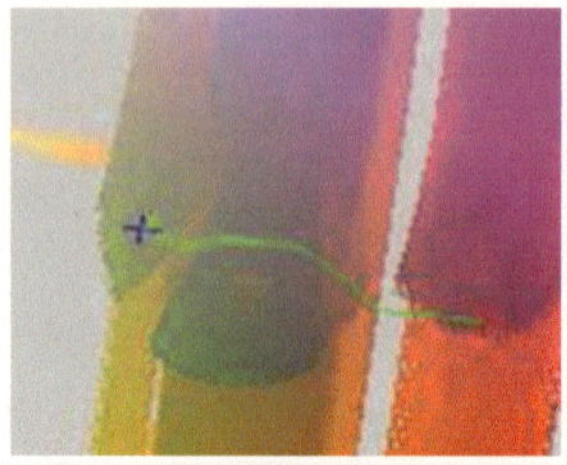	12.267
4	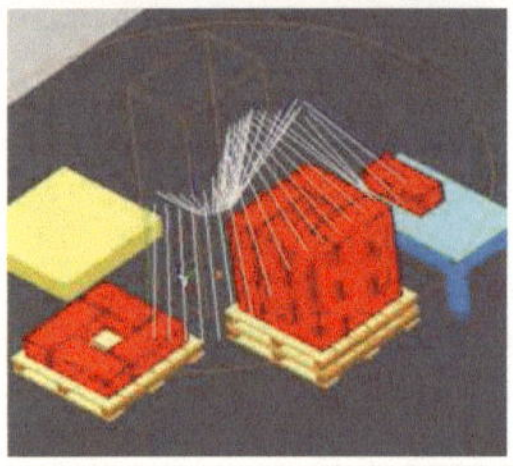		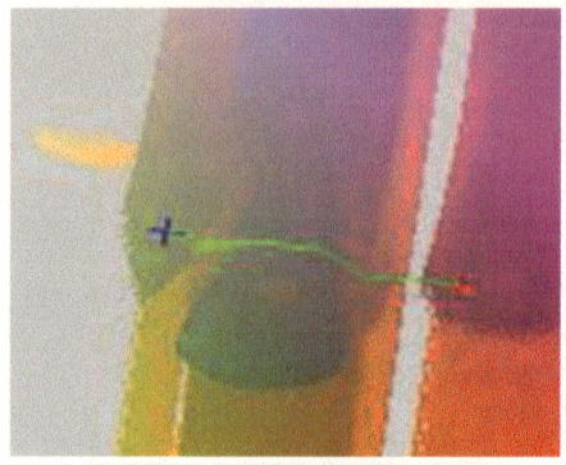	9.734

Table 2. Palletizing Task Simulation and Generation of Optimal Trajectory using A* Algorithm

In this progress plane, the objects surrounding the robot were scanned and the outline of a section was generated. The left side of the Fig.10.(b) describes the specified palletizing task layout. The outline, including its interior, could be considered an obstacle. In this study, the outline was acquired by using an end effecter of the robot, and the free-movement and obstacle zones in the C-space were generated as shown at the right side of Fig.10. To help distinguish the 3D shape of C-space, various brightness and color are used. This figure is necessary to generate the optimal path using the A* algorithm described in the next chapter.

4.2 Application of the A* Algorithm for Trajectory Generation

The A* method is a thorough, robust planning technique that determines either the minimum cost path or whether no safe path exists. By exploring a map, the A* algorithm generates nodes that are used to recode the current status. This technique is used to find the optimal path between the gripping point (starting point) and the place's down point (end point). The original A* technique is outlined below. To begin, a 2D rectangular grid was produced in which the cells were either safe or forbidden. The planning began at the starting point, and the cells adjacent to this cell were probed. On the basis of a cost function, the cell with the minimum cost was explored next. The cost function refers to the summation of costs, which required one to move from the starting node to the current node, and the "estimated" cost, which required one to move from the current node to the goal (a lineal distance). Based on this algorithm, palletizing simulation is performed in the 3D space and Table.2 is the results of the simulation.

5. Consideration of the Real Size of the Robot for Trajectory Generation

5.1 Modified Slice Plane (with horizontal thickness)

One of the disadvantages of the A* algorithm is the required computing time. The aforementioned approach considers the robot arm as a bar. Hence, the computing time load is relatively low. A real industrial robot, however, has an original volume, and these factors

have to be applied to the A* algorithm. The next step was to consider the real volume of the robot when it scans obstacles and generates C-obstacles.
To do this, the slice planes were redefined because it was assumed that the original slice plane had no thickness but that the modified slice plane had a thickness and that the factor that changed the scanning point of an obstacle of each angle was a group of both sides of the boundary of the modified slice plane (Fig. 11.). The thickness of the plane was determined individually by the thickness of the robot arm, including its gripper and load.

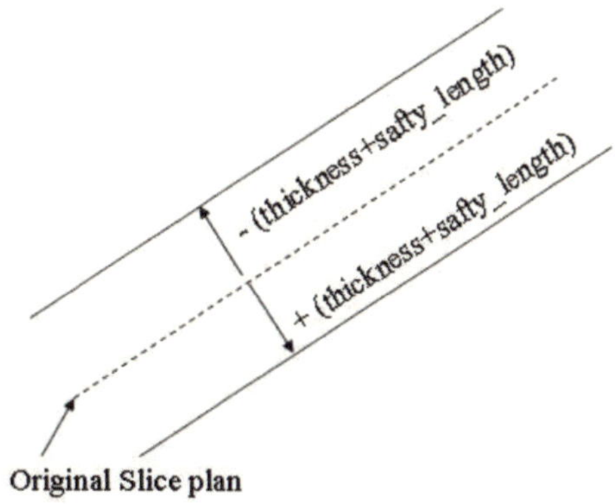

Fig. 11. Modified slice plane

5.2 Convex List and Graham's Algorithm

Fig. 12. shows the scanning points that used the modified slice plane.
The proposed system used factors of convex list points of objects and the sum of half of the thickness and a safe distance. As shown in Fig.12, the convex list was generated using the inside apexes of objects and intersection points. If the number of intersection points was less than two, the slice plane is regarded as meeting with one apex or edge.

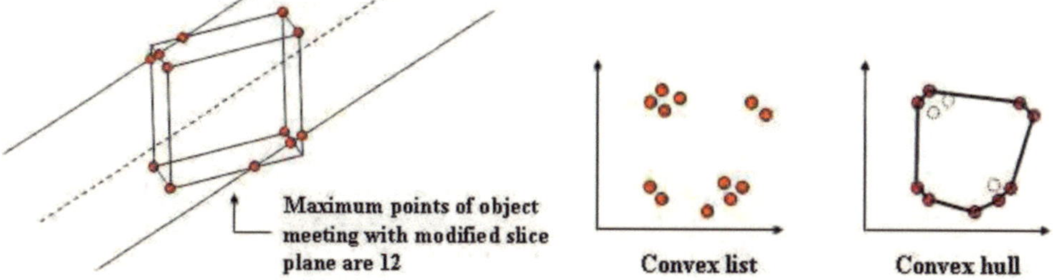

Fig. 12. Convex list generation

Finally, Graham's algorithm was used to generate the convex hull. This hull was used as the new boundary of the object when the modified slice plane was applied.

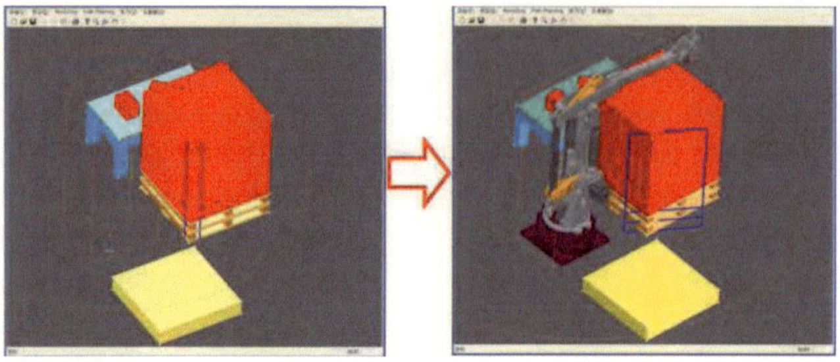

Fig. 13. Modified slice plane

Fig. 13 describes the effect of the modified slice plane. As shown in the figure, the slice plane became larger.

5.3 Consideration of Vertical Thickness

The previous chapter showed the horizontal thickness of a real robot and proposed the modified slice plane that was used to generate the obstacle area of an object. As a next step, the vertical thickness of the robot was considered. Fig. 14. illustrates outlined margin of robot manipulator and its realization on the proposed simulator.

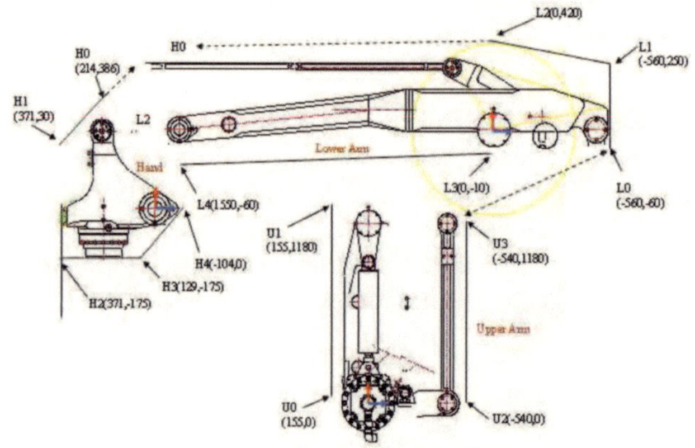

(a) Outlined Margin of Robot Manipulator

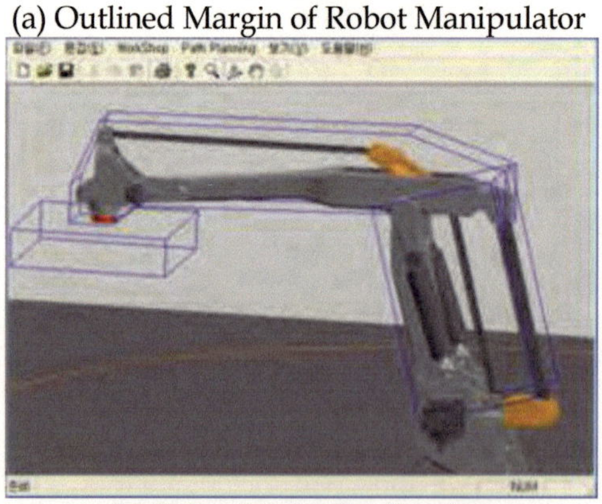

(b) Robot Model Realization

Fig. 14. Boundary line of the target robot system

These assumptions of the boundary of the gripper and its load (box) consider the total volume of the robot, including the robot arm, the gripper, and its load. Hence, when the modified slice plane (vertical thickness of the robot, gripper, and its load) is applied, the designed simulator is considered the vertical thickness of the robot arm, including the gripper and its load, simultaneously.

5.4 Consideration of the Performance of the A* Algorithm Using the Modified Slice Plane

If the robot body is a line, the computing time is very short and is therefore not an issue. When the modified slice plane was applied, however, the computing time was substantially increased. The possible explanation for this could be that the results were duplicated at the intersection points in each step and were added to the computation load of Graham's algorithm for the generation of the convex list. Fig. 15. shows an illustration of this simulation.

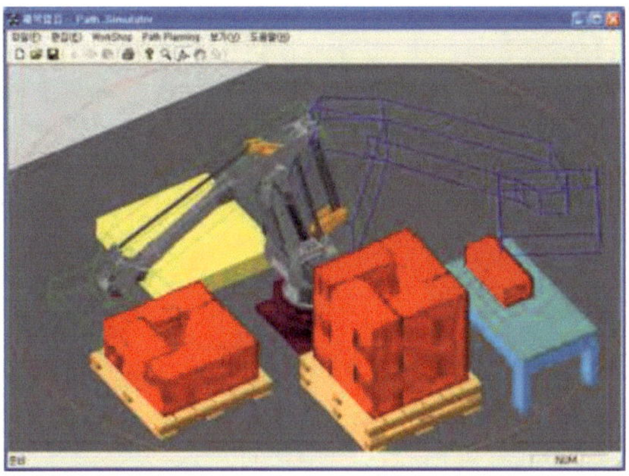

Fig. 15. Simulation of the A* algorithm using the modified slice plane

6. The Overlap Method to Generate the Palletizing Trajectory

The computing load is a critical problem in the area of software development. The purpose of this study, as described in the introduction, was to develop an OLP (offline programming) simulator specific to palletizing automation. As shown in Table 2, if the real size of a palletizing robot is considered to generate the optimized trajectory, an A* algorithm is a relatively expensive method. To use this algorithm, the C-space has to be generated, but this requires a large amount of computing load.

To focus on the characteristics of the palletizing task, a new strategy devoted to the generation of the set of boundaries (convex) of the obstacles was proposed. As shown in Fig. 16., the proposed method overlaps the scanned images of each box at one plane and obtains the outer line of the overlapped image. This method used the total traveling distance from the pickup point of the boxes to the place-down point via the outer line of the overlapped area.

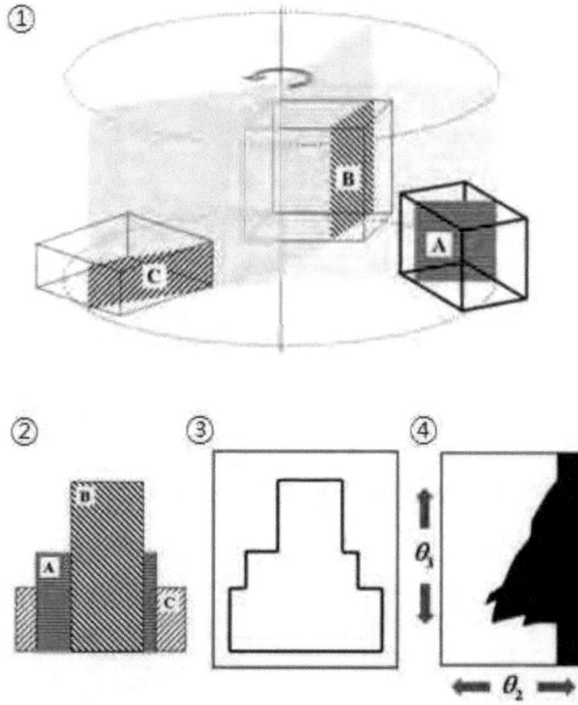

Fig. 16. Procedure of Overlap method

The following equation was used in this study to optimize this distance:

$$
\begin{aligned}
T_{opt} &= A[abs\{(P_{via} - P_{pick-up})_{\theta_2}\} + abs\{(P_{place-down} - P_{via})_{\theta_2}\}] \\
&+ B[abs\{(P_{via} - P_{pick-up})_{\theta_3}\} + abs\{(P_{place-down} - P_{via})_{\theta_3}\}]
\end{aligned} \tag{3}
$$

where T is the distance that the robot must negotiate to palletize one box, which is the absolute summation of the distance from the pickup point to the outer line of the overlapped area and the distance from the outer line to the place-down point (Fig. 17.).

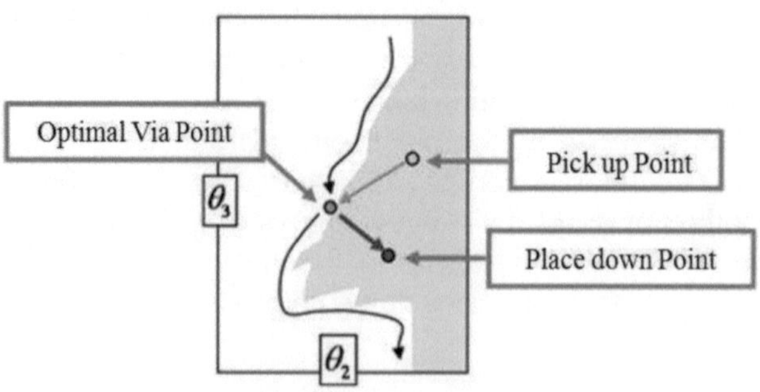

Fig. 17. Determination of θ_2 and θ_3 to generate trajectory

The robot path, however, is not composed of 3 points only (a place-down point, an optimal via point, and a place-down point). Therefore, this algorithm is expended to find an extra via point that would travel the whole path, from the start to the end point. (Fig.18.)

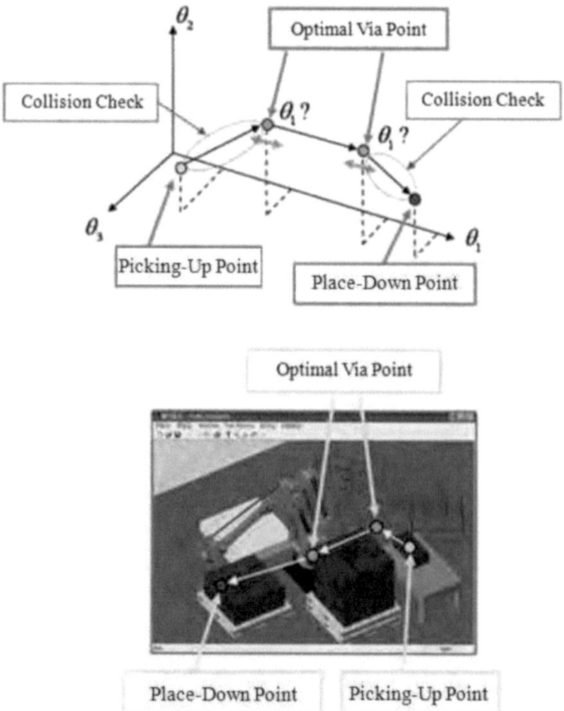

Fig. 18. Determination of θ_1 to generate the optimal via point

To do this, the aforementioned optimal 1 via point is used as a 1st optimal via point. If the gripper of the robot reaches this point, a collision between the gripper and the obstacle can be avoided by changing θ_1. The definition of the collision or gap between the robot and the obstacle is decided beforehand (user-defined setting of the designed OLP S/W - "safe distance"). Through this treatment, the intermediate via points are decided. Finally, the total travel points are composed of [picking-up point] → 1st optimal via point, [via(θ_{11}, θ_{12}, θ_{13})] → [···] → nth via point, [via($\theta_{n1}, \theta_{n2}, \theta_{n3}$)] → final optimal via point, [via(θ_{f1}, θ_{f2}, θ_{f3})] → [place-down point]. Here, i and j of θ_{ij} means ith generated via point of jth joint of robot manipulator.

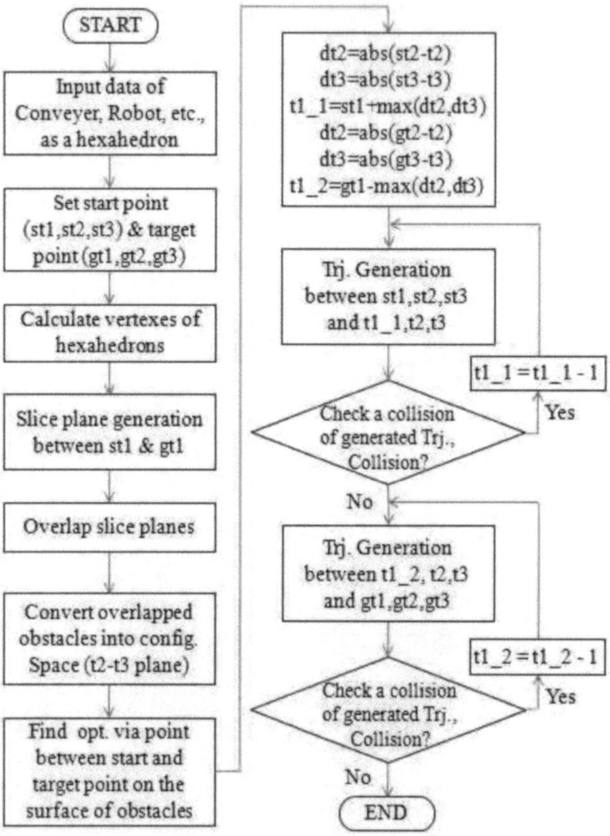

Fig. 19. Basic algorithm of the overlap method

- st1, st2, st3: , , of the starting point
- gt1, gt2, gt3: , , of a goal point
- t2, t3: , of an optimal path point
- t1_(i): of an optimal path point (ith iteration)

This method deals with every surrounding obstacle of the robot in every unit step of the process ("unit step" means one cycle of pick-and-place task). As the shapes of the obstacles that surround the robot are changed at every step, this approach has the advantage of being able to calculate the pick-and-place path. Fig. 19. shows the detailed algorithm of the overlap method.

7. Conclusions and Considerations

To prove the efficiency of the proposed methodology, all type of trajectory generation method described in this chapter is simulated and its results are compared.

Step	Time (Sec.)		
	Line(A*)	Volume(A*)	Overlap Method
1	0.014051	2.380194	0.412106
2	0.01543	2.066007	0.427587
3	0.01558	1.857863	0.415959
4	0.017952	2.318304	0.43101
5	0.014286	2.265576	0.411975
6	0.016003	2.213547	0.422834
7	0.013669	1.360996	0.443131
8	0.014206	2.20403	0.440602
9	0.016555	1.328561	0.454023
10	0.015094	1.298407	0.438623
11	0.017387	1.660548	0.466195
12	0.01523	1.298854	0.4273
13	0.01889	0.886562	0.46002
14	0.01344	1.159015	0.428835
15	0.016348	1.091212	0.43728
16	0.01413	1.301314	0.424192
17	0.017107	0.386479	0.464915
18	0.016205	0.429178	0.47078
19	0.017836	0.361664	0.484123
20	0.014301	0.389356	0.439179
21	0.020215	0.278059	0.491373
22	0.014119	0.408476	0.441098
23	0.018028	0.288906	0.466881
24	0.014902	0.295928	0.439116

Table 3. Elapsed Time of Each Method (Box, 24ea)

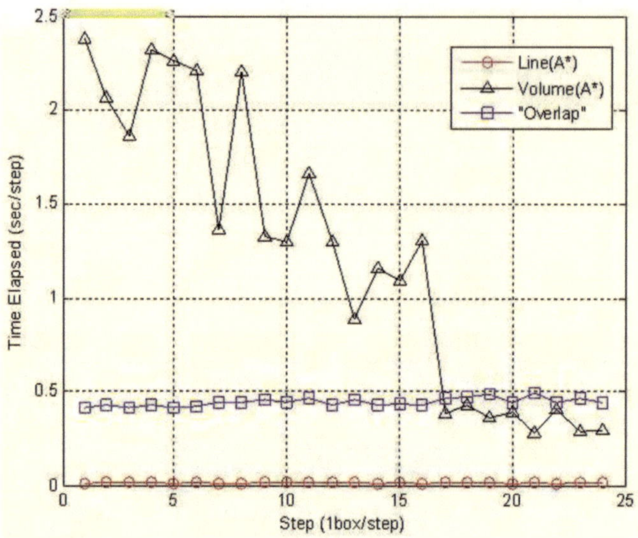

Fig. 20. Elapsed time of the 1-step palletizing task

As shown in the third column of Table 3 and Fig. 20., with the A* algorithm that considered the volume of the robot, the computing time of each step was remarkably different depending on the situation which is encountered at every step. The overlap method, however, produces fast and stable computation results, regardless of the place-down position and configurations of surrounding obstacles.

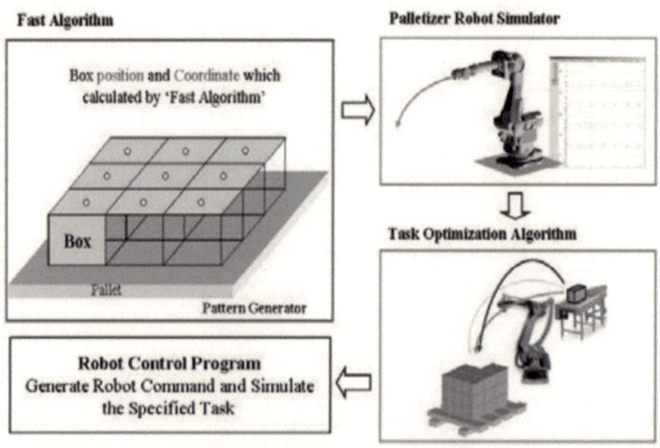

Fig. 21. Schematic diagram of the proposed palletizing OLP Simulation S/W

This chapter focuses on the realization of palletizing OLP Simulation S/W, which is a newly adapted method of treating and generating the path of the robot palletizing task efficiently. Fig. 21. shows the organization of this study. First, the box-loading patterns are generated on the pallet. This pattern was generated through the Fast algorithm that is handled in this chapter. The generated boxes were given a unique position value indicating the location and posture on the pallet where each box would be placed (i.e., the place-down point). Finally,

the robot moves the loads from pick-up points to place-down points through the trajectory generated by the overlap method.
Finally, Fig. 22. shows the total simulator that uses the proposed algorithms. The user defines the size of the box, the pallet, and the position of the working component on the workspace, and the simulator generates the optimized box-loading pattern, motion simulation of palletizing robot through the optimized trajectory and its task result as shown in enlarged image of Fig. 22.

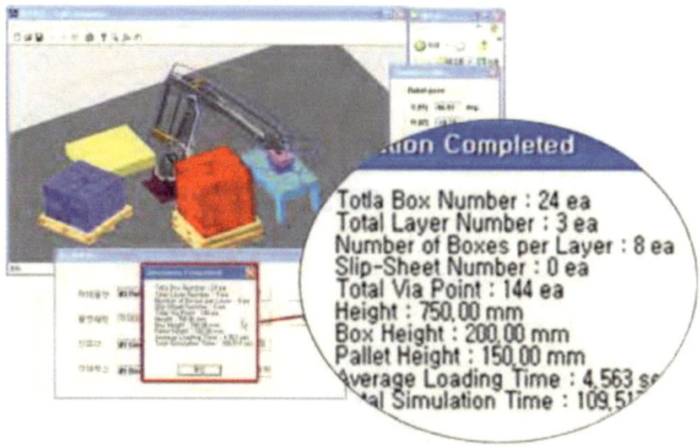

Fig. 22. Task results of the proposed OLP Simulation S/W

For application in the real robot system, however, the accuracy problem involving the synchronization of the robot simulator with the real robot, or the "calibration" problem, is an absolutely important issue. The calibration of a proposed palletizing S/W is suggested for future research. Provisionally, "three-point calibration" was considered to check and compensate for the errors between the workspace and the installed robot system.

8. References

Debanik Roy (2005). Study on the Configuration Space Based Algorithmic Path Planning of Industrial Robots in an Unstructured Congested Three-Dimensional Space: An Approach Using Visibility Map, *Journal of Intelligent and Robotics Systems*, Vol. 43, No. 2-4, Aug 2005, pp. 111-145, 0921-0296

Harold J. Steudel (1979). Generating pallet loading patterns: A special case of the two-dimensional cutting stock problem, *Management Science*, Vol. 25, No. 10, Oct 1979, pp. 997-1004

J. H. Kim; J. S. Choi; H. Y. Kang; D. W. Kim & S. M. Yang (1994). Collision-Free Path Planning of Articulated Robot using Configuration Space, *Transactions of the KSAE*, Vol. 2, No. 6, Nov 1994, pp. 57-65, 1225-6382

John J. Craig (2004). *Introduction to Robotics – Mechanics and Control, 3rd Edition*, Pearson Education Int., 978-0201543612

Michael A. Hernan I (2000). *An Introduction to Automated Palletizing*, Anderson Technical Services, Inc.

Pettersson, M.; Olvander, J. & Andersson, H. (2007). Application Adapted Performance Optimization for Industrial Robots. *IEEE International Symposium on Industrial Electronics*, pp. 2047-2052, 978-1-4244-0755-2, Jun 2007

Ronald Graham (1972). An Efficient Algorithm for Determining the Convex Hull of a Finite Point Set, *Info. Proc. Letters 1*, pp. 132-133, Jan 1972

Seung-Nam Yu; Heu-Kwon Yoon; Sung-Jin Lim; Young-Hoon Song & Chang-Soo Han (2005). The development of Robot Palletizing S/W using Fast Algorithm and 3-D Robot Simulator, *Proceedings of Korean Society of Mechanical Engineers*, pp. 1663-1668

T. Lozano-Perez (1987). A Simple Motion Planning Algorithm for General Robot Manipulators, *IEEE Journal of Robotics and Automation*, Vol. RA-3, No. 3, Jun 1987, pp. 224-238, 0882-4967

Warren, C.W. (1993). Fast Path Planning Using Modified A* Method, *Proceedings of the 1993 IEEE International Conference on Robotics and Automation*, pp. 662-667, 0-8186-3450-2, Atlanta, GA, USA, May 1993,IEEE Comput. Soc. Press

Xiaojun Wu; Qing Li & Heng, K.H. (2005). A New Algorithm for Construction of Discretized Configuration Space Obstacle and Collision Detection of Manipulators, *Proceedings of 12th Int. Conf. on Advanced Robotics*, pp. 90-95, 0-7803-9178-0, Jul 2005

Young-Gun G & Maing-Kyu Kang (2001). A fast algorithm for two-dimensional pallet loading problems of large size, *European Journal of Operational Research*, Vol. 134, No. 1, pp. 193-202,

Zhao, C.S.; Farooq, M. & Bayoumi, M.M. (1995). Analytical solution for configuration space obstacle computation and representation, *Proceedings of the 1995 IEEE IECON 21st International Conference*, pp. 1278-1283, 0-7803-3026-9, Orlando, FL, USA, Nov 1995, IEEE

18

Implementation of an automatic measurements system for LED dies on wafer

[1]Hsien-Huang P. Wu, [1] Jing-Guang Yang, [1]Ming-Mao Hsu and [2]Soon-Lin Chen, [2]Ping-Kuo Weng and [2]Ying-Yih Wu

[1]Department of Electrical Engineering National Yunlin University of Science and Technology #123 University Rd. Section 3 Yunlin, Douliou 640 Taiwan, ROC e-mail: wuhp@yuntech.edu.tw

[2] Materials RD Center, Solid-State Devices Materials Section Chung-Shan Institute of Science and Technology (CSIST)

P.O. Box 90008-8-7 Lung-Tan, Tao-Yuan 325 Taiwan, ROC

Abstract

High brightness light-emitting diode (HB-LED) has now become one of the most popular lighting device in our daily life. As the LED industry becomes prosperous, techniques for improving production efficiency become more and more important. In this paper, a system integration method was proposed and successfully realized to implement an automatic measurements and grading (AMG) system for the LED dies after the wafer has been scribed and broken. It can be used to greatly improve the speed, efficiency and accuracy in the testing process. This suggested approach combines machine vision, optical measurement instruments, and mechanical technology to create an affordable, flexible, and highly efficient LED measurement and grading system. System architecture and details on each subsystem were described and performance was evaluated. The average speed of measurement is 3.5 LED dies per second based on repeated testings and evaluations. The experimental results demonstrate the effectiveness and robustness of the proposed system. We hope the results presented in this paper can help the LED manufacturer to make more informed decisions on the design or purchase of the AMG machine.

1. Introduction

The incandescent electric light is almost everywhere in our daily lives. However, it has not changed very much since first invented more than 100 years ago. Another popular lighting technologies based on fluorescent and compact fluorescent were developed about 30 years ago. A new type of lamp invented recently, HB-LED (High Brightness), appears to be the first truly innovative electric light. These new semiconductor-based lights promise not only to replace incandescent and florescent lights in almost every application but also create some brand new applications that were not possible with conventional lamps. After slowly developing over many years in niche applications, the costs are dropping and volumes are

increasing at rapid rates. New visible colors and wavelengths, even in infrared and ultraviolet regions, are emerging as promising new products.
HB-LEDs are finding new applications every day and the future product trends are in its favor. Beginning in very specific applications, HB-LEDs are now entering many general products such as televisions, PC displays, LCD backlight application, digital cameras, cell phones, traffic lights, billboards, signboard, and indoor/outdoor general-purpose lighting. In the automotive market, HB-LED are replacing tail lights, interior lights, and soon, even headlights. Entire buildings are being lit externally in different colors schemes depending on the time of day and turning light into a type of variable illumination "paint". Rapid growth is also in many other markets including wireless, optical and telecommunication. Volumes for HB-LEDs are reaching tens of millions per month and triggering new economies of scale in manufacturing that are dropping costs substantially.
The production of LED, which is shown in Fig. 1, can be separated into four sections: epitaxy (top-stream), processing/fabrication (mid-stream),

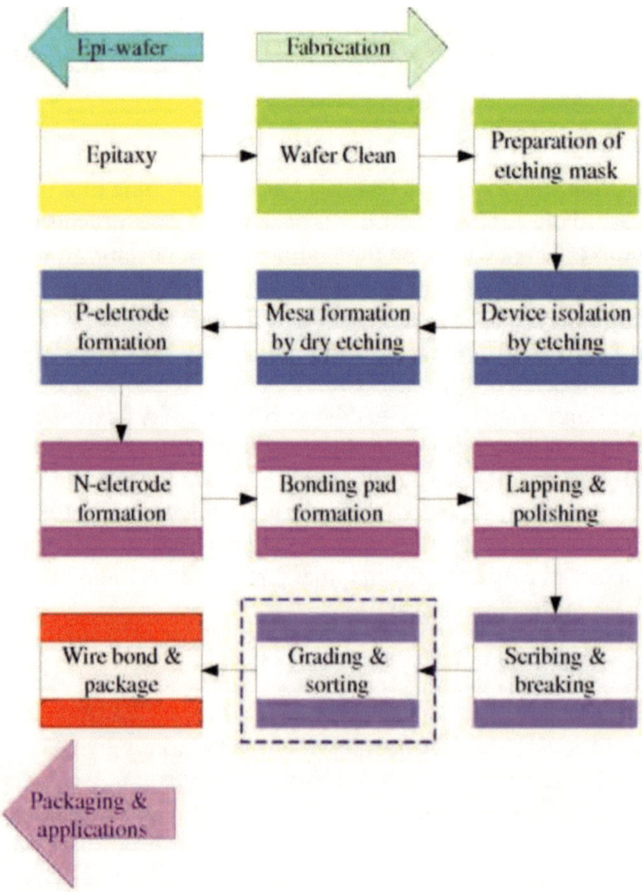

Fig. 1. The flow of LED production process and the block of grading developed in this paper.

packaging/modules (bottom-stream), and lamp design & applications (lighting companies). With each new stage of growth comes new challenges for manufacturers in each section trying to manage the growth. Apparently, automation in the process of manufacturing is an indispensable tool to increase the production rate. In this paper, we will focus on the development of an automated measurement and grading (AMG) system for the HB-LED dies in the fabrication (mid-stream) section based on machine vision. This system belongs to the "grading and sorting" module of the fabrication process (midstream) in the Fig. 1. Since the LED will be measured after the wafer was scribed and broken into individual dies, their relative positions are irregular and current AMG machine for wafer before breaking can not be used.

Machine vision has been successfully employed in many fields and various kinds of applications. For example, apple grading [1] and seeds refined grading [2] in agriculture, sheet-metal forming defect detection in automobile industry [3], computer-aided diagnosis and archiving in medical imaging [4][5], web inspection in textile industry [6], bottle inspection [7][8], and defect detection in TFT-LCD industry [9][10][11]. While many studies have been published in so many diverse fields, little information is available on the automated measurements of LED dies. The only literature related to the LED production was conducted by Huang [12] on the LED micro structure inspection. To date, while there exists LED prober on the market [13][14], no studies have been shown in the literature to measure and grade the HB-LED dies based on the machine vision technique. The purpose of this paper is to detail an approach on how to design a system that can measure and grade the LED dies automatically by integrating the mechanical, optical, electronic, and machine vision modules. The remainder of this paper is organized as follows. Section 2 presents the hardware modules and architecture used in the proposed system. In section 3, the software used to integrate the hardware modules is described. In section 4, experimental results are presented that confirm the performance of the system implemented. Section 5 summarizes the conclusions of the paper.

2. Architecture of the Hardware System

Measurement system based on the machine vision involves the harmonious integration of elements of the following areas of study: mechanical handling, lighting, optics, sensors, electronics (digital, analog and video), signal processing, image processing, digital systems architecture, software, industrial engineering, human-computer interfacing, control systems, and manufacturing. Successful integration of these mechanical, optical, electronic, and software subsystems is essential for automatic inspection and measurement of natural objects and materials. In this section, the hardware components used in the design will be described.

The LED measurement and grading needs to be conducted by two consecutive phases. The first phase employs machine vision system to identify and record the position of each individual *LED die* on the wafer. The second phase uses these position data to move each individual LED die to the place where the probe of the optical system is located above. The wafer is then moved upward and the LED die is stimulated to turn on by the touch of the probe and its electro- and optic properties are measured. These measurement data can be used for grading and sorting of the LED dies. Videos with a Readme file to illustrate the action of the proposed system in phase II can be accessed at our web site [15].

To achieve the requirements mentioned above, the automated measurement and grading system is formed by an industrial PC, machine vision system, mechanical system and electrical-optical measurement system. The system allows accurate measurement of each LED die while the LED wafer is moved on the carriage of the move-table continuously. The overall architecture of the LED measurement and grading system is presented in Fig. 2 where the bottom part contains the mechanical (X-Y-Z table) and machine vision systems for and die position measurement and placement. The top part contains the probe and the optical and electrical measurement system. The entire hardware system can be roughly divided into three main modules and will be described below.

2.1 Machine Vision System

The vision system contains a black and white area scan CCD camera (STC-130BJ) with 510 *x* 492 picture elements and an Angelo RTV-24 image acquisition board. The acquisition board provides image sequence with 640 *x* 480 resolution. The purpose of the vision system is to automatically identify the center location of each LED *die* based on visual information. In order to obtain accurate position estimation, MML (Machine Micro Lens) telecentric lenses is used to provide a constant perspective angle across the field. The combination of the CCD sensor and the lens generates images with an approximate resolution of $1.67\mu m \times 1.61\mu m$ per pixel. This pixel size is precise enough for our targeted LED with a *die* size of $300\mu m \times 300\mu m$. Because the wafer is flat and contains specular surface, coaxial lighting with diffuse type of source (model 3AM-LV-15R,LED red color) is used to obtain reliable images without shadows and specular reflection. The configuration of imaging unit is illustrated in Fig. 3 where a beam splitter is used to direct light from the diffuse source toward the subject at an orthogonal angle. The beam splitter also allows the radiance from the wafer surface to reach the CCD camera for image formation. Several example images acquired by this unit is illustrated in Fig. 4.

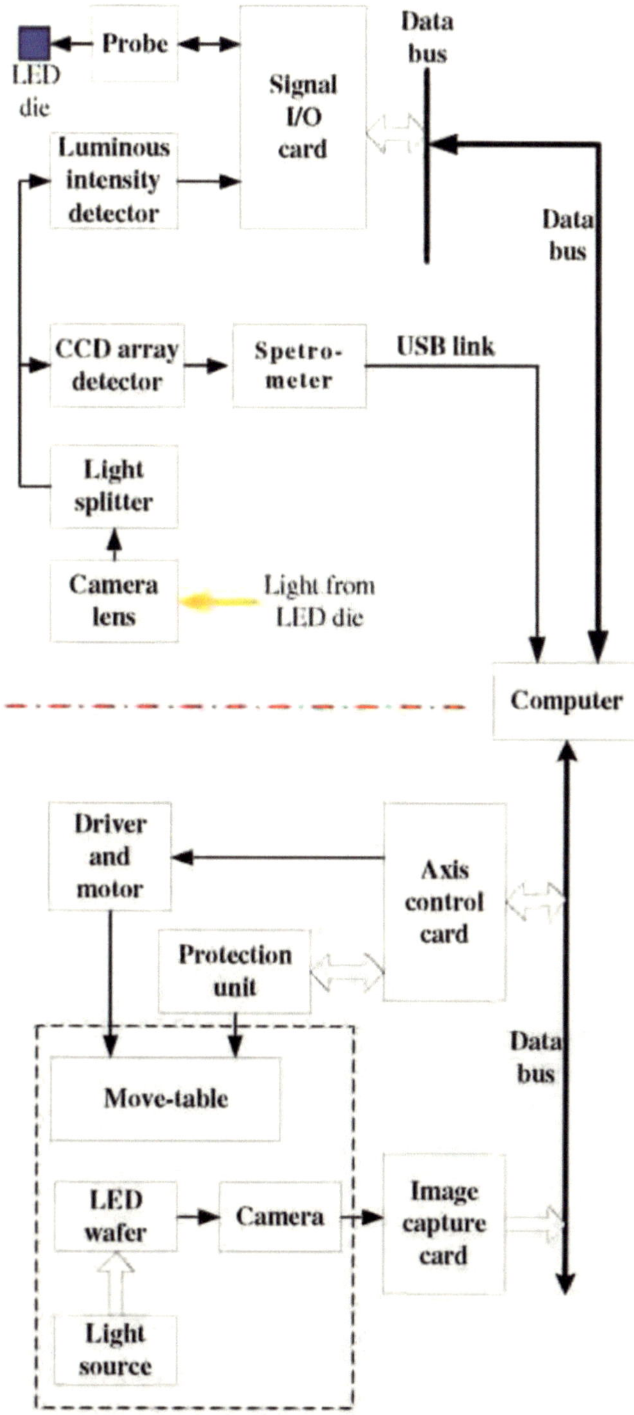

Fig. 2. The overall architecture showing the all hardware modules used in system.

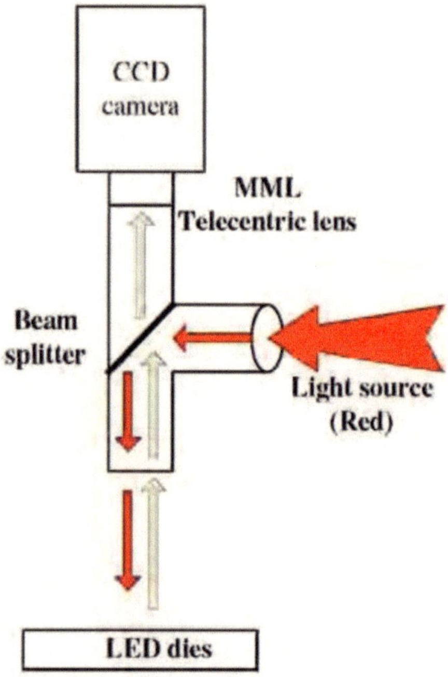

Fig. 3. Coaxial lighting with diffuse type of source is used for reliable images for specular surface of the wafer.

2.2 Mechanical system

Because a very small field of view is used to capture high resolution image, only a small area of the LED wafer can be observed by the camera in each acquisition. To cover all the LED dies, the wafer needs to be moved around systematically so that all the dies can be found and located. A 4-axis mechanical move-table with proper (stepping or servo) motor

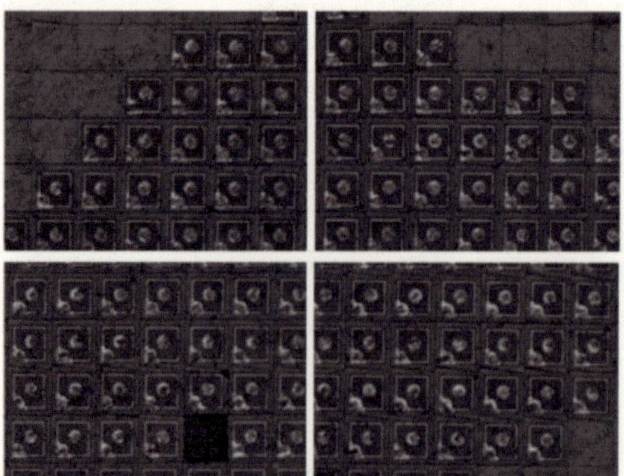

Fig. 4. Four real image frames of the LED dies on the wafer captured by the camera.

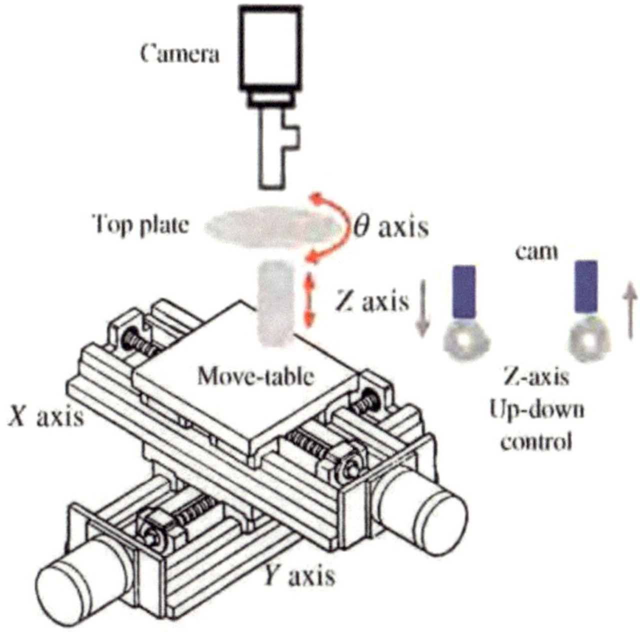

Fig. 5. The sketch of a 4-axis move-table, where a motor-cam combination translates a rotation into a up-down movement in Z-axis.

can be used to achieve position control for image taking. The standard table includes motor mounting plates, couplings, lead screws, large base and top plate, limit switch, etc. The sketch of a 4-axis move-table is shown in Fig. 5 to illustrate the model of the mechanical system. In our implementation, the X-axis is on top of the Y-axis and these two axes are used for wafer position control. The Z-axis is for the control of the up (in measurement status) and down (in motion status) of the wafer by using a motor to control the cam which translates a rotation into a up-down movement (see Fig. 5). The θ-axis is used to adjust the angle of the wafer so that the LED dies on the same row can be aligned with the X-axis. This will assure straight movement of the wafer along the X-direction for all the rows of the LED dies. A 4-axis high speed stepping motor card is therefore needed to meet the above requirements.

This 4-axis structure is mounted on a carriage that is driven by a ballscrew mechanism. The ball-screw mechanism is also called a lead screw. When the motor turns the shaft of the ball screw, the carriage will move horizontally along the length of the ball-screw shaft. The length of the screw lead per motor-rotation is $2mm$, and it takes 800 motor steps to complete one

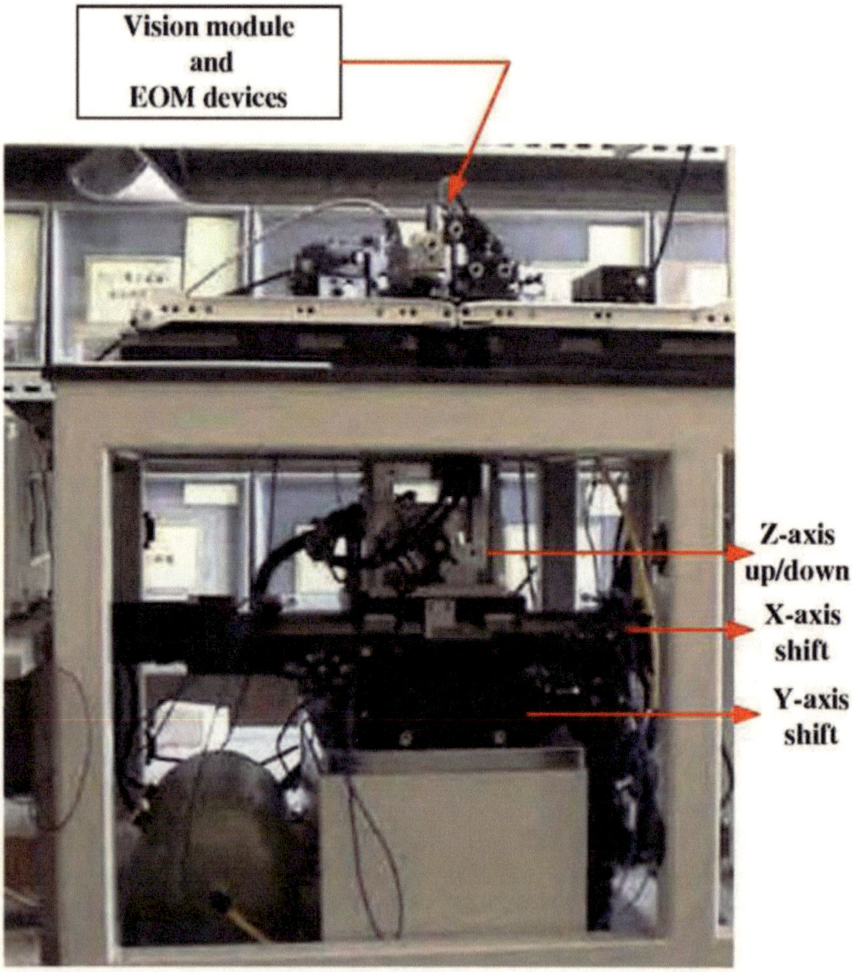

Fig. 6. The finished 4-axis mechanical move-table.

rotation. Therefore, the resolution of each motor step for both X-axis and Y-axis tables is $2.5\mu m$. The finished mechanical move-table in our system is presented in Fig. 6. In order to guarantee the accuracy of the positional measurement, linear optical encoders with appropriate travel and measurement range are installed in the X and Y axes of the move-table. The encoder includes main grating scale and optical encoder head and the encoder head consists of LED light source, optical detector module, and secondary phase-grating. The installation of the encoder help to maintain the precision of the mechanical and control system.

In the process of die position measurement, the center of each die is estimated and recorded in the unit of motor step. For each image captured, the top-left corner is designated as the reference point for the current frame, and its position (in units of motor steps) is saved as (r_x, r_y) in the coordinate of move-table (or simply the *table coordinate*). Furthermore, the conversion ratios between image coordinate (in pixel) and table coordinate (in motor step) are ρ_x and ρ_y for X-axis and Y-axis respectively. Given the resolution of the motor step ($2.5\mu m \times 2.5\mu m$) and the image pixel ($1.67\mu m \times 1.61\mu m$) used in our current implementation, these two conversion ratios are:

$$\varphi_x = \frac{1.67\mu m}{2.5\mu m} = 0.668 \quad (1)$$

$$\varphi_y = \frac{1.61\mu m}{2.5\mu m} = 0.644 \quad (2)$$

where one pixel in the image corresponds to φ_x motor steps in the horizontal direction and φ_y motor steps in the vertical direction. These two conversion ratios are needed when the distance or position measurement based on pixels in the image coordinate are converted to the motor steps for the table coordinate.

2.3 Electrical-Optical measurement (EOM) system

After positions of all the LED dies were found by vision and mechanical systems, the wafer will be ready for the measurements of the electrical and optical parameters. The hardware modules and their real images used in the parameter measurements are shown in the top part of Fig. 2 and Fig. 7 respectively. There are significant differences between LEDs and other light sources which made it necessary to introduce new quantities (CIE 127) for their characterization with precisely defined measurement conditions. The photometric detectors that we used in these modules have proven to meet the standard CIE 127-1997 [16]. The EOM system will send out a bias voltage and read back the response (voltage and current) through the probe to check the electrical parameters of the die. Simultaneously, the LED die is turned on by the bias voltage and its light will be picked up and sent to the intensity detector and Spectrophotometer [17] (see Fig. 7) for the measurements of luminous intensity, dominated wavelength, chromaticity coordinates, color temperature, and purity. Note that the numbers (1 and 2) shown in Fig. 7 are used to illustrate the connection of the optical devices: components

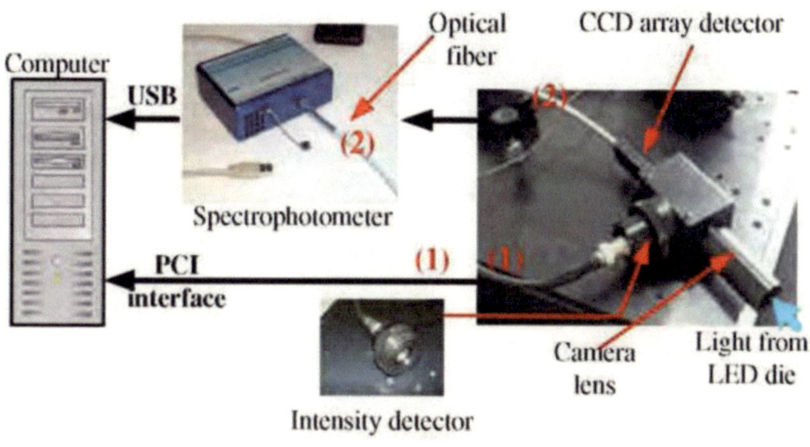

Fig. 7. The connection of the devices used in the optical measurements.

with the same number are connected together. The installation of the EOM module on the mechanical system is illustrated in Fig. 8.

The modules in the EOM system are all off-the-shelf devices and can be replaced as soon as better models are available. Note that the Technical Committees TC2 of the CIE is preparing a revision of CIE 127 [18]. The main reason is that CIE 127 was focused on luminous intensity measurements and did not cover sufficiently the measurement of total luminous flux and color of LEDs, which are now becoming very important for solid state lighting and signaling applications. Another reason is that LED technologies are changing rapidly and some measurement methods must be updated for varieties of new LEDs. Our choice of not constrained to a certain EOM components makes the design of the AMG system become very flexible.

3. Software System

The software system for the measurements and grading of the LED dies employs two separate phases in the whole process: one is for die positioning and the other is for die measurement.

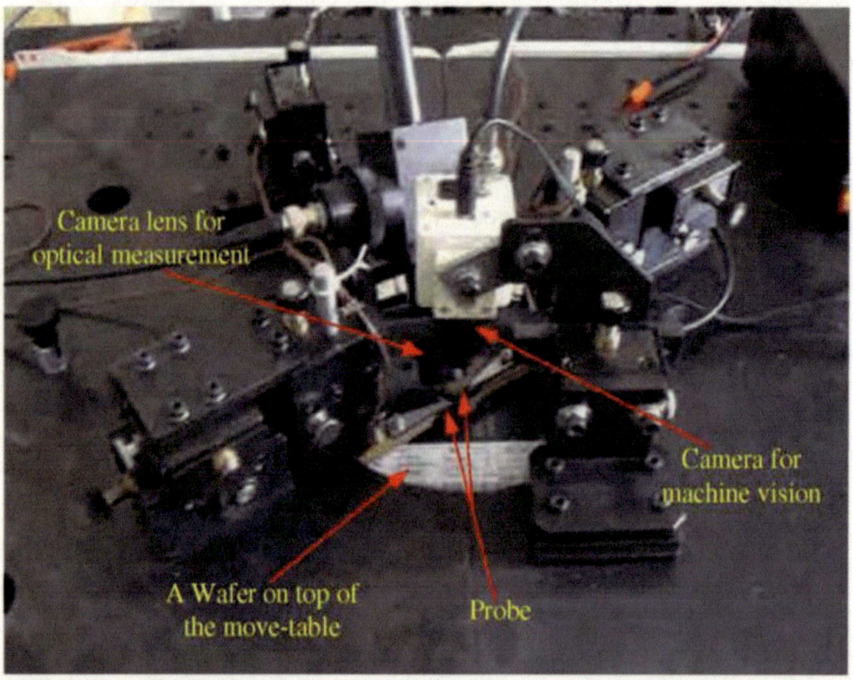

Fig. 8. The structure and position of the probe relative to the other devices.

3.1 Phase I: Positioning of each individual die

Before the dies on the wafer can be probed to measure the electrical and optical parameters, their positions must be found first. This is done by moving the whole wafer around such that all the dies can be captured by the camera of the vision system and their positions be

obtained. In order to finish the position estimation of the dies as fast as possible, the scanning route should be continuous, as shown in Fig. 9(a). The wafer will be moved continuously in the X-axis direction (left or right) and images of multiple rows of the dies are taken until the (left of right) boundary of the wafer is met (this will be called a *rows-scan* in the following discussion). It then goes downward (Y-axis) and executes another rows-scan in the reverse direction. The number of rows-scan performed was saved in variable i. This process is repeated until the end of the wafer, as illustrated in Fig. 9(b).
The LED wafer will be placed manually on top of the move-table and one calibrated line on the table is used to align the row of the dies on wafer with the scan line of the camera by rotating the θ-axis of the moving table. After the wafer is placed properly, the user should

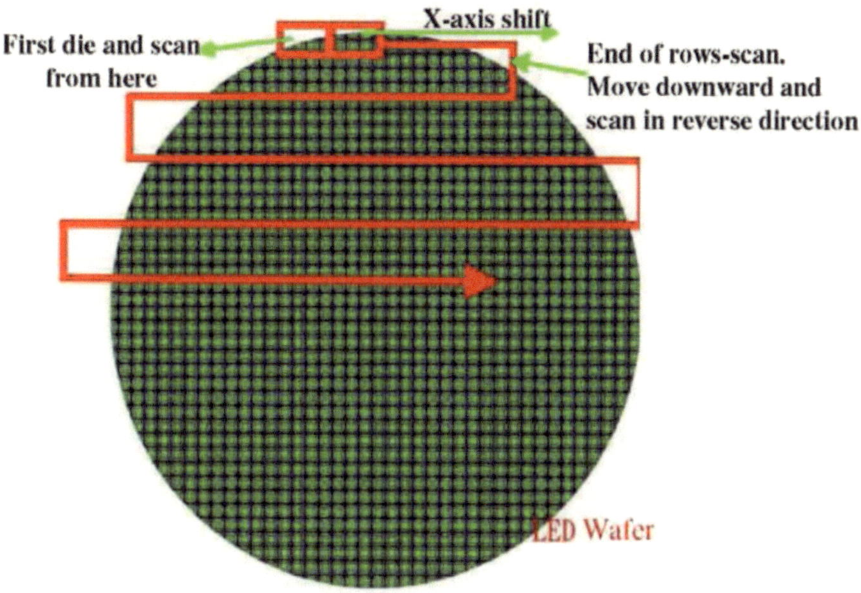

(a)

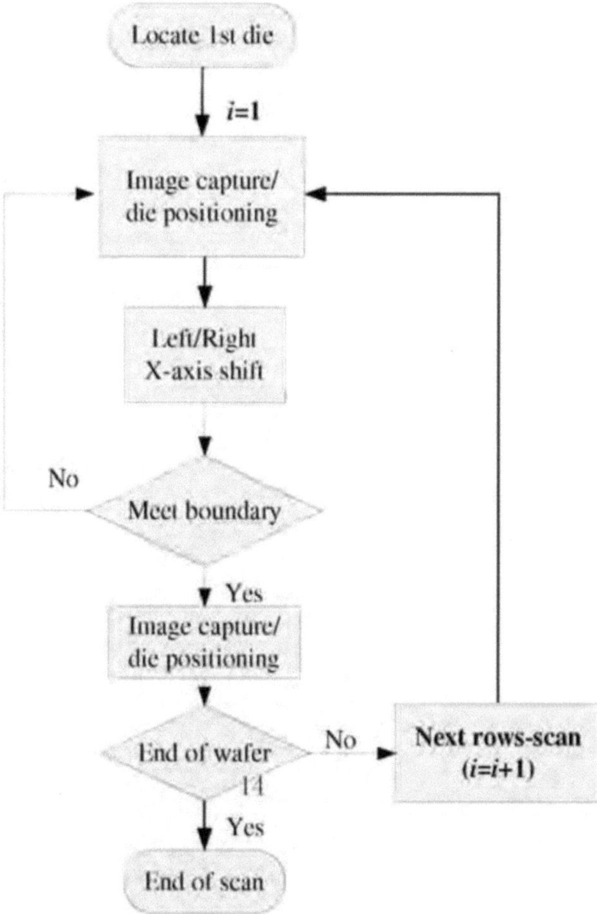

(b)

Fig. 9. The scanning route for the image taking of the dies.

start the scanning by moving the table using keyboard or joystick so that the first die on top-left corner of the wafer can be seen by the camera (this can be observed from the monitor). Since the home position of the move-table is located at the bottom-right corner, the scanning will be started by moving the table to the top-left.

3.1.1 Calculation of the centers of the dies

Given each frame captured, positions of the dies in image coordinates inside the FOV (Fig. 10(a)) needs to be calculated. Since the metal electrode of the LED usually shows strong contrast compared to the background under red light source with coaxial illumination, the die block can easily be located by proper thresholding. For each acquired image, the estimation of each die position is done by the following processing steps:

(1)Thresholding the image automatically (by Otsu's algorithm [19]) to generate a corresponding binary image where die area appears as white (Fig. 10(a)).

(2)Applying horizontal (Y) projection and vertical (X) projection respectively on the binary image (Fig. 10(b)) to generate two arrays of data $P_H(y)$ and $P_V(x)$. Since the rows of the dies on wafer have been aligned with the scan line of the camera, the projections from different groups (i.e. successive columns or successive rows) will not overlap. Given an $M \times N$ (M pixels by N lines) binary image, this can be done by accumulating (projecting) all the white pixels on each row in the image to form a vector P_H, and collecting (projecting) all the white pixels on each column in the image to form a vector P_V.

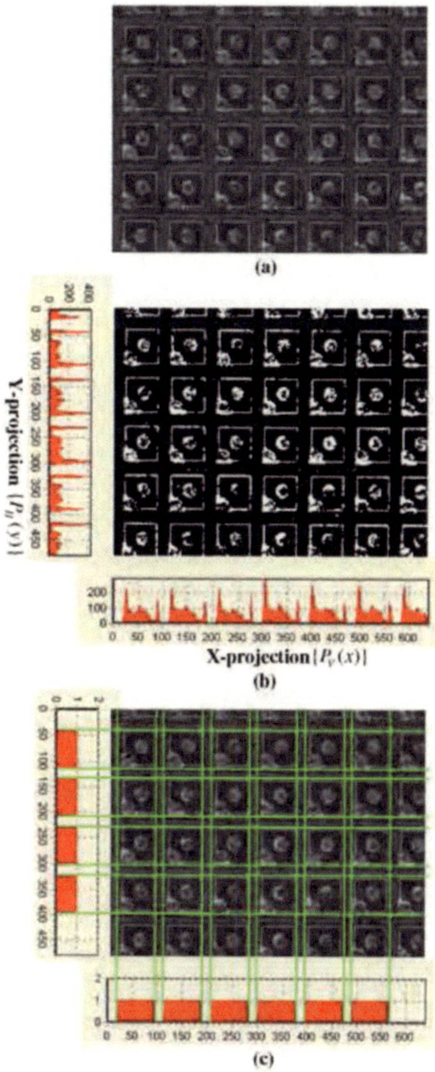

Fig. 10. The extraction of LED die, block based on the projection method. (a)The thresholded input image, (b)Projection to the X and Y directions, (c)Identify the nonzero region of the projection data to form the *final blocks of projection* and then back-projected. Blocks formed by intersection of the back-projection are the positions of the dies.

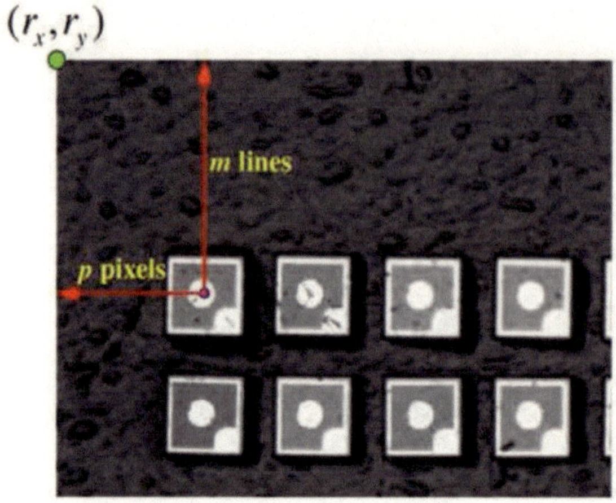

Fig. 11. The relationship between image coordinate and table coordinate.

Assuming the binary image is $f(x, y)$ with $f(x, y) = 1$ for white pixel and $f(x, y) = 0$ for black pixel, the projections are defined as:

$$P_H(y) = \sum_{x=1}^{M} f(x, y) \quad \text{and} \quad P_V(x) = \sum_{y=1}^{N} f(x, y) \tag{3}$$

(3)Identifying the nonzero regions of the projection in these two arrays and thresholding them to form the *final blocks of projection* (Fig. 10(c)). The blocks which touch the image boundary are removed. However, these removed blocks will appear as complete blocks in the next captured frame and be positioned, as illustrated in Fig. 12.

(4)Back-projecting from the *final blocks of projection* to obtain the intersection region (Fig. 10(c)).

(5)Marking the intersection regions as die blocks.

(6)Calculating the center of each die block in image coordinates (p, m). Steps (1)-(6) will be repeated until images of all the LED dies are acquired. There are two ways to generate position data for each individual dies from the information of image coordinates obtained in the above process. One is to mosaic all the images captured into one global frame and the image coordinates of all the die blocks are re-positioned relative to the first die block. The image coordinates are then converted to the coordinates of the X-Y table. The other simpler approach is to convert the image coordinate into table coordinate in step (6) as soon as all the die blocks in one frame were located. That is, the position data are saved in the table coordinate directly using the unit of motor steps. The conversion from image coordinate, (p, m), to number of motor steps in X-axis and Y-axis (table coordinate), (s_x, s_y), can be computed by

$$(s_x, s_y) = (r_x, r_y) + (\rho_x p, \rho_y m) \tag{4}$$

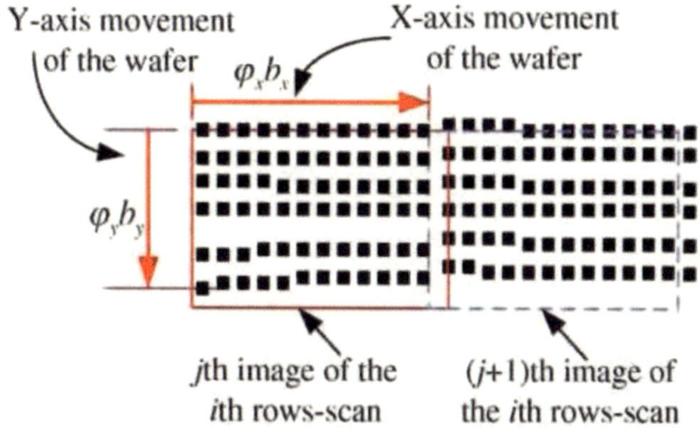

Fig. 12. The distance of movements in horizontal (x) and vertical (y) directions for each image capturing of the wafer. Each image captured will contain several rows and columns of the dies on the wafer.

where (r_x, r_y) is the top-left corner of the current image frame expressed in the table coordinate (Fig. 11), and φ_x and φ_y are the conversion ratios defined in (1) and (2). Note that the distances for the table movements in both X and Y directions are determined by the field of view (FOV) of the camera, and determination of their exact values will be described below. In the step (3) above, the coordinate of the right boundary of the last complete block will be the beginning position of the next frame, as shown in Fig. 12. This position data in image coordinate, (b_x), will be converted to table coordinate as the new r_x by

$$r_x \Leftarrow r_x + \rho_x b_x \tag{5}$$

and sent to the mechanical system. This value is then used as the next position for the following frame capture in the X-direction. Furthermore, while doing the rows-scan and estimating all the positions of the dies frame by frame, the highest Y-axis value of the die is also updated and recorded in b_y. After the boundary of the wafer is reached, the value of b_y is converted to table coordinate as new r_y by

$$r_y \Leftarrow r_y + \rho_y b_y \tag{6}$$

and then send to the mechanical system for movement of the wafer on top of the move-table in the Y-direction to the next (r_x, r_y) position for a new rows-scan.

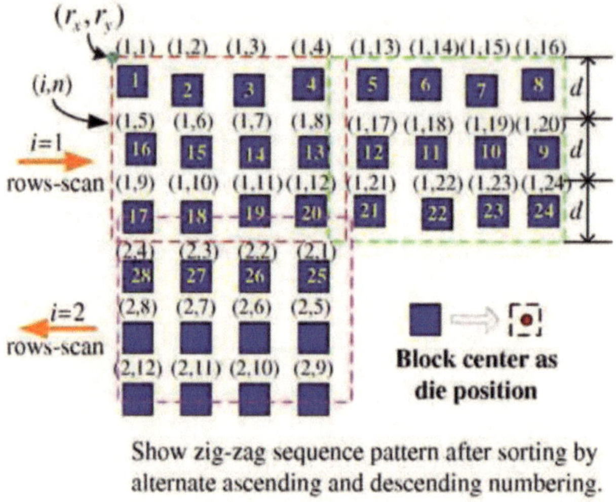

Fig. 13. Proper ordering of the dies is necessary for fast movement and processing in measurement phase.

3.2 Phase I: Reordering of the dies

In order to measure each die sequentially by the EOM system in Phase II, the position data of the dies in each individual row should be sent to the mechanical system continuously, as indicated by the numerical order inside each die in Fig. 13. However, since one whole row of the dies can not be captured in a single image frame, this ordering can not be obtained directly in the above processes. This problem can be solved by storing three types of information in the process of positioning the dies and execute a reordering by sorting after all the positions of dies in the wafer were obtained.

The first type of information to record is the position data of each die in table coordinate, that is (s_x, s_y) in the unit of motor step defined in Eq. (4). The second type of information recorded is the number of rows detected in each individual rows-scan (R_s^i) and the average die position of the Y-axis for each row ($C_k^i : k^{th}$ row in the ith rows-scan). The former information can be obtained from the number of blocks of projection in step (3), and the latter data can be calculated from the average of the centers of the dies in each row.

The third type of information recorded is the initial order of the dies inside each frame, which contains a pair of number (i, n), as shown in Fig. 13. The first number, i, represents the order of the rows-scan performed, and the second number n is used to identify each individual die for all the dies located in that particular rows-scan. The dies in the same frame are numbered sequentially from left-to-right and then top-to-bottom. The dies captured in the subsequent frame are then continuously numbered, as demonstrated in Fig. 13. The total number of dies in all the captured frames for the ith rows-scan is saved as D^i.

Given the example in Fig. 13, there are three rows of dies ($R_s^1 = 3$) and two image frames captured in the first ($i = 1$) rows-scan, and the distance between two successive rows is approximately d (in the unit of motor step). Since all the above position data and related information are saved in database (Paradox in the implementation) and can be accessed by SQL, the reordering can be achieved by two sorting processes based on SQL command. The general idea is to sort the dies group-by-group where each group contains all the dies in the rows acquired in one rows-scan. The dies in the group are first sorted in the ascending order based on the values of the Y (table) coordinate s_y. The same group will then be sorted row-by-row based on the values of the X (table) coordinate s_x in the alternate order of ascending and descending. The final order, which is numbered in a zigzag sequence pattern, will be like the number inside each die block shown in Fig. 13. Assuming there are N_{rs} rows-scan obtained for the wafer under measurement, the sorting can then be accomplished by the following algorithm.

Parameters (s_x, s_y), R^i_s, C^i_k, D^i, and N_{rs} defined above will be used in the description of the algorithm. Note that C^i_k contains the average die position of the Y-axis for each row in the kth row of the ith rows-scan, and d holds the average distance between two successive dies in the Y-axis.

Algorithm of the reordering

$R_t = 0$ {Total number of rows sorted at the end of each rows-scan}

CurrentOrd = 1 {Current position to insert for storing the sorted dies}

$D_t = 0$ { Total number of dies on the wafer}

for $i = 0$ to N_{rs} **do**

$D_t = D_t + D^i$ {D^i: total number of dies in the ith rows-scan}

end for

Allocate an array struct *SortDie()* of size D_t {to store reordered die positions}

for $i = 0$ to N_{rs} **do**

Sorting all the dies with the same i in the (i, n) numbering based on the s_y in ascending order. {by SQL command}

for $k = 1$ to R^i_s **do**

1. Pick the die whose s_y value falls in the range of $(C^i_k - s/2, C^i_k + d/2)$ by SQL command
2. Count the number of dies obtained above and save it in *NumDieInRow*.
3. Allocate an array *DieForSort* of size *NumDieInRow* to save the position data (s_x, s_y) of these dies

if $(R_t + k)$ is odd **then**

Sort array *DieForSort* based on the s_x values in ascending order

else {$(R_t + k)$ is even}

Sort array *DieForSort* based on the s_x values in descending order

end if

Copy the contents of the sorted array *DieForSort* to array *SortDie* start at position *CurrentOrd*

CurrentOrd = *CurrentOrd* + *NumDieInRow* + 1

Free the memory of array *DieForSort*

end for

$R_t = R_t + R^i_t$

end for

After the reordering is accomplished, the position data of the dies are now in the correct zigzag sequence. These position data in the unit of motor step can now be sent to the mechanical system for the smooth control of the table movement in the measurement phase.

3.3 Phase II: Measurements and grading

Because positions of the electrodes on the die relative to the center of die block are known, the probes can be installed properly before the measurement. Given the database obtained in the first phase for the sorted locations of each die block, the things left to be done in phase II is to move all the LED dies one-by-one (using the X-axis and Y-axis tables control) so that each die can be positioned below the probes (Fig. 8). Then the wafer will be pushed upward (using the cam control on the Z-axis) for the electrodes of the die to touch the probes and the LED be turned on. The current and voltage values are transferred to the signal I/O card through the probe for estimation of the electrical parameters. Simultaneously, the light it emits is picked up by the camera lens and distributed by the beam splitter to the luminous intensity detector and spectrophotometer for optical parameter estimation (Fig. 7). The electrical and optical parameters obtained will be graded and saved in the database for future use.

In order to obtain stable measurement, the wafer surface (Fig. 14(a)) must be kept flat and soft enough so that the probe can touch the die under constant and suitable pressure. This condition can be achieved by placing the wafer on top of a vacuum holding plate. The vacuum holding plate is built with a thin plate made by porous ceramic which is enclosed in a adjustable holder (Fig. 14(b)) and then connected to a vacuum pump. In most cases, the porosity of the ceramic can be varied from 20% to 60% by volume. The porous structures are always interconnected in which liquid and gas can flow through them with low pressure drop. When the wafer is sitting on top of the plate, its flatness can be maintained by the proper vacuum pressure generated from the pump and passed through the porous ceramic inside the holding plate, as shown in Fig. 14(c). The vacuum pressure is continuously monitored and the pump will only be activated when the pressure is not enough.

4. Experimental results and discussion

The wafer (Fig. 14(a)) with its LED die images shown in Fig. 4 was used for our system development and measurement testing. There are approximate 14000 dies on the wafer and the system only took an average of five minutes to acquire 815 frames and finished the positioning process in Phase I. However, the measurement in Phase II will take roughly about 67 minutes. This corresponds to an average speed of 3.5 LED dies per second for the measurement and grading process. The dies on the wafer can then be graded based on the measurement data, and the position data will be displayed on the screen for monitoring (Fig. 15). Furthermore, these position and graded data in the database will also be used in the sorting process for the gripper in the sorter to distribute the LED dies into different container. This sorting process is not covered in this paper.

Given the results in Fig. 15, it has been verified visually that all the dies are correctly identified and their electrical and optical parameters are saved

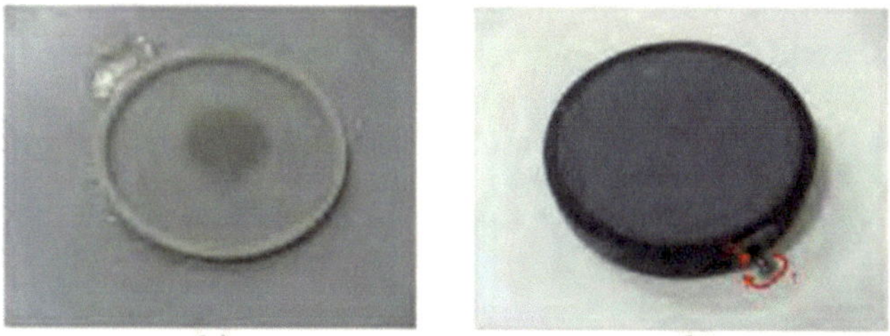

(a) (b)

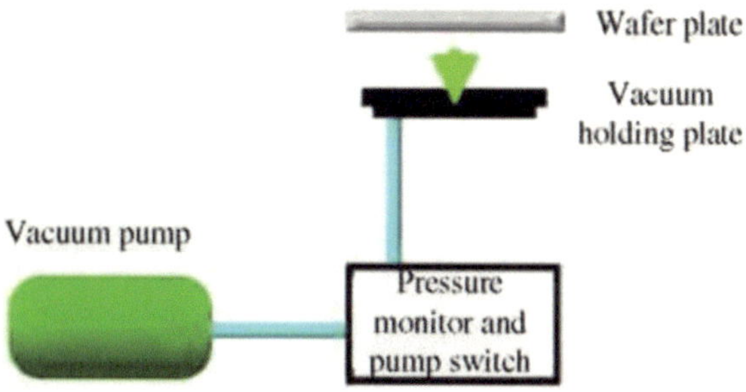

(c)

Fig. 14. (a)Led die wafer placed on a plate to be measured. (b)Vacuum holding plate built with porous ceramic. (c)The system setup for maintaining the flatness of the wafer surface.

and graded in the corresponding database. The white areas represent either empty or the faulty LEDs. Because the wafer is very expensive (~US$1000), we only used one wafer for all the measurements and repeated testing. Under frequent contacts with the probe in the common laboratory environment, quality of the LED dies were gradually degraded. This will be likely to ruin the LED wafer and generate those empty zones. When the system is put to use in the controlled measurement condition, the whole system will be installed in a clean room and the die will only be touched once by the probe. As one can expect, the quality will be much better than that we show here in Fig. 15.

Because "ground truth" of the position data for each individual die can not be obtained, effectiveness of the system was evaluated by thorough manual verification for the dies that have no response in the probing phase. There are two possible reasons for the "no response" condition: one is caused by the wrong position data of the die and the other is caused by a faulty die. This verification process was conducted by searching the database for the dies with no optical and electrical measurements, and then these dies were probed again one-by-one by moving the X-Y table using the position data obtained in the Phase I. We found that all these dies can be positioned and probed properly and their failure to respond were all caused by their fault-iness. This verification process proved that the position data for each die can be correctly found and each individual die can be properly probed for measurement.

While mechanical accuracy of the positional measurement based on the motor step can be secured by the linear optical encoder, we found that flatness of the wafer sometimes could be the source of the problem. In order to make sure the wafer is flat enough in the process of position estimation and measurement, porous ceramics has been used and the results are satisfactory.

5. Conclusion

This paper has presented an integrated design scheme for the implementation of an AMG system for LED dies on the wafer. The system was built by using off-the-self components which makes it very cost-effective. The measurement speed of 3.5 dies per second achieved is limited not by the mechanical system and vision system but by the response time of the LED and the measurement

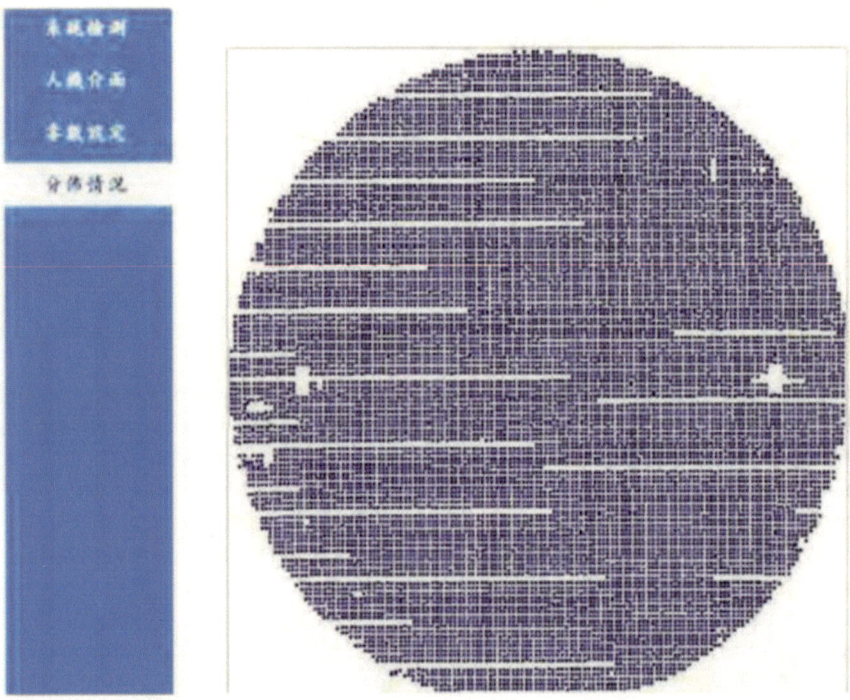

Fig. 15. The position data of the LED die detected in the measurement and grading process.

instrument. However, the current performance is enough in the field of LED inspection and therefore three prototypes of the AMG system have been made for the field tests. The methodology adopted in the proposed system is to utilize the off-the-self hardware devices and components so that the system can be setup economically and rapidly. Moreover, as soon as better devices are available, the system can be updated immediately to maintain its competitiveness.

While the vision system for estimation of the die positions is designed for a specific pattern of the LED die, the modular design of the system and simple structure of the die make the change of the vision software can easily be done for different pattern. One example to demonstrate another pattern type of the LED is shown in Fig. 16, where the regularity of the pattern can be used to accurately detect each die position. The results of this research might help the LED industry to make more informed decisions on the purchase or design of the AMG machine. Since there is no literature available on the design of the LED measurement and grading system, we hope the contents presented here can draw the attention and interests to further the development on this important topic. Video to show the proposed system in action can be accessed in our web site at http://web.ee.yuntech.edu.tw/lab/videolab/main/download.asp.

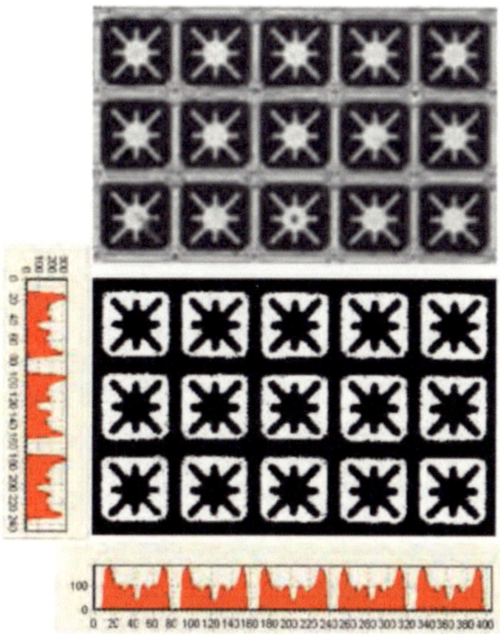

Fig. 16. Image to show another pattern type of the LED dies. Its regularity makes the position estimation of the die simple and accurate.

6. Acknowledgement

The authors would like to thank the support of the CSIST and the National Science Council, Taiwan, R.O.C., under grants NSC95-2221-E-224-029.

7. References

D. Unay, B. Gosselin, "Artificial neural network-based segmentation and apple grading by machine vision," IEEE International Conference on Image Processing, Vol. 2, pp. 630-633, Sept. 2005.

Y. Xun, J. Zhang, W. Li, and W. Cai, "Automatic System of Seeds Refined Grading Based on Machine Vision," *The Sixth World Congress on Intelligent Control and Automation,* Vol. 2, pp. 9686-9689, June 2006.

F. Gayubo, J.L. Gonzalez, E. de la Fuente, F. Miguel, and J.R. Peran, "On-line machine vision system for detect split defects in sheet-metal forming processes," *18th International Conference on Pattern Recogni*-tion,Vol. 1, pp. 723-726, Aug. 2006.

I.C. Sluimer, A.M.R. Schilham, M. Prokop, and B. van Ginneken,"Computer analysis of computed tomography scans of the lung: a survey," *IEEE Transactions on Medical Imaging,* 2006, vol. 25, pp. 385-405.

H. P. Wu, "Patient information extraction in digitized X-ray imagery," *Image and Vision Computing.* Vol. 22, Issue 3, pp 215-226, 2004.

J.H. Sorebo, R.D. Lorenz, "Web inspection using gradient-indexed optics," *IEEE Transactions on Industry Applications,* Vol. 41, No.6, pp. 1476 - 1482,2005.

F. Duan, Y. Wang, and H. Liu, "A real-time machine vision system for bottle finish inspection," *ICARCV 2004 8th Control, Automation, Robotics and Vision Conference,* Vol. 2, pp. 842-846, Dec. 2004.

F. Shafait, S.M. Imran, S. Klette-Matzat, "Fault detection and localization in empty water bottles through machine vision," *E-Tech* 2004, pp. 30-34, July 2004.

H.D. Lin, C. H. Chien, "Automated Detection of Color Non-Uniformity Defects in TFT-LCD," *International Joint Conference on Neural Networks,* pp. 1405 - 1412, July 2006.

C.L. Changm, H.H. Chang, C.P. Hsu, "An intelligent defect inspection technique for color filter," *IEEE International Conference on Mecha-tronics,* pp. 933-936, July 2005.

J.S. Ryu, J.H. Oh, J.G. Kim, T.M. Koo, and K.H. Park, "TFT- LCD panel Blob-Mura inspection using the correlation of wavelet coefficients," *IEEE Region 10 Conference TENCON,* Vol. A, pp. 219-222, Nov. 2004.

Y.C. Huang, K.C. Chuang, M.S. Lin, and C.F. Chen, "Inspecting LED micro structure by piezo servo system," *IEEE International Conference on Mechatronics,* pp. 151-156, July 2005.

LEP series of LED prober from QMC/Korea with http://www.iqmc.co.krproducts/shop/product_view.htm?ID=0300312.

IPT 6000 LED prober from FitTech Co./Taiwan with http://www.fittech.com.tw/ProductDetail.aspx?UID=40.

http://web.ee.yuntech.edu.tw/lab/videolab/main/download.asp

CIE (Commission Internationale De L'Eclairage or International Commission on Illumination) Publication 127, *Measurement of LEDs* (1997).

CS-100 Spectrophotometer based on Photodiode-array, made by SUN-WAVE OPTO. from Taiwan.

TC2-45 Measurement of LEDs - Revision of CIE 127 and TC2-46 CIE/ISO standards on LED intensity measurements.

N. Otsu, "A threshold selection method from gray-level histogram," *IEEE Trans. on System, Man and Cybernetics,* Vol. SMC-9, pp. 62-66, 1979.

19

Advanced Manufacturing Technology Projects Justification

Josef Hynek and Václav Janeček
University of Hradec Králové, Faculty of Informatics and Management
Czech Republic

1. Introduction

Manufacturing companies worldwide are pressurized to undergo a transformation processes in order to compete more effectively and under these circumstances advanced manufacturing technology (AMT) is considered to be a very important tool improving their ability to succeed with their products on extremely competitive international markets. It is widely believed that AMT has a great potential to provide the respective companies by a whole variety of tangible as well as intangible benefits and the reduction of production cost, increased volume of production, improved quality as well as better safety at work are usually amongst the most mentioned ones. On the other hand it is also generally understood that the adoption of AMT requires a high level of initial investment and also the level of risk associated with the implementation of the AMT project is higher especially when the particular company lacks relevant experience. Moreover the payback period of advanced manufacturing technology investment is as a rule longer than the payback period of rather traditional and usually less expensive technology. That is why the process of adoption and utilization of advanced manufacturing technology has been carefully studied and examined in last two decades and numerous studies were published in order to provide some guidelines for managers of manufacturing companies with the view of helping them to make good and well-founded decisions.
We also strongly believe that it is important to study the respective processes when the crucial decisions about AMT projects justification resulting into their practical implementation or on the contrary their rejection are made. The deep comprehension of the fundamentals of these processes allows us to derive the appropriate pieces of knowledge that could turn out to be helpful to technology specialists. We will present selected results of two extensive surveys targeted on adoption and utilization of advanced manufacturing technology that were carried out recently in the Czech Republic in this chapter. We will focus on the phase of advanced manufacturing technology project economic justification and findings ascertained in the Czech Republic will be compared with the outcomes of analogous surveys that were carried out earlier in the United Kingdom and the United States of America. We will demonstrate there are many problems of advanced manufacturing technology projects justification that we have in common in all the above

mentioned countries and we believe that technology specialists as well as managers worldwide could learn from issues presented and discussed here.
Based on our results we suppose that technology specialists empowered in advance by broader insight of what kind of difficulties to anticipate they should be able to prepare their AMT projects accordingly and to improve their chance to get the management approval for the project financing and its implementation.

2. Problem definition

We have already pointed out that advanced manufacturing technology is rather expensive and the relevant project is associated with a higher degree of risk. Therefore the proper and sound justification of the investment decision is required. If the project is incorrectly undervalued and it does not get through the justification process, the company will miss the opportunity to derive potential benefits and its competitiveness might be jeopardized. On contrary, if the project is overvalued because of technology enthusiasm or because of the other reasons, it will be implemented and then it is likely that it will not meet the initial expectations. It will cause a disappointment and furthermore, it will complicate the justification process for further AMT projects that will be perceived through biased lens as the former experience was not a positive one. Whatever the motives are, we can see that the both problems, underestimation as well as overestimation of AMT projects, are terribly wrong and unfortunate. That is why the appropriate methods used for AMT projects justification and their proper utilization are extremely important.
It is widely accepted that there are three general groups of investment appraisal techniques - economic approach, analytic approach and the strategic approach. The economic justification approach seems to be very natural and straightforward one and perhaps that is why it is so wide-spread in relevant companies worldwide. AMT investment has to be financially sound and viable because such a project competes for limited resources with many other projects. Therefore various financial and accounting justification techniques such as payback period (PP), return on investment (ROI), net present value (NPV), and internal rate of return (IRR) are frequently used by managers in order to assess the economic aspects of the project. However, many researchers argue that these methods support decisions that are sensible when viewed in isolation and they do not always indicate the best action when we take into account the whole organizational context (Chan et al., 2001). Furthermore, these methods could be misleading when employing too short payback periods or too high discount rates, neglecting various benefits of the new AMT system or being unable to quantify them properly in financial terms. To overcome the problems inherent in using purely economic appraisal techniques, analytic and strategic appraisal approaches have been promoted.
The analytic justification approaches are predominantly quantitative but more complex than the economic techniques. It is believed that especially when intangible benefits are taken into account, these techniques can be far more appropriate by being more realistic, offering better reflection of reality and taking more factors into consideration (Meredith & Suresh, 1986). Various scoring and ranking models could be used including some traditional optimization techniques as well as risk analysis approaches. It is clear that the transformation process from the decision problem to the particular model involves a great deal of simplification and many important factors could be easily overlooked. Furthermore,

models involving various weights of individual factors are rather vulnerable to bias brought along with subjective judgments.

The strategic justification approaches tend to be less technical that economic and analytic methods, but it should be stressed that they are quite often used in combination with them. The main advantage of the strategic approaches is their direct linkage to the goals of the company. Criteria such as meeting the business objectives, comparison with competitors, the retention or attainment of competitive advantage and industry leadership might be utilized as suitable factors for the relevant decision making processes where AMT projects are scrutinized. Of course, it would be unwise to assign too much importance to strategic justification methods and to overlook the economic and tactical impact of the project. That is why recent studies have promoted hybrid approaches based on suitable combination of economic, analytic and strategic appraisal techniques (see (Raafat, 2002)).

We will focus on the economic justification techniques in the rest of this chapter. These techniques seem to be widely used in manufacturing companies worldwide when the decision concerning AMT investments should be made. It is quite natural because the cost of such project is usually well known (although it could be very easily underestimated too) and it is necessary to cover the cost by relevant revenues and various benefits. We will show some typical problems related to the utilization of economic justification techniques and we would like to stress that some researchers have even claimed that these techniques are inappropriate for evaluating AMT projects (Bucher & Lee, 2000).

3. Literature review

There are many interesting papers describing various issues of AMT projects justification from different points of view. Perhaps the easiest way to get quickly oriented in the field is to start with a comprehensive bibliography on justification of AMT (Raafat, 2002) that cites over two hundred articles from a variety of published sources. Chan et al. (2001) concisely reviewed various approaches used in the process of investment appraisal of AMT and concluded that improved approach that would integrate the currently used evaluation approaches was needed. Abdel-Kader & Dugdale (1998) wrote an interesting paper reporting the results of a survey investigation into the investment decision making practices of large UK companies and their study focused especially on investments in AMT. On the other hand, Ariss, Raghunathan & Kunnathar (2000) published their findings concerned factors affecting the adoption of AMT in small manufacturing firms in the United States. Hofmann & Orr (2005) presented the results of their postal survey that was conducted amongst German manufactures and one part of their questionnaire was devoted to the assessment of AMT proposal too. Finally, we have decided to put forward the paper written by Small (2006) that summarizes the results of investigation on the justification of investments in AMT at US manufacturing plants.

We proudly acknowledge that the biggest motivation to start our own investigations in the field of AMT in the Czech Republic came from the work of Lefley & Wharton (1993), Lefley (1994), and Lefley & Sarkis (1997). These authors examined carefully the investment appraisal processes in the United Kingdom and the United States of America. They carried out extensive surveys both in the UK and the USA in order to learn more about current practices in respect of capital investment in AMT projects, to identify if there were perceived difficulties in appraising these projects and to elicit the opinions of senior executives on the various issues related to AMT projects evaluation. Among other things they found out that AMT projects were evaluated by the simplest financial criteria that seem to be unsuitable in this respect. Moreover, they realized that financial directors do have many difficulties when assessing various benefits of AMT projects, and finally, that investment into AMT could be easily influenced by business culture where managers are under pressure to produce short-term results.

The first study in this field in the Czech Republic (Lefley et al., 2004) revealed that despite of many differences ascertained especially in the extent as well as the level of evaluated and implemented technology, where Czech manufacturing companies lagged behind their western competitors, there were many problems that were common for managers from all the three surveyed countries. These results fostered our interest to conduct the second survey in the Czech Republic in 2005 in order to identify the relevant changes in the results that were expected due to the quickly transforming Czech economy and its openness. And finally, we undertook the last survey in the Czech Republic in 2008 and we were interested in evaluation of AMT benefits this time. The results of this investigation are being carefully analyzed, processed statistically and we plan that we will be able to publish them later this year. However, the survey results described below have been derived from the first and the second survey only.

4. Survey methodology

To keep in line with the earlier UK and US surveys which were used as a basis for comparison we have decided to employ the same questionnaire as Lefley & Wharton (1993) utilized earlier for their investigations. We translated their original English questionnaire into Czech language and verified its localization by means of a pilot survey.

The original questionnaire comprised of three sections. Questions in the first part were intended to establish the level of implementation of AMT that had been achieved to date. Three levels of AMT were identified which correspond to the levels of sophistication proposed by Dornan (1987) and Meredith & Suresh (1986). Level 1 systems cover stand-alone projects e.g. robots, NC machines, CAD etc. Level 2 systems are linked systems e.g. linking together of a number of CNC machines, CAD/CAM etc., and Level 3 systems are fully integrated systems including computer integrated manufacturing (CIM) and flexible manufacturing systems (FMS).

In part number two of the survey the respondents were asked which techniques and criteria were used in capital project appraisal and what methods, if any, were used to measure and take into account project risk. Information was obtained about the measures used to assess the performance of senior executives as it appears that management in general is reluctant to make long-term risky investments (such as those in AMT) and prefers to invest in short-term projects that show early profits and low risk (Lefley, 1994).

The third part of the survey was designed to explore opinions about the need for AMT investment, the efficacy of the investment criteria used and the extent to which other factors and considerations had a bearing on capital investment decisions.

We added one more additional section to the questionnaire that was used in the Czech Republic in 2005. It was devoted to the utilization of EVA (economic value added) indicator in our companies as there were some suggestions that there might be a relationship between utilization of this concept and investment behavior of manufacturing companies.

To assure a straightforward comparison of collected data in different countries we carefully followed the methodology used by our predecessors. The survey was aimed at those companies who, it was believed, would have had some experience in the appraisal of AMT projects and that the person who was asked to complete the questionnaire should have had a significant contribution to make in final investment decision. A number of databases were reviewed (with the main stress on data acquired from EDB and Czech business register) to identify the largest manufacturing companies. As we wanted to restrict the survey to 'large' Czech manufacturing organizations, we finally chose sample size of 416 firms in 1999. Within our last survey we have decided to include also the middle sized Czech manufacturing firms and so we have increased the sample to 1030 in 2005.

Our first postal survey started at the end of 1998 and of the 416 questionnaires sent out 92 was returned giving a response rate of 22.12%. A usable sample of 79 completed questionnaires with a response rate of 19.0% was considered to be reasonable under the existing circumstances.

The second postal survey has been conducted from January till April 2005 and 1030 questionnaires were sent out and 135 have returned, 3 of them were unusable. We can see that the rate of response is 12.8% only which is significantly lower rate that the one we achieved in 1999. The reason that we did not reach comparable numbers with our former survey could be explained by the fact that in our current survey the middle sized firms were addressed too.

This article deals with the selected results derived from the first three parts of our questionnaire only and due to limited space we cannot dwell on the other issues here. Readers who are interested in further details are advised to look at (Hynek & Janeček, 2007) or (Hynek & Janeček, 2008).

5. Survey results and discussion

The main part of this section will be devoted to economic justification of AMT projects, but we believe that the facts we will present here should be perceived in a broader context. That is why we will outline basic facts concerning the experience of Czech companies in the area of AMT projects evaluation as well as the levels of implemented technology that were achieved by surveyed companies. Furthermore, selected personal opinions of managers will be put forward in order to show some important problems and difficulties that could significantly influence the chance of AMT projects to pass successfully through the evaluation process.

5.1 Appraisal experience and level of AMT

First of all, from the point of view of further discussions concerning AMT projects justification it could be worthwhile to learn more about the experience of Czech manufacturing companies in the area of AMT projects evaluation. We can see from table 1 that 82.3 % in year 1999 and 78.3 % in year 2005 of Czech manufacturing companies claimed they had evaluated AMT projects. These numbers are significantly lower than results described by Lefley and Sarkis (1997) who reported that 99.3 % of UK and 96.7 of US companies stated that they had evaluated AMT projects over the past ten years. It is clear that Czech managers are less experienced in this respect. Moreover, we have to take into account the time difference among the surveys.

Furthermore, 84.8 % of Czech manufacturing companies in year 1999 and 92.3 % in year 2005 stated that they expect to consider such projects within the next ten years. Once again, comparing these findings with 97.1 % of respondents in UK and 99.2 % in US (Lefley & Sarkis, 1997), there is a significant difference here despite the fact that the latter result ascertained in the Czech Republic might be considered as a positive signal evidencing the raising awareness of the importance of AMT projects amongst Czech managers.

AMT projects evaluated	1999	2005
Number of companies	65	101
Percentage	82.3 %	78.3 %
Total number of companies	79	132

Table 1. Companies that had evaluated AMT investment proposals

Secondly, we were interested in the level of manufacturing technology that was taken into consideration and consequently the level of technology that was actually implemented in the surveyed companies. There were some thoughts that massive foreign investment into transforming and quickly developing Czech economy during last two decades could accelerate the processes of adoption of advanced technology in manufacturing companies. The respective results are summarized in the table number 2 and 3 below.

% number of companies that evaluated AMT project at:	1999 [%]	2005 [%]
Level 1 (stand alone projects)	57.0	40.4
Level 2 (linked systems)	35.4	41.3
Level 3 (fully integrated systems)	15.2	18.3

Table 2. Level of evaluation of AMT projects

% number of companies that implemented AMT project at:	1999 [%]	2005 [%]
Level 1 (stand alone projects)	51.6	45.0
Level 2 (linked systems)	33.9	36.9
Level 3 (fully integrated systems)	14.5	18.0

Table 3. Level of implementation of AMT projects

It is clear that many projects that were originally planned on a higher level were unable to reach the stage of practical implementation and only the restricted version of the project (on a less sophisticated level of technology) was carried out. There is an obvious positive tendency that we can see in the table number 2 as the percentage of Czech manufacturing companies that evaluated the higher level AMT project proposals have been increased in 2005. The same is true for the implementation stage but comparing these results internationally we have to admit significant differences in respect of stages reached by UK, US and Czech manufacturing companies in relation to the evaluation and implementation of AMT projects. For example, taking into account the results of British and US surveys (Lefley & Sarkis, 1997) it is unmistakable that significantly greater number of UK (55.1%) and US companies (50.9 %) had evaluated the most sophisticated projects (on the third level) while the Czech companies have in majority only the first and the second level experience (we can see from table 2 that only 18.3 % of companies reached the third level technology evaluation experience).

Moreover, based on the results shown in table number 3 it is evident that Czech manufacturing companies are lagging behind their British and American competitors in the adoption of advanced manufacturing technology. The contrast is especially visible when focusing on the most advanced fully integrated systems (only 18.0 % of Czech firms implemented them comparing to the 43.0 % in the UK and 43.4 % in the USA). Moreover, as we can see in table number 3 the situation in the Czech Republic has not changed very much between 1999 and 2005 and therefore the gap is still huge (Hynek & Janeček, 2007). It is obvious that the high level of foreign direct investment in the Czech Republic did not fetch along anticipated acceleration of advanced technology adoption in manufacturing companies and the achieved levels of AMT implementations are lower than those previously observed in the UK and USA. Unfortunately, as we will discuss below, we have found that reasons for this unfavorable position of Czech manufacturing companies does not lie with lack of investment money only but it might be deeply rooted in management attitudes too.

5.2 AMT projects justification

Our findings that were described in the previous section clearly demonstrated that the level of AMT evaluation as well as its utilization in the Czech Republic is lower than the levels observed earlier in the UK and the USA. Furthermore, we have indicated that the process of AMT adoption might be influenced by management attitudes towards technology investment in general and, of course, the particular evaluation and justification approaches chosen by the relevant decision makers could be seen as a direct and straightforward way of influencing the outcome of the AMT projects evaluation processes. Some researchers and technology promoters expressed their concerns over conventional appraisal techniques such as payback, return on investment, or net present value, claiming that these techniques are inadequate and biased against technological investment in general (see, for example, Chan et al, 1999). Their criticism is based on assumption that while the cost of the proposed AMT project is in general easily quantifiable, there are many benefits that are very often difficult to estimate. Moreover, as AMT projects tend to be of long-term nature and sometimes even full deployment of particular AMT project requires substantial time period, the profits cannot be expected in short time and that is why the decisions on these projects require a long-term perspective. Subsequently, if the chosen appraisal method is well known for favoring short term profits, the relevant investment decision that is based on such method is easily predictable.

Table 4 shows financial criteria used to assess AMT projects by financial directors of Czech manufacturing companies. It is obvious that more than 60 % of Czech managers employ the simple non-discounted cash flow payback period (non-DCF Payback) as the criterion to decide whether to finance such a project or not (see table 4 for more details) and more than 70 % of them use discounted version of payback (DCF Payback). And it is exactly payback criterion that is often criticized and attacked for its inappropriateness regarding AMT projects. Naturally, this criterion prioritizes projects capable of early repayment of initial expenses while as a rule capital intensive AMT projects tend to be slow in generating positive net cash flows. Indeed, many argue that the utilization of the payback method virtually guarantees the rejection of projects such as AMT (Lefley et al, 2004). On the other hand it has to be stressed, that the problem is not caused by the criterion itself, but it arises when a short payback period is requested by the company management. As we can see in

(DeRuntz and Turner, 2003), while the western companies generally accept a payback period of 1 to 5 years as a reasonable amount of time to recover the initial cost, the Japanese companies are much more flexible in this respect as they use the payback method more as a performance measure than as a rigid financial criterion that must be met.

Financial appraisal criteria used	1999 [%]	2005 [%]
IRR/yield	31.1	35.5
NPV	45.9	38.7
DCF Payback	71.6	76.6
Other DCF	5.4	10.5
Non-DCF Payback	63.5	62.1
ARR	35.1	23.4
Other non-DCF	1.4	2.4

Table 4. Financial appraisal criteria

It has been anticipated that many companies would use more than one criterion and that is why we have made inquiries regarding the number of financial appraisal criteria being used and their importance. The results are summarized in tables 5 and 6. It should be noted that percentages given in table 6 below add up to more than 100 % because some respondents gave equal first ranking to more than one technique.

Number of methods used	1999 [%]	2005 [%]
1	23.0	22.6
2	32.4	33.1
3	20.3	25.0
4 or more	24.3	19.3

Table 5. Number of different financial appraisal methods used

Criteria ranked first or first equal	1999 [%]	2005 [%]
IRR/yield	5.4	9.7
NPV	28.4	17.7
DCF Payback	51.4	58.1
Other DCF	1.8	4.0
Non-DCF Payback	43.2	62.1
ARR	13.5	8.9
Other non-DCF	0.0	1.6

Table 6. Percentage of companies ranking criteria first

It is definitely a positive ascertainment that more than 40 % of financial directors use more than two financial criteria when assessing an AMT project proposal. On the other hand it is evident that every fifth company relies on single criterion only and here of course the important issue is which criterion is employed in these cases. We can see from table 6 that Czech managers without any doubt prefer both versions of payback criterion. Moreover, within our last survey the above mentioned and criticized simple non-discounted cash flow payback period (non-DCF PB) has been ranked as the most important one in the Czech Republic (62.1 % in 2005) while discounted version of this criterion came second (58.1 %). It should be emphasized that there is a huge gap afterwards as the third most important criterion (net present value) maintained its position from 1999 but it was ranked as the most important criterion by 17.7 % of Czech managers only in 2005.

Comparing these results with earlier ascertainments of Lefley and Sarkis (1997) we could find out that non-discounted cash flow payback period (non-DCF PB) was ranked as the number one criterion in the United Kingdom (38.5 %). American managers inclined to use more sophisticated methods that make allowance for the time value of money and that is why DCF Payback (ranked first by 33.3 % of managers) was closely followed by internal rate of return (IRR) that was preferred by 28.2 % of US managers. From this point of view it is quite interesting that IRR is rather popular amongst British managers too (28.0 %), while only 5.4 % of Czech managers in 1999 and 9.7 % in 2005 marked it as the most important criterion.

It should be noted that the higher number of methods and techniques used within AMT projects evaluation process should be facilitated by various pieces of software. Therefore, companies were asked if spreadsheet packages, dedicated software or other computer aids were used in the process of evaluating advanced manufacturing technology investment proposals and the results are shown in table 7. We can see that a very high proportion of Czech companies use spreadsheet software (75.7 % in 1999 and even 89.4 % in 2005). Approximately one out of six managers employs some dedicated computer software, while other computer aids were reported to be used semi-occasionally.

Computer aids	1999 [%]	2005 [%]
Spreadsheets	75.7	89.4
Dedicated software	16.2	18.3
Other computer aids	4.1	4.8

Table 7. Use of computer aids

Of course, it was anticipated that conventional criteria are widely used and therefore the respondents were asked to indicate, based on their own experience and judgment, whether or not they agreed with the statement that, "conventional appraisal methods such as Payback, NPV and IRR favored short term projects". According to (Lefley & Sarkis, 1997) more than 70 % of companies in the UK and USA agreed with the statement, while significantly fewer in the Czech Republic (55.6 % in 1999 and 53.2 % in 2005) were of the same opinion. The relatively low proportion of Czech managers who thought conventional techniques favor short term investments seems to support the above mentioned views that conventional financial appraisal methods do not automatically favor short-term projects and that these criteria could be used for AMT project proposal evaluation too. Of course, these techniques should be used wisely because when short payback periods or unjustifiably high discount rates are used then a short-term bias can easily occur.

In this respect it could be interesting to find out if there is a tendency to set up some tight hurdle rates for AMT projects justification in companies. The respondents were asked to express their level of agreement with the relevant statement and their responses are summarized in table 8.

We can see that nearly every second manager agreed with the statement and admitted that there is a tendency to set up very tight hurdle rates which could indicate rather disadvantageous starting position for AMT projects. High hurdle rates in combination with the above mentioned traditional appraisal methods could easily result in the AMT project rejection. On the other hand we have to say that in many cases the high hurdle rates are used by managers in order to make appropriate adjustment for a higher degree of risk and uncertainty that relates to AMT projects and it is rather typical approach taken by many companies worldwide when evaluating more risky investment project. Hence such behavior should not be automatically perceived as deliberate intention to discriminate against AMT projects especially when the particular company lacks experience with the project proposal that is under consideration.

There is a tendency to set too high a hurdle rate for AMT projects.	1999 [%]	2005 [%]
Agree	48.0	48.8
Disagree	52.0	51.2

Table 8. There is a tendency to set too high hurdle rates for AMT projects

Some researchers as well as practitioners advocate for exploitation of non-financial criteria and rather strategically oriented criteria believing that there is too much importance attached to conventional techniques. That is why the respondents were asked to express, based on their own experience and judgment, whether or not they agreed with the statement that, "too much importance is attached to conventional techniques". Their responses are presented in table 9.

Too much importance is attached to conventional techniques	1999 [%]	2005 [%]
Agree	51.4	44.4
Disagree	48.6	55.6

Table 9. Too much importance is attached to conventional techniques

It is clear that Czech managers do not feel like having a serious problem with conventional appraisal techniques utilization and their views are perfectly conformable with the opinions of British and US managers where also slightly less than five out of ten managers agreed with the above presented statement that too much importance is attached to conventional appraisal techniques.

It was also noted (Hynek & Janeček, 2007) that a high proportion of companies in all three countries (83.4 % on an average) referred back for re-appraisal those proposals that had failed the initial financial appraisal (the results concerning the situation in the Czech Republic are displayed at table 10). Of course, the introduction of a referral process into the investment justification procedure creates further opportunity for managers to examine the whole proposal carefully once more, to take into account strategic considerations, re-assess and quantify potential benefits or even adjust financial criteria that has to be fulfilled (for example, by reduction of required payback period, or by lowering the pertinent discount rates). On the other hand, it is the very same moment when exactly opposite measures and actions could be taken and there is a large space in which the accept/reject decision could be manipulated. It could be anticipated that in these cases the formal appraisal procedure transforms itself into a ritual where the final decision is based on other influences, which might be of a political, rather than an economic nature. In this context we should put and understand the interesting fact that more than eight out of ten respondents confirmed the referral procedure.

Project proposals re-evaluated	1999 [%]	2005 [%]
Agree	89.2	81.5
Disagree	10.8	18.5

Table 10. Percentage of proposals re-appraised

Having admitted that there might be some political influence in the referral procedure and in the projects evaluation process in general it is natural to ask to which extent do senior executives use their dominant role based on their formal as well as informal authority in order to affect the relevant decisions related to AMT investment in both directions. That is the reason why the respondents were asked to express their level of agreement with the statement that more importance is attached to the experienced judgment of senior management than to financial indicators. The results are shown in table 11 and we can see that slightly over fifty percent of Czech respondents agreed with the statement in 1999 (51.9 %) and their number declined further in 2005 (45.7 %). Nevertheless, it should be stressed that the number of managers who agreed with the statement is relatively high overall and it is clear the concerns expressed by some researchers as well as practitioners seems to be legitimate.

More importance is attached to the experienced judgment of senior management than to financial indicators	1999 [%]	2005 [%]
Agree	51.9	45.7
Disagree	48.1	54.3

Table 11. More importance is attached to the experienced judgment of senior management than to financial indicators

To conclude this section we would like to stress that despite the mentioned criticism the traditional financial appraisal techniques play important role in the process of AMT projects evaluation and justification. We have shown that managers prefer the simplest techniques like payback period that are very easy to understood and interpret, but we have also mentioned that there is a danger of bias towards projects delivering short-term profits when these techniques are used mechanically, shortsightedly, and without broader impact considerations. Furthermore, the risk that AMT projects would be disadvantaged by utilization of these simple techniques could be moderated by utilization of several methods and we have shown that more than eight out of ten projects are re-evaluated if they failed to pass through the initial financial evaluation process.

5.3 Personal opinions of managers

We have already mentioned that the process of AMT justification might be seriously influenced by management attitudes towards technology investment in general. AMT is often considered as one of critical factors that plays important role in the process of acquiring competitive advantage. It seems to be a widely accepted opinion but do managers really think so? We wanted to verify this ascertainment and the respondents were asked to indicate based on their own experience and judgment, whether or not they agreed with the statement that non-investment in AMT was a high risk strategy. Responses to this statement were summarized in table number 12 and we can immediately see there that surprisingly large proportion of executives in the Czech Republic (33.3% in 1999 and 30.7 % in 2005) disagreed that non-investment in AMT is a high-risk strategy.

Non-investment in AMT is a high-risk strategy	1999 [%]	2005 [%]
Agree	66.7	69.3
Disagree	33.3	30.7

Table 12. Non-investment in AMT is a high-risk strategy

Comparing these findings with the results of Lefley and Sarkis (1997) who reported more than the decade ago than 74.8 % in the UK and 81.9 % in the US agreed with the statement that non-investment in AMT is a high-risk strategy, it is clear that significantly higher proportion of Czech managers do not consider AMT as strategically important investment. It is a rather surprising ascertainment taking into account that Czech manufacturing companies after transformation of our economy had to find new market opportunities for their products. Many of them oriented themselves mainly on strongly competitive markets in Western Europe, many others were sold to foreign investors and it was anticipated that new owners would bring new technologies too. It is difficult to generalize, but as we concluded in (Hynek & Janeček, 2006a) it is likely that many companies have apparently decided to rely on skilful and relatively cheap labor force and that is why the relevant companies seems to be somewhat slow in AMT adoption. Moreover, we are afraid that in today's mutually interlinked and quickly changing global world the exploitation of such strategy sounds like a rather shortsighted decision.

Obviously it is not easy to change management attitudes towards AMT investment and perception of its importance from day to day. Fortunately enough, there are some other issues we should pay our attention too and we think that there might be some space where improvement of the current state of art is more feasible. Moreover, we will show that while there are significant differences in perception of the strategic importance of AMT investment in general between managers working under conditions of transforming Central European economy and managers representing two of the most developed countries in the world, there are some problems they have in common too. For example, we have learned that many AMT projects are likely to be rejected just because the lack of understanding of what the contribution of new technology really is.

We could see in table number 13 that more than 60 % of Czech executives agree with the statement that it is difficult to assess all potential benefits of AMT investments (67.1 % in 1999 and 60.3 % in 2005). The level of agreement with the relevant statement was even higher in the UK (81.6 %) while 63.9 % of American managers shared the view (Lefley & Sarkis, 1997).

Thinking about reasons that we can see three possible explanations of this unfavorable situation. First of all, there are some benefits where managers seems to be unable to foresee and to assess their impact and magnitude there because of lack of experience, lack of relevant input data etc. Secondly, the company is not sure whether some particular benefit will be realized at all and thus the benefit falls into this category and stays there without any attempt to quantify it. And thirdly, it is often believed that brand new technology will bring along some new benefits and completely unexpected benefits that are impossible to predict before the technology reach the stage of regular utilization. While the first problem seems to be based on lack of experience and administrative-technical reasons, the other two explanations seems to be much more of a speculative nature.

AMT investments are difficult to assess because they have non-quantifiable benefits	1999 [%]	2005 [%]
Agree	67.1	60.3
Disagree	32.9	39.7

Table 13. AMT investments are difficult to assess

However, whatever reason applies it helps to create the feeling that there are some non-quantifiable benefits that were not taken into account. And we will demonstrate that there is a problem related to proper assessment of non-quantifiable benefits and their expression in financial terms which means that these benefits will not be taken into relevant economic calculations.

We can see in table 14 that large majority of Czech managers (90.1 % in 1999 and 81.7 % in 2005) agreed with the statement that not all potential benefits of AMT are taken into account because they are difficult to quantify in financial terms. It should be noted that these numbers are in compliance with the earlier findings of (Lefley & Sarkis, 1997) who reported that 80.9 % of British managers agreed with the statetems and 81.2 % of American managers did so. It is important to repeat here that the respondents of our surveys were financial directors and decision makers of surveyed manufacturing companies. Recalling back this fact we can see that the situation is very serious and some measures should be taken in order to make sure that AMT project proposals have a fair chance to get through the justification process and to get the pertinent investment approval.

Not all potential benefits of AMT are taken into account because they are difficult to quantify in financial terms	1999 [%]	2005 [%]
Agree	90.1	81.7
Disagree	9.9	18.3

Table 14. Not all benefits are taken into account

According to Primrose (1991) people advocating investment in AMT have made considerable efforts to identify the company-wide benefits which it can produce. The problem is that they describe these benefits always in general terms, such as the following: increased flexibility of production, better-quality products, improved documentation, ability to respond to market needs, need to keep up with competition, improved company image, better management control, obtaining experience of new technology, etc. Managers usually start with the belief that a particular aspect of AMT could be used in their department and they would select an application which was aimed at improving operating efficiency. Having defined the required specification, they try to justify the expenditure afterwards. And now it is necessary to identify the benefits. The nature of intangible benefits is such that they do not have to appear in the department where the investment is made, but occur elsewhere in the company. In addition, the relationship between cause and effect is indirect, so that their magnitude has to be estimated rather than directly calculated. In fact there are two distinct problems and these must be dealt with separately. First of all the form in which the benefit is quantified, and secondly estimating the magnitude of the benefit (see (Primrose, 1991) for more details).

6. Conclusions

The presented selected results of two AMT surveys focused on the specific issues of advanced manufacturing technology justification that were carried out in the Czech Republic demonstrate that the economic justification of the relevant projects is definitely not an easy process. Moreover, there are many problems that seem to be common for managers in Central Europe who has to face the conditions of transforming economy and managers from technologically most developed economies in the world.

First of all, our results clearly demonstrate that Czech manufacturing companies are lagging behind their British and American competitors in the adoption of AMT and the optimistic prognoses that the high level of foreign direct investment will bring along acceleration of AMT adoption as well as the latest technology has not been proved yet.

We have also shown some pieces of evidence that AMT projects might be very easily knowingly as well as unknowingly disadvantaged because of a whole spectrum of reasons. Based on our results it is clear that managers exploit rather unsuitable financial criteria, too much importance is given to the simplest methods that clearly prioritize short-term outcomes and thus short-term projects. British and American managers seem to be more aware of this fact and perhaps it is the reason why they tend to utilize more sophisticated criteria and greater number of criteria in general than managers in the Czech Republic do. However, we have stressed that the problems could be avoided if the criteria are used wisely and we have mentioned as an example the difference between payback period utilization in western companies on one side, and Japanese companies on the other one.

We have seen that more than eight of out ten AMT projects are re-evaluated if they failed the initial financial appraisal. As the result of this phenomenon many projects are carried out only partially. It could be the restricted version of the original project that lacks the originally intended level of integration, or it could be done at the expense of the originally planned level of technology used. In both cases there is a danger that restricted version of the originally planned AMT project will be unable to deliver originally planned benefits and the project will not live up the expectations. Furthermore, we have pointed out that introduction of the referral process establishes ground for various influences that might be of a political rather than economic nature.

Finally, we have examined management attitudes towards AMT projects. We have realized that comparing our results with the outcomes of earlier survey conducted in the UK and USA, significantly higher proportion of Czech managers do not consider AMT as strategically important investment. On the other hand, there are some serious issues that significantly influence the process of AMT adoption and these issues are common for the managers from all three surveyed countries. First of all, two thirds of managers agreed with the statement that AMT investments are difficult to assess because they have non-quantifiable benefits. Secondly, over eighty percent of respondents supported the view that not all potential benefits of AMT are taken into account because they are difficult to quantify in financial terms. Putting these ascertainments in other words we can see that there is a clear lack of understanding of what the contribution of the proposed AMT project really is. Moreover, managers are fully aware of the fact that some benefits are not taken into their calculations because they are unable to estimate them and express them in financial terms. We have already expressed (Hynek & Janeček, 2006b) our view that there is an important space and great opportunity right here that should be taken by technology specialists. They should be able to identify, describe and explain the complex benefits of a particular AMT project and hereby prepare better background material for financial executives. Their involvement in this phase could assure that various tangible as well as intangible benefits will be taken into consideration, properly assessed and consequently expressed in financial terms. Of course, this task could be fulfilled only by technology experts who are able to see the particular technology not simply from technological point of view. Their knowledge and broader understanding of technology benefits for the company as a whole could considerably improve the chances of AMT projects to get the management approval.

Of course, it should be also accentuated that economic approach to AMT projects justification is widely used but it is not the single approach and we recommend employing strategic and analytical approaches too. These approaches do have their own drawbacks too and therefore wise combination of different approaches should be encouraged in order to make sure that AMT project proposals are assessed properly. This is the only way providing enough opportunities to avoid later disappointment.

7. Acknowledgement

This research has been supported by the Grant Agency of the Czech Republic project No. 402/07/1495.

8. References

Abdel-Kader, M. G. & Dugdale, D. (1998). Investment in Advanced manufacturing technology: a study of practice in large U.K. companies. *Management Accounting Research*, No. 9, pp. 261-284, ISSN 1044-5005.

Ariss, S. S.; Raghunathan, T. S. & Kunnathar, A. (2000). Factors Affecting the Adoption of Advanced Manufacturing Technology in Small Firms. *SAM Advanced Management Journal*, Vol. 65, No. 2, Spring 2000, pp. 14-29, ISSN 0749-7075.

Bucher, P.G. & Lee, G.L. (2000). Competitiveness Strategies and AMT Investment Decisions. *Integrated Manufacturing Systems*, No. 11/5, pp. 340-347, ISSN 0957-6061.

Chan, F.T.S.; Chan, M.H.; Lau, H. & Ip, R.W.L. (2001). Investment Appraisal Techniques for Advanced Manufacturing Technology (AMT): A Literature Review. *Integrated Manufacturing Systems*, No. 12/1, pp. 35-47, ISSN 0957-6061.

DeRuntz, B. D. & Turner, R. M. (2003). Organizational Considerations for Advanced Manufacturing Technology. *International Journal of Production Economics*, Vol. 2002, No. 79, pp. 197-208, ISSN 0925-5273.

Dornan, S. B. (1987). Cells and Systems: Justifying the Investment. *Production*, February 1987, pp. 30-35.

Hofmann, C. & Orr, S. (2005). Advanced Manufacturing Technology Adoption - the German Experience. *Technovation*, Vol. 25, No. 7, pp. 711-724, ISSN 0166-4972.

Hynek, J. & Janeček, V. (2005). Adoption of Advanced Manufacturing Technology - New Trends in the Czech Republic. In: *Proceedings of the IEEE 9th International Conference on Intelligent Engineering Systems*. IEEE, Piscataway, NJ, 2005, pp. 75-78, ISBN 0-7803-9474-7.

Hynek, J. & Janeček, V. (2006a). Information Gap between Technology Specialists and Decision Makers. In: *Proceedings of IEEE 3rd International Conference on Mechatronics*, IEEE, Piscataway, NJ, pp. 61-64, ISBN 1-4244-9712-6.

Hynek, J. & Janeček, V (2006b). Problems of Advanced Manufacturing Technology Projects Approval. In: *Proceedings of the 5th WSEAS Int. Conf. on System Science and Simulation in Engineering (ICOSSE'06)*, Tenerife, Canary Islands, Spain, WSEAS Press, pp. 412-416, ISBN 960-8457-57-2, ISSN 1790-5117.

Hynek, J. & Janeček, V. (2007). Advanced Manufacturing Technology Projects Justification. In: *Proceedings of the 4th IEEE International Conference on Mechatronics*, University of Kumamoto, Japan, 2007, pp. 277-282, ISBN 1-4244-1184-X.

Hynek, J. & Janeček, V. (2008). Economic Justification of Advanced Manufacturing Technology, In: *Proceedings of the 2nd WSEAS Int. Conf. on Management, Marketing and Finances (MMF'08)*, Harvard University, Cambridge, Massachusetts, WSEAS Press, 2008, pp. 103-108, ISSN 1790-5117.

Lefley, F. & Sarkis, V. (1997). Short-termism and the appraisal of AMT capital projects in the US and UK. *International Journal of Production Research*, Vol. 35, No. 2, pp. 341-368, ISSN 0020-7543.

Lefley, F. & Wharton, F. (1993). Advanced Manufacturing Technology Appraisal: A Survey of U.K. Manufacturing Companies. *Proceedings of the 4th Int. Production Management Conference: Management and New Production Systems*, London Business School, 1993, pp. 369–381.

Lefley, F. (1994). Capital investment appraisal of advanced manufacturing technology. *International Journal of Production Research*, Vol. 32, No. 12, pp. 2751-2776, ISSN 0020-7543.

Lefley, F.; Wharton, F.; Hájek, L.; Hynek, J. & Janeček, V. (2004). Manufacturing investments in the Czech Republic: An international comparison. *International journal of Production Economics*. Vol. 88, No. 1, pp. 1-14, ISSN 0925-5273.

Meredith, J. R. & Suresh, N. (1986). Justification techniques for advanced manufacturing technologies. *International Journal of Production Research*, Vol. 4, No. 5, pp. 1043-1058, ISSN 0020-7543.

Primrose, P. L. (1991). Investment in Manufacturing Technology. Chapman&Hall, London 1991, ISBN 0412409208.

Raafat, F. (2002). A comprehensive bibliography on justification of advanced manufacturing systems. *International Journal of Production Economics*, Vol. 2002, No. 79, pp. 197-208, ISSN 0925-5273.

Small, M. H. (2006). Justifying Investment in Advanced Manufacturing Technology: a Portfolio Analysis. *Industrial Management & Data Systems*, Vol. 106, No. 4, pp. 485-508, ISSN 0263-5577.

20

Installation of Mechatronics Education Using the MindStorms for Dept. of Mechanical Engineering, O.N.C.T

Tatsushi Tokuyasu
Oita National College of Technology
Japan

1. Introduction

In the late 1950s Japanese economy were growing remarkable. In order to sustain this economic growth and foster human resources who can provide advances in technology and science, the government established particular kind of educational institution, National College of Technology, in 1962. The educational system of National College of Technology is positioned between industrial high school and technological university, because the age range of student is from fifteen to twenty. Basically, the first grade students are the graduates of junior high school so that students are not sufficiently grown to understand special subjects and mechanisms of most advanced technologies. There are 55 scholastic institutions throughout Japan at the present day, so that one or two schools have been established in each prefecture. Each school is allowed to construct department and to have own educational curriculum, and most of schools have own advanced course (http://www.kosen-k.go.jp/english/index.html).

Educational programs for early year are mainly composed of general studies such as mathematics, foreign and international languages, social studies, and physical education, and so on. Meanwhile special subjects including graduation research are implemented in higher-grade educational program. The purpose of graduation research is to foster the ability as engineer, and then final-year students engage in researching at their interesting laboratory for one year, in which lots of scientific papers have been published every year.

Mechatronics is one of the major technical studies and consists of wide variety of technologies, such as mechanics, electronics, programming, and robotics, and so on, so that it is difficult to compose an educational program of mechatronics. In fact, many textbooks for mechatronics covering a broad range of specific subject have been published. To select a text book suit for levels of student is also difficult. Then, the author assumed that practical work more effect for mechatronics installation education than classroom lecture using textbooks.

Mind Storms, a set of small plastic and various shaped blocks, motors and several different types of sensors, produced by LEGO Group, has been already known as one of good educational toolkits for mechatronics installation(Komatsu et al., 2000)(Nakashima et al.,

2001)(Inagaki et al., 2001). A variety of of guide books for MindStorms have been published[5]. By using a toolkit of MindStorms, trainers can easily get trainees started on learning the basic factors of manufacturing. In the department of mechanical engineering of Oita National College of Technology, mechatronics education begins from the first grade as a part of practical training program. This chapter introduces a method of mechatronics installation by using Mind Storms have implemented in the department of mechanical engineering of Oita National College of Technology (http://www.oita-ct.ac.jp/).

2. Orientation

The mechatronics installation education presented in this chapter have been implemented as a part of practical training subject from 2006 against the first grade students belonging to department of mechanical engineering. The practical training of the first grade student is composed of the following six courses; lathe turning machine (eighteen hours), drafting (three hours), hand finishing (fifteen hours), mechatronics installation (eighteen hours), assembly of a micro-car (twelve hours), and fishing lure manufacturing (six hours). Because of a number of classmates, forty students take part in the course every year, students are divided into four groups. Each group participates in these courses as shown in Table 1. Total learning time of the practical training is twenty-four, and only six times of the total number is assigned to the mechatronics installation course.
In order to make students comprehensively learn a subject of mechatronics, MindStorms has been adopted as educational tool of this mechatronics installation course. MindStorms is an integrated educational toolkit for mechatronics learning. A toolkit contains variously-shaped blocks, a RCX controller, a Robolab programming software, and some kinds of sensors. It enables students experimentally learn the factors necessary to make a controllable robot. A lot of installation books of MindStorms have been published(Sato, 2000)(Eto, 1999).

Group	1	2	3	4	5	6	7	8	9	10	11	12	13	14	15	16	17	18	19	20	21	22	23	24
A	L						D	H					M						A					F
B	A				F		L						D	H					M					
C	M						A				F		L						D	H				
D	D	H					M						A				F		L					
Month	4	5			6				7			9	10			11				12		1		2
Day	27	11	18	25	1	8	22	29	6	13	21	7	5	19	26	2	9	16	30	14	21	18	25	1

L: Lathe Turning Machine M: Mechatronics Installation
D: Drafting A: Assembly of A Micro-Car
H: Hand Finishing F: Fishing Lure Manufacturing

Table 1. Time table of practical training though the year

3. Method

The subject of practical training has been held for three hours one day a week. Especially, as shown in Table 1, the mechatronics course consists of six lessons. The followings describe the educational contents undertaken in each week.

3.1 Installation

At the first session of this mechatronics course, firstly basic concepts of mechatronics are introduced into the students, for example, a history of robotics, fundamental construction of mechatronics devices, name derivation of mechatronics, and so on. A car wiper drive system, moving from side to side, is one of familiar mechatronics devices even young students, and to think this mechanism is employed as the first problem of this course. Students firstly sketch a mechanism which enable wiper blades to move from side to side under a bonnet of car based on their imagination. Next, they try to shape the sketch by using MindStorms. Fig. 1 shows a piece of work a student made, where four linkage blocks and two gears are used and it achieved to imitate the behaviour of wiper blades. A technical point is to change the direction of motion from rotational movement of gear to linear reciprocating motion in right-and-left of wiper blades.

Fig. 1. A car wiper drive mechanism made from MindStorms

3.2 Duplication of mechanical module

Several kinds of mechanical modules, such as lever slider, one-way ratchet, spur wheel, geneva drive, eccentric crank, etc, produced by SHINKO engineering research corporation, had been introduced in order to educate dynamics of mechanism. Though a subject of dynamics of mechanism begins in the fourth grade, these modules are suited to installation tools of dynamics of mechanism because students can experimentally observe behaviours of gears, linkages, and other mechanical components. For example, in the case of eccentric crank module as shown in Fig. 2, the rotational motion driven from handle operation is transformed into the translational motion of the work. After observing these mechanical modules, students try to duplicate their favorite mechanical module by using MindStorms. In order to completely build the mechanical module, firstly proper block selection has to be done under considering overall stiffness of module. Some of students wrote a conceptual drawing by using selecting blocks. In the phase of assembling, fitness between other shaped blocks due to their different thickness. Most of students selected the eccentric crank and had spent about two hours. Seventy percentage of students could duplicate the function of eccentric crank by using MindStorms.

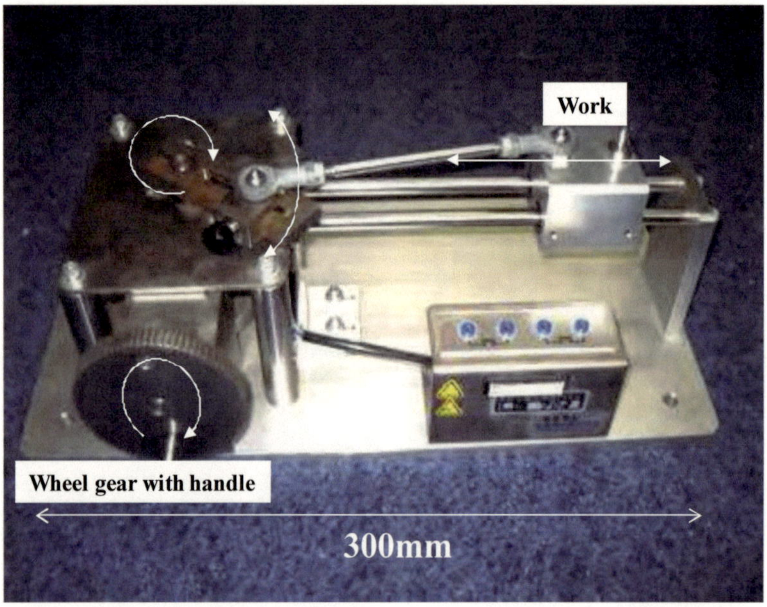

Fig. 2. Mechanical module of eccentric crank

3.3 Introduction for Sequential Control System

The third session begins with introduction of sequential control that is used in a variety of devices supporting our dairy comfort life, such as auto vender, laundry machine, and some kind of audio devices, etc. In order to educate sequential control method and mechanisms necessary for running an automated production line, a purchasable miniature model which imitates an automated production line system is adopted. The department had installed an automation education line system few years ago. Fig. 3 shows each work cells of the installed miniature automation line system. This production line makes two simple components to put together on a floating pallet as shown in Fig. 4. The automated sequential tasks of each work cells are followings; the endeffector of the first work cell grasps a male component from the rotary table and puts it on the floating pallet on the belt conveyer. The second work cell has a suction type end effecter which vacuums a female component up and puts it on the male component fixed on the floating pallet, and then two different types of components are completely combined on the pallet. At last, the third cell takes the complete product.

Each of works' movements gathered attention of students. They carefully observed mechanisms of rotary table of the first work and/or movements of each end effecters. A control panel of this automation line system is allocated under the table. Because the control panel administrates all of sensors, actuators, and work processes, each of indication lamps keep lighting-up during the automation system is running. The observation of control panel enables students to learn the roles of controller to manage actuators and sensors.

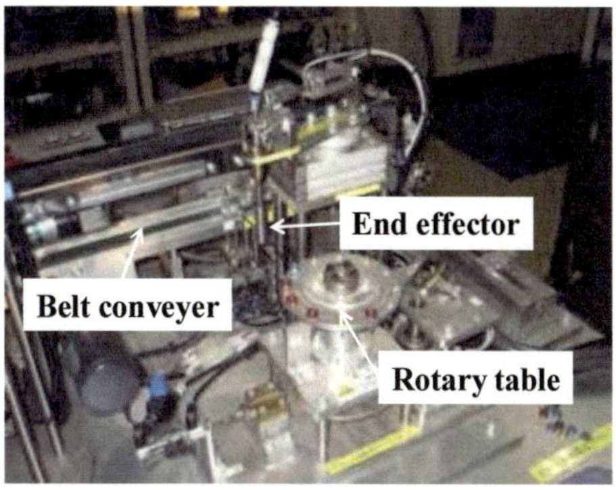

(a) First work cell

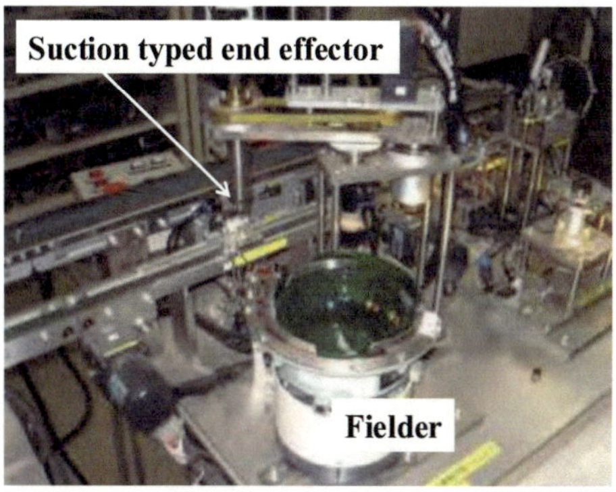

(b) Second work cell

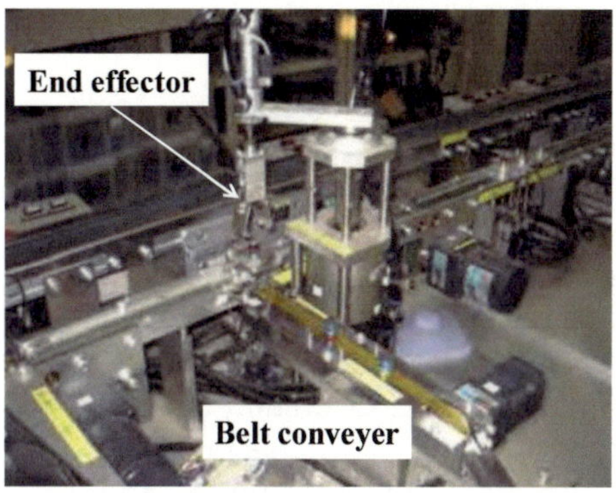

(c) Third work cell

Fig. 3. Main work cells of the miniature automation production line

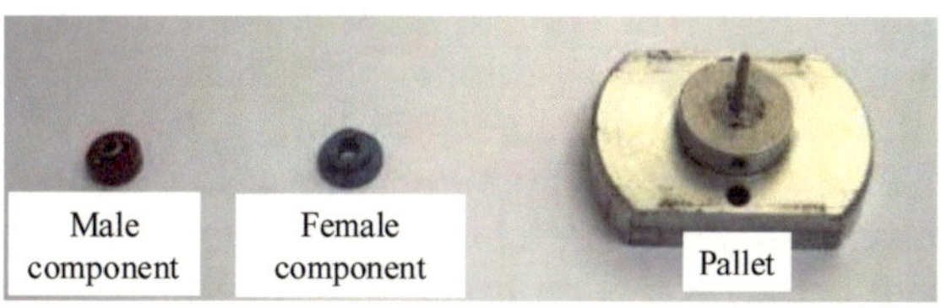

Fig. 4. A pair of components and pallet used in the miniature automated production line

3.4 Assemble a multi-legs robot

In the fourth session a guide book for MindStorms written in Japanese (Sato. 2000) is used as a manual to assemble a mobile robot. This book shows how to assemble several types of robot, such as multi-legs walking robot and wheel type mobile robot and so on. Fig. 5 shows a multi-legs walking robot named Musimusi No.5, which can be built by reference to the study guide. It averagely took about an hour and a half to complete this robot.

Fig. 5. Six legs autonomous robot, Musimusi No. 5

3.5 Programming

RCX is a delicate controller for MindStorms which has a 8 bit micro processor. It has three input ports and three output ports. Fig. 6 shows a RCX connecting with a motor, a lamp, a touch sensor, and a light sensor. A GUI based programming software, Robolab, is prepared as a regular programming software of MindStroms. Fig. 7 shows a programming environment of Robolab, where only users align command icons and connect them with wires in order to control some motors and actuators of MindStorms. All control command icon is prepared in the function panel. Operation commands for icons like wires can be selected on the tool panel as shown in Fig. 7. This programming environment of Robolab is well-suited to the beginners because of its friendliness, and makes it possible to code some high-level programming techniques such as conditional branching and infinite loop, etc.
Since this mechatronics course has to be implemented in short space of time, teachable programming techniques are restricted. Fortunately some fundamental programs are previously installed in software for self-education then there is no difficulties to introduce the basic function of Robolab and make students to understand a signal flow of program. After students understand and run the programs by using RCX, they try to make the programs working out several instructions in order to cultivate their programming ability. The following itemized instructions is an example of the training, and its corresponding program is shown in Fig. 7.

I. When the touch sensor mounted at port 1 is pressed, then both of motors mounted at port A and C rotate in left direction for three seconds.
II. When the touch sensor mounted at port 1 is pressed again, both of motors mounted at port A and C rotate in right direction.
III. When the touch sensor mounted at port 1 is pressed, then the rotating two motors stop.

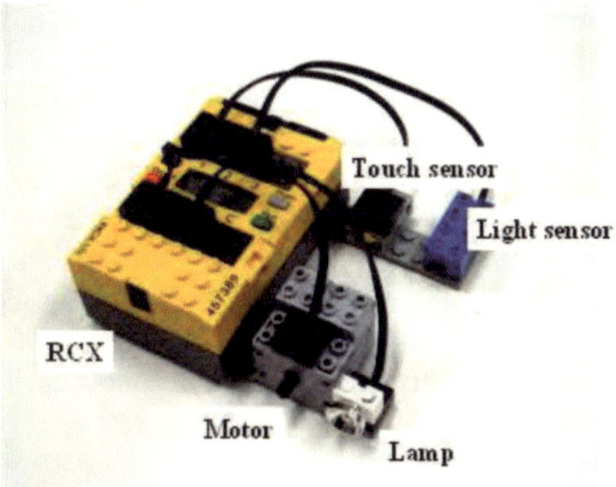

Fig. 6. A configuration of RCX, which connects with actuators and sensors.

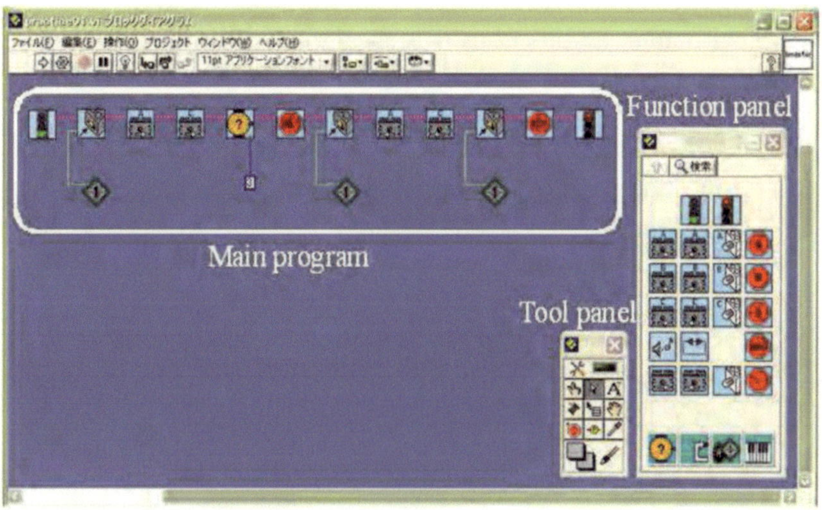

Fig. 7. Graphical user interface of Robolab

3.6 Obstacle course

Through the fourth and fifth sessions, students experienced a basic method to assemble a movable robot and learned how to code a program of ROBOLAB which controls the sensors and motors. A research factor of robot engineering is involved in the final session, where students try to make a mobile robot getting up to the goal with avoiding and/or overriding some obstacles on the field.

Only the final session, students are divided into five teams, then each team consists of two or three students. This team formation aims to avoid the restriction due to a number of components of MindStorms, and to make up for each other's deficiencies. A problem field is informed to student after the fifth session, and then students begin to think a strategy where they proposed shape and mechanism of robot and program code are proposed with each other within one week. In the final session students can use total 150 minutes for assembling and coding their robot and program, and the remaining time is used for competition. Of course there is no problem to exchange of opinions with other teams, and change their strategy they thought. Before the competition, they explain their strategy and the function of the developed robot.

The teams of the first group challenge a field of obstacle course shown in Fig. 8(a), where two obstacles are fixed between the start area and the goal area. A robot has to recognize the obstacles by using using touch sensors and change the direction of movement to avoid the obstacles.

Fig. 8(b) shows the second group's field of obstacle course, where an object which is twenty centimetres square by ten centimetres height is fixed in the centre of the field. A robot has to have functions that enables the robot to go strait and go around the obstacle in light direction. Only two rules a robot has to follow are that a robot must not contact the obstacle and take over the outer square line.

The third group's students try to make a robot that is able to go up the stairs. Fig. 8(c) shows a scheme of the field of obstacle course. The height of each step is five centimetres and and the width is fifteen centimetres. This stair-like obstacle is fixed at one meter distance from the start line, so the robot has to not only go straight and climb the stairs.

In the fourth problem three hurdles made of rubber grips are used as the obstacles. Fig. 8(d) shows a scheme of the fourth field of obstacle course. The hurdles are assigned fifteen centimetres apart and these height are one, two, and three centimetres from the floor. Additionally, hurdle's stiffness escalates afterward.

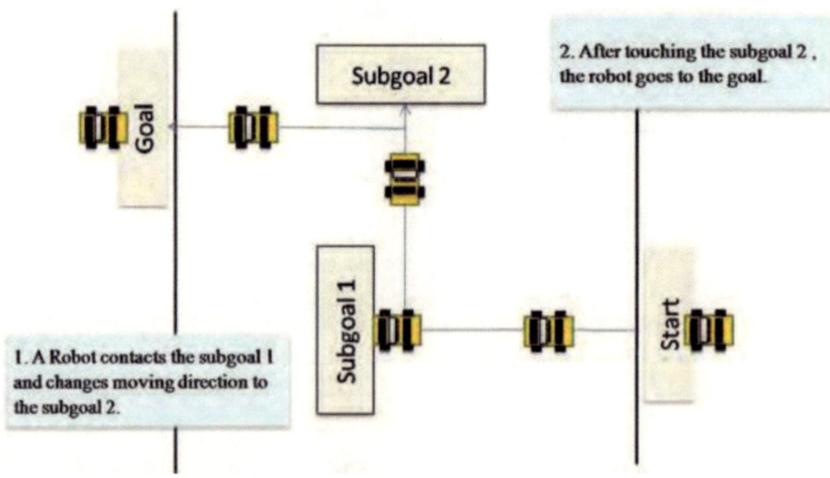

Fig. 8.(a) Scheme of the first obstacle course

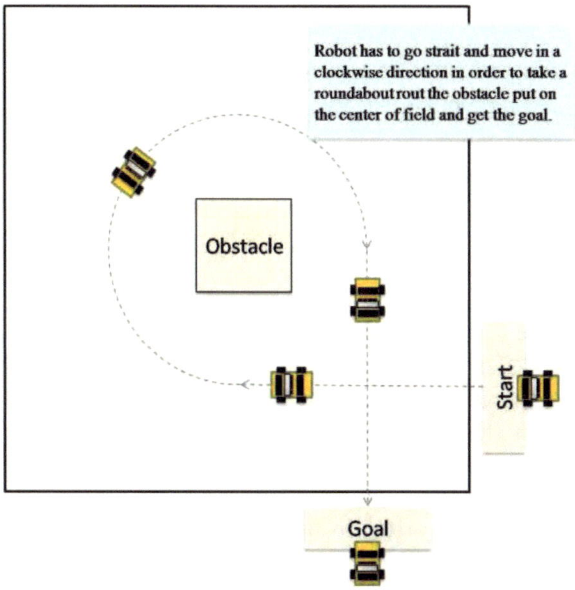

Fig. 8.(b) Scheme of the second obstacle course

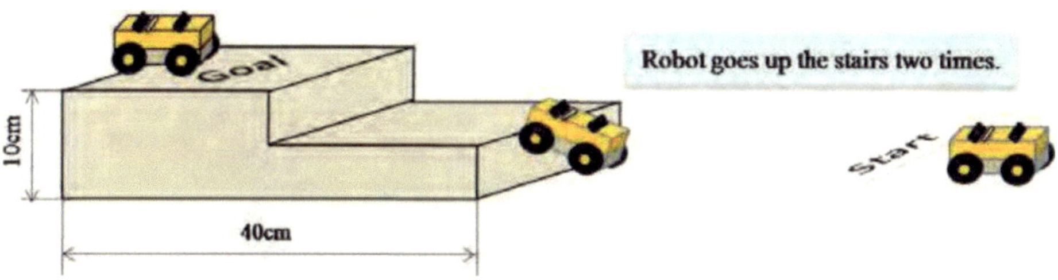

Fig. 8.(c) Scheme of the third obstacle course

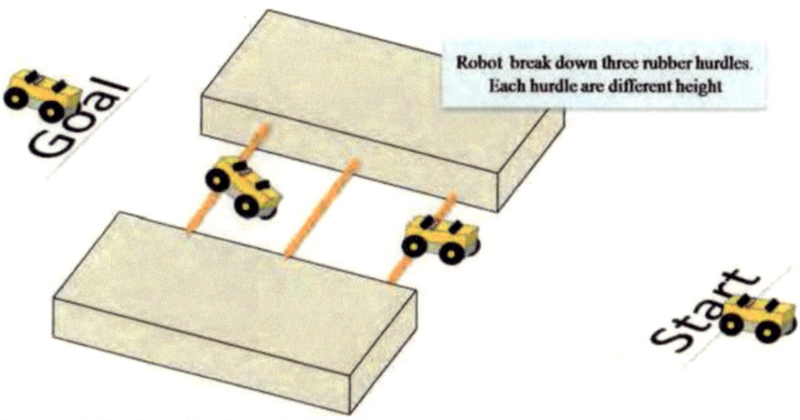

Fig. 8.(d) Scheme of the fourth obstacle course

4. Result

This mechatronics installation course aims to make students to briefly know not only basic concepts of mechatronics but also difficulties of manufacturing. A toolkit of Mindstorms has an infinite of design freedom, so that this toolkit is able to foster students' creativity and design talent. Most of students had worked at this mechatronics installation course in earnest and developed interests in mechatronics. After learning basic techniques to assemble a movable robot and write a programming code, students addressed the challenge of obstacle course at the final session.

Fig. 9 shows two robots that could accomplish to clear the first obstacle course as shown in Fig. 8 (a). One is a type of autonomous robot that mounts RCX on its body and employed a rear-wheel-drive system. Students assumed that the functions required for a robot are to recognize accurately the surface of obstacle's wall, to keep going straight, and to change the direction of movement orthogonally. Firstly the robot moves toward the obstacle 1. After recognizing the surface of obstacle 1 by using the touch sensor fixed in front of body, the robot once goes back to turn a right and goes forward to the obstacle 2. After that, the robot goes to the goal in the same matter used to avoidance the obstacle 1.

On the contrary the robot shown in Fig. 9 (b) is manually controlled by the students. They used a RCX as a manual controller so that no sensors are mounted on the robot. The direction of movement is changed by differently adjusting the motor powers. The control signal is transmitted to the motor according to the timing of the student pushes a touch sensor mounted on RCX.

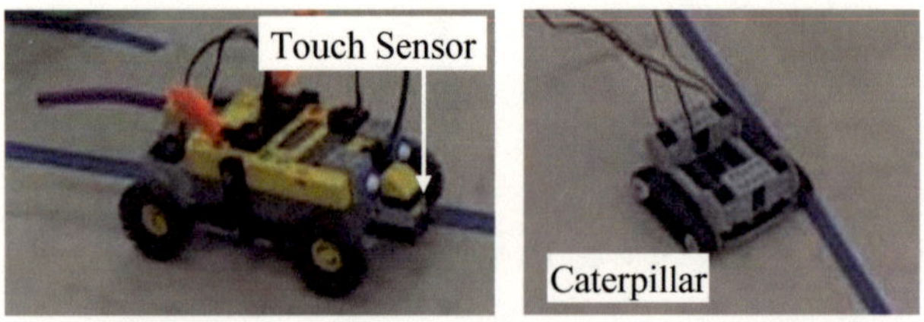

(a) Autonomous robot (b) Manual control robot

Fig. 9. Robots for the first obstacle avoidance problems

Fig. 10. An experimental scene of the second obstacle course

Fig. 10 shows a scene of the competition of the second team. Unexpectedly only one robot shown in Fig. 10 could clear this obstacle course. Most of robot could not take a roundabout route within the outer square line. The team employed a four wheel drive system, however different sized wheels used in left and right sides in order to reduce turning radius of the robot. This team spent most of time to adjust their program of RCX by repeating test run, because rotational speeds of left and right motors have to be asymmetrically controlled even to keep going straight.

Five teams addressed the third problem shown in Fig. 8 (c). Most of teams aimed to climb the stairs by driving front wheels as shown in Fig. 11(a), however, no team could climb even the first step by using this way, they ended in failure due to the weight of RCX. Only one robot could complete the stair-like obstacle course. Fig. 11(b) shows an overview of the robot and describes its body's degrees of freedoms, where the lack gear moves along the joint q1 in order to lift up the front caterpillar. The q2 shows the angle of gradient for the front caterpillar. A remote control system is adopted in order to trim the weight of robot.

Along the following five steps, shown in Fig. 9(c), the robot climbed up stairs in the experiment; (1) the robot moves toward the stairs with fixing the angle of front caterpillar and quit going forward in front of the stairs. (2) Using the movable joints q1 and q2, the robot puts a part of front caterpillar on the first step. (3) The robot got back to the beginning posture as well as (1) on the first step. (4) The robot repeated the previous three behaviours once more to get the top of the stairs. (5) The robot arrived at the goal at last.

Fig 11.(a) An example of robot ended in failure for the third obstacle course

Fig. 11.(b) An overview of the robot completed the third obstacle course

Fig. 11.(c) Experimental scenes going up the stairs in the third obstacle course

Fig. 12.(a) A tank type robot with double arms for the forth obstacle course

Fig. 12.(b) A long leg type robot for the forth obstacle course

Fig. 12 shows the robots could clear the forth obstacle course shown in Fig. 8(d). The robot shown in Fig. 12(a) rotates the front arm along the direction of q1 for the three purposes of the followings; (a) lifting the body up, (b) breaking down the rubber hurdles, and (c) going forward. The tires equipped at the end of arms does not rotate to go forward. As the result, this robot became entangled with rubbers several times, however, it finally could ride over all hurdles.

The robot, as shown in Fig. 12(b), only goes straight by rotating swastika shaped legs. These legs always lift the body higher than the rubber hurdles and enabled the robot to get the goal without the legs hitch the rubber hurdles.

According to the result of questionnaire about this mechatronics course, the following answers are received; all students had the interests in mechatronics, 68% of students were highly interested in the obstacle course conducted in the final session, and 17% of students were especially concerned with RCX programming. The others concerned with making the mechanical modules by using MindStorms, the miniature automation line system, and the assembling a multi-legged robot at the same rate.

4. Discussion and Conclusion

The author constructed an installation course of mechatronics and conducted on the students of department of mechanical engineering, Oita national college of technology. The course is composed of six sessions and is aiming to grow up the mechanical engineers who can adapt quickly to changes in industrial society. Then, the education programs of computer technology and information processing are more emphasized in this course. Certainly the specific subjects involved with mechatronics are constructed as a part of curriculum in the older grades, however there is some difficulties to make students of department of mechanical engineering to have interests in electronics and/or information science. Viewed in this light, it is better to begin mechatronics education with undergoing experiments like this course since they were in early grade.

This course employed the obstacle course as the final project. Working at making a robot for obstacle environment provides a research factor for students. They have an opportunity to discuss about their strategy to make a robot and they can experience concurrent engineering between assembly of robot and design of the controller. Meanwhile, development of a line tracing robot based on MindStorms has commonly used as a training issue in mechatronics education. A line tracing robot generally consists of mechanisms, light sensors, and motors. Additionally, a repeatable structure program is necessary. Most of mechatronics factors are contained in making a line tracing robot, however, there are lots of information about line tracing robot based on MindStorms and Robolab on the Internet. Certainly, difficulties of line tracing robot can be increased by changing the line width and/or the path of line. The reason why obstacle course is employed as the final project of the mechatronics course is that nobody knows an appropriate solution corresponding to obstacle environments.

Students worked in a team to address each given environment. Obstacle environment had to be different corresponding to students' growth through the year, so that the evaluation method for students' grade must not focus on the result of the final problem. In this course the grade of student is evaluated from their submitted reports for every session, where their activeness and written description of their impressions for each sessions become mainly evaluation object.

It was difficult to keep enough time to make students to learn programming techniques of Robolab such as branch connection and conditional statement, so that some students could not know convenience of programming for controlling actuators. Increment of the programming session make it possible to enhance the quality of robot and raise success rate of the final project. Temporal distribution of every kind of mechatronics contents is the most important problem.

5. References

Jin SATO. (2000). *TETSUJIN Technique for MindStorms of Jin Sato*, Ohmsha, ISBN 4-274-08682-8, Japan.

Jiro Eto, Yuki SHIRAKAWA, Tetsuro MAKISE, Jin SATO, Daisuke Kurabayashi, and Go FURUKAWA. (1999). *LEGO MINDSTORMS Perfect guide*, Shueisha, ISBN 978-4-88135-769-9, Japan.

Chieko KOMATSU, Toshikazu MINOSHIMA, and Takafumi MATSUMARU. (2000). *Effeciency of experimental study on Mechatronics by using the LEGO MindStorm*, Proceedings of Robotics and Mechatronics, Kumamoato prefecture, Japan, May and 2000, 1A1-81-128, (In Japanese).

Tomoyuki NAKASHIMA, Hiyoshi HAGIWARA, and Takafunmi MATSUMARU. (2001). *Learning by Experience System on Mechatronics using LEGO MindStorms*, Proceedings of Robotics and Mechatronics 2A1-A2, Kagawa prefecture, Japan, June and 2001

Eiichi INAGAKI, Yoshiaki SAWA, and Hiroyuki Okamura. (2001). *Practical Education by using LEGO MindStorms at a Lecture Room*, Proceedings of Robotics and Mechatronics 2P1-A2, Kagawa prefecture, Japan, June and 2001